TK
7888.3
B7
v.1

DESIGN AUTOMATION OF
DIGITAL SYSTEMS

DESIGN AUTOMATION

volume one
THEORY AND TECHNIQUES

CONTRIBUTORS

Ralph J. Preiss
Melvin A. Breuer
Benson H. Scheff
Stephen P. Young
Uno R. Kodres
Maurice Hanan
Jerome M. Kurtzberg
Sheldon B. Akers

Edited by
MELVIN A. BREUER
University of Southern California

OF DIGITAL SYSTEMS

PRENTICE-HALL, INC., Englewood Cliffs, New Jersey

DESIGN AUTOMATION OF DIGITAL SYSTEMS

volume one
THEORY AND TECHNIQUES

Edited by
Melvin A. Breuer

© 1972 by Prentice-Hall, Inc.,
Englewood Cliffs, New Jersey

All rights reserved. No part of this book
may be reproduced in any form
or by any means without permission
in writing from the publisher.

10 9 8 7 6 5 4 3 2 1

ISBN: 0-13-199893-5

Library of Congress Catalog Number: 79-39507
Printed in the United States of America

Prentice-Hall International, Inc., *London*
Prentice-Hall of Australia, Pty. Ltd., *Sydney*
Prentice-Hall of Canada, Ltd., *Toronto*
Prentice-Hall of India Private Limited, *New Delhi*
Prentice-Hall of Japan, Inc., *Tokyo*

CONTENTS

PREFACE, ix

chapter one
INTRODUCTION
1

    1.1 Design Automation Objectives and Usage, *2*

    1.2 Evolution, *4*

    1.3 Designing a Computer System, *10*

chapter two
LOGIC SYNTHESIS
21

2.1 Introduction, *21*

2.2 Notation, Definitions, and Basic Results, *26*

2.3 Covering a Single-outputs Switching Function, *33*

2.4 Covering a Multiple-output Switching Function, *50*

2.5 Synthesis of Constrained NAND Networks Via Factorization, *59*

2.6 Circuit Modification Techniques, *77*

2.7 Data Structures, *85*

2.8 Miscellaneous Topics, *87*

chapter three
GATE-LEVEL LOGIC SIMULATION
101

3.1 Introduction, *101*

3.2 The Logic Simulation Process, *103*

3.3 Developing the Logic Model, *113*

3.4 Exercising the Model, *131*

3.5 Fault Simulation, *143*

3.6 Special Simulation Techniques, *150*

3.7 Simulator Example, *155*

3.8 Simulator Selection Criteria, *164*

chapter four
PARTITIONING AND CARD SELECTION
173

4.1 Introduction, *173*

4.2 The Influence of Building Materials and Techniques, *174*

4.3 Graph Theoretic Representation of the Logic Design, *176*

4.4 The Standard Library of Replaceable Units, *181*

4.5 The Partitioning Problem, *187*

4.6 The Selection Problem, *202*

4.7 Unsolved Problems and Conclusions, *209*

chapter five
PLACEMENT TECHNIQUES
213

5.1 Introduction, *213*

5.2 Placement Problem Definition and Graph Models, *214*

5.3 Interconnection Rules, *219*

5.4 Associated Mathematical Problems, *221*

5.5 Constructive Initial-placement Methods, *226*

5.6 Iterative Placement-improvement, *245*

5.7 Branch-and-bound, *264*

5.8 Conclusions, *271*

chapter six
ROUTING
283

    6.1 Introduction, *283*

    6.2 Wire-list Determination, *287*

    6.3 Layering, *295*

    6.4 Ordering, *308*

    6.5 Wire Layout, *312*

    6.6 Final Comments, *327*

chapter seven
FAULT TEST GENERATION
335

    7.1 Introduction, *335*

    7.2 Definitions, *337*

    7.3 Review of Classical Approaches, *342*

    7.4 Test Generation Models For Combinational Logic, *345*

    7.5 Test Generation Models For Sequential Logic, *376*

    7.6 Automated Test Application, *388*

PREFACE

The design and development of digital systems can be partially automated by using digital computers as design tools. To accomplish this, effective techniques must be established for each of the various steps of design, evaluation, manufacture, and maintenance. In the process of automated development of such systems, digital computers have been utilized both as controlling devices and ancillary aids.

The rapid growth of design automation has been evidenced by the investments of large digital system manufacturers, as well as by the relatively significant commitments made by small companies and some software organizations.

Unfortunately, the growth of the automation technology has not been matched by the existence of adequate and easily obtainable literature. This difficulty has been compounded by the fact that until recently, few techniques related to design automation were taught at the university level. As a result, there is considerable ignorance of many existing techniques, even among practitioners. The user can find only sparse justification to permit him to evaluate, with confidence, the adequacy of various techniques. Thus, work is frequently duplicated and, as new people enter the field, they often "re-invent the wheel."

Recognizing this problem, the Executive Committee of the IEEE Computer Group created an ad hoc committee of the Design Automation Committee to write a substantive book on design automation which would provide: 1) a definitive introduction to the technical aspects of digital system design and development for individuals with either hardware or software oriented backgrounds, 2) a guide to the selection of techniques for inclusion in a design automation system, and 3) a foundation from which others could develop new and better design automation techniques.

This volume and a subsequent one are the results of this effort which first began late in 1968. This volume deals with four aspects of the logic of a system; namely, synthesis, simulation, testing, and physical implementation. The latter area includes partitioning, placement, and routing. The subsequent volume will deal with system level simulation, simulation and synthesis at the register transfer level, file maintenance, and interactive systems.

These two volumes collectively describe practical techniques which have either been successfully employed, or have been considered to be useful alternatives in a design automation system. The constraint of practicality often eliminates both total enumerative and elegant theoretical techniques. Enumerative techniques, though effective since all digital systems are finite, are usually computationally infeasible due to high costs. On the other hand, the implied benefits of many theoretical solutions may not be achievable in actual practice.

We realize that what is and is not practical varies with the user, the number of parameters in the problem, and time. Due to the rapid technological advances, solutions to problems posed today may not be acceptable solutions for these same problems tomorrow. Here we discuss techniques in which the cost of developing and executing the design automation system are reasonable, and in which acceptable results are obtainable.

This work is suitable not only for those who are familiar with some aspects of design automation, either from a hardware or software viewpoint, and wish to learn about other areas, but also for those just entering the design automation field. It will be of particular interest to students who desire to learn aspects of computer design which are different from those taught in the traditional courses on logic design and switching theory. The reader need only have a basic knowledge of digital computers and logical design.

We believe this work will also have strong appeal to the applied combinatorial analyst, since we cover aspects of many key subjects in this field such as discrete optimization theory, branch and bound techniques, integer programming, graph theory, the covering problem, the traveling salesman problem, the assignment problem, and the quadratic assignment problem.

The subject matter has been divided into two volumes, the first being somewhat hardware oriented, the second software oriented.

Volume 1 deals with the problems of logic synthesis, simulation, testing, and physical implementation.

In chapter one we introduce the general area of computer aided design of digital systems, discuss the motivation for and goals of computer aided design, and outline the historic development of this growing area. Finally, the chronological steps involved in the realization of a new digital system are outlined, and we indicate where and how design automation can interact in each step of the design and manufacturing process.

The classical problem of logic synthesis is discussed in chapter two, where techniques are presented for logic simplifications, factorization, and conversion between different families of logic elements. Practical restrictions related to fan-in and fan-out constraints are considered.

Chapter three deals with the simulation of a digital system at the logic or gate level. Here, the goals of logic simulation, simulation languages, techniques, and systems are discussed.

The next three chapters deal with three major problems related to the physical construction of a digital system.

Chapter four relates to various partitioning and assignment problems associated with component layout, as well as to problems of evaluating the effectiveness of standard circuit modules.

Chapter five presents a detailed analysis of various placement algorithms dealing with the problems of initial placement and placement improvement. These techniques are applicable to a wide range of backplane functional units, e.g., components, modules, and boards.

The interconnection or routing problem is treated in chapter six. The discussion covers both discrete and etched multi-layer connection techniques, including the related problems of ordering of interconnections, path selection, pin selection, and layer selection.

Finally, in chapter seven we cover the problem of generating fault diagnostic and detection tests for combinational and sequential logic circuits.

As programming languages have evolved from machine code to assembly languages to machine independent languages, so digital design languages have evolved from circuit descriptions to Boolean expressions to register transfer languages and finally to system level languages. Volume II will deal primarily with new design techniques at the higher levels of description as well as associated support systems. More specifically, this work will cover material on the following subjects:

1. Descriptive languages for digital devices at the system or functional level, and related simulation techniques used in evaluating system performance.
2. Register transfer languages.
3. Synthesis of digital devices from their register transfer level description.
4. Simulation of systems at their register transfer level.
5. Data-base systems for use in an integrated design automation system.
6. Interactive graphic systems and their role in computer-aided design.

As a final note I would like to thank my co-authors who have contributed so much of their time and effort in the preparation of this book.

MELVIN A. BREUER

chapter one

INTRODUCTION

RALPH J. PREISS

IBM Systems Development Division
Poughkeepsie, New York

This chapter presents an overview of the purpose and evolution of design automation and its application to the computer design process. As an overview of a field involving many manufacturers, universities, and research foundations, the reader should not assume that any process described is practiced by any particular organization. Rather, the reader should realize that this overview is a synthesis of the many practices that the author has observed, read about, or discussed with other design automation practitioners.

The chapter is divided into three sections. Section 1.1 reviews the objectives of design automation and the attempts that have been made to achieve them. Section 1.2 covers the history of design automation as an evolutionary discipline. Section 1.3 provides a summary of the steps involved in the design of a digital system, and it emphasizes the designer's responsibilities and interplay with a design automation system.

1.1 DESIGN AUTOMATION OBJECTIVES AND USAGE

Design automation, as used here, is a goal rather than an immediately realizable objective. It is the art of utilizing digital computers to help generate, check, and record the data and the documents that constitute the design of a digital system [1].

The objectives of design automation are cost- and time-reduction between start of design and completion of fabrication. These objectives are accomplished by relieving design engineers of repetitive, time-consuming manual tasks such as:

1. generating detailed design information and documenting it on what is here called *systems logic pages*;
2. controlling changes to systems logic pages;
3. checking systems design for electrical, logical, and physical compatibility;
4. preparing wiring lists, cable lists, location charts, and other manufacturing data, such as the bills of material, and tests.

A wide variety of topics are covered by the design automation umbrella. The only unifying theme is the elimination of repetitive manual operations or computations in the design of digital systems. Excluded from consideration are the computations for the design of components, circuits, mechanical structures, and cooling. Design automation, therefore, limits itself to filling the gap between the systems specifications and the manufacturing data. It is involved with converting the systems specifications into logic hardware, packaging the hardware into mechanical structures, and describing this process for fabrication.

If the design process could be fully automated, the conceptual data that describe a proposed digital system in a higher-level language could be converted directly into the mass of detailed data, i.e., part numbers, assembly sequences, cabling information, etc., necessary to manufacture the machine. Failing complete automation, design automation can be used as a data base from which to extract design information on demand, calculate (with specialized design mechanization aids) additional details, and then return the new data to the data base. This permits creative manual intervention and allows additions and deletions to be made as specifications change.

The data base in which the design is described may contain computer-generated as well as hand-generated data. This includes, for each logic function, the part number code implementing it, its physical placement in the machine frame, test points, cabling and internal wiring information, cross-referencing data, notes, and design change activity information. Supplementary information, such as the exact location and types of bends or the lengths of cables and interconnecting wires, may be omitted from the main documents and relegated to secondary documents. Similarly, the detailed description of subassemblies, such as discrete components mounted on cards, may be relegated to secondary documents. However, each of these secondary documents, once entered into the design automation system data base in computer-readable form, is seldom retranscribed but is utilized by the system in subsequent calculations.

While the basic goals of design automation tend to be the same from one manufacturer to another, the automating philosophies tend to be quite different. On the one extreme, a manufacturer considers the accuracy of his field documents paramount and therefore uses the machine-readable data base and the same documents in design and manufacture. On the other extreme, a manufacturer may permit differences between the field and the design and the manufacturing documents. The latter approach could best be described as having multiple data bases. It is characterized by multiple transcriptions to or from computer-readable form to take advantage of the computer for certain calculations, such as logic simulation (requiring logic data), wiring (requiring interconnection data), or load checking (requiring electrical data). Errors that crop up through transcription mistakes are detected by procedural means. The field documents are usually draftsman-produced, and they may or may not fully reflect the latest design changes.

Design automation programs are generally not shared among manufacturers, since the programs are quite heavily dependent upon design philosophies, circuit technology used, and the computer available to the design engineer to help him with his work. (The use of higher-level languages usable with different computers has been quite discouraging. These tended to be inefficient: fine for experimental purposes, but too expensive for production use.) Furthermore, design automation is especially advantageous if standardized packaging and logic are used and if a standard control system, unique to individual manufacturers, is set up. These aspects are usually proprietary since they are tied into providing an economic advantage of one design/manufacturing process over another.

Besides automating design checking and record-keeping, a design automation system serves another important function: effective change control. By reducing design and engineering-change time, it permits logic changes to be made late in the development process, assures accurate manufacturing instructions and test procedures, and instills a confidence that manual procedures could not hope to attain.

1.2 EVOLUTION

In the mid-1950s, the first generation (tube generation) of stored-program computers was in its prime. At the same time, transistor technology was getting to be understood, and uniform circuit characteristics could be assured in large batches of manufactured components. It was obvious that the new (second generation) computers being designed would utilize this new technology.

While transistor circuits were smaller, faster, and cheaper than their tube counterparts, their use implied that the new generation of computers would have to be redesigned from the frame up. In addition, the small size of the new components meant the introduction of new problems in packaging and interconnection. Where a single 7-tube circuit might require some 100 cubic inches of space, perhaps ten transistor circuits might require no more space. The smaller components permitted an increase in circuit speeds, but inductive noise was increased at the same time because of the closeness of the interconnecting wires.

Also, at this time, computer manufacturers were faced with the prospect of meeting contractual deadlines for military computers using transistors. These deadlines forced computer manufacturers to look for new ways to design computers in order to reduce their risks.

1.2.1. Efforts to Automate Design Procedures

Around 1955, without a single bit of literature appearing in any of the journals, the idea of design automation caught on, and the major computer manufacturers were all in the computer-aids-to-design, or design-mechanization, or design-automation business.

In a typical company, study of the design procedures showed that a logic designer sketched out his ideas in rough form and passed them on to draftsmen who put them into more readable form. Additional information would be added to a copy of the draftsman's drawings until enough information was contained so that the wiring and cabling layout could be developed. This process in itself would add additional information to the master diagram. Each time an addition or a change was made, the history would be recorded, and the appropriate designer would initial the drawings after review.

With the advent of transistor circuits, the wiring layout designer's job became more difficult. He had to make sure that two adjacent wires didn't travel too many inches in parallel because of "noise buildup." He constituted a vital link and he needed assistance in the design and build process. The computer could be used to lay out the wiring grids, take into account capacitive loading and interwire noise, and print the diagrams of these wires. This approach could provide a more direct involvement of the designer in the final hardware. It could also aid in the manufacturing process by calculating the path and the sequence that the wire-wrap people should follow. Furthermore, these calculations could save days in the schedule.

The computer designers also realized that they could substantially improve design turnaround if the logic diagrams were maintained on magnetic tape files and the computer could print these diagrams for every engineering change. It would only be necessary to keypunch a few cards describing a change, update the master file, and a printout of the modified design could be obtained in a few minutes.

1.2.2. First Uses of Design Automation

One of the first design automation systems described in the literature was presented at the 1956 Western Joint Computer Conference by Cray and Kisch [2]. They described a three-phase program which started with the checking of logic equations for logical, clerical, and timing errors. Further checking could be done by counting the number of inputs and driven outputs per circuit (i.e., fan-in/fan-out checking). The Boolean equations could also be sorted into various categories, and printed outputs could be provided to those who needed them. In the second phase, the simulation phase, the designer could apply input values to his Boolean equations from a simulated switch panel and watch the results as though he had built the machine and were testing it out. Finally, in the third phase, he could package his design into assemblies, compute the interconnections between chassis, provide lists of the origin and destination points, and designate the color and the lengths of the wires. One interesting observation which also might indicate the pioneering nature of this article is that it cited no references. Another early article by Kloomok, Case, and Graff, describing a similar but more elaborate system, was presented at the Eastern Joint Computer Conference in 1958 [3]. After 1960, many manufacturers were busy discussing their systems [4], [5], [6], [7], [8], [9]. Figure 1-1 depicts the major problems facing computer designers and design automation. What follows here is a narrative account of the design-automation evolution.

Let us return to the development of design automation systems. Initially, the input for the wiring layout was from system logic blueprints. From these, the coordinates of the series of points which were to be connected by wires were keypunched. The computer program consisted of the wiring rules and legal wiring channels. After reading the keypunched coordinates, the computer produced a listing that indicated the length of each wire, the sequence in which each should be installed, and the route each wire should take (Fig. 1-1, items 1 and 2).

Problem	Solution	Problem

1. Capacitive loading and interwire noise in new circuit families ⟶ 2. Minimize noise generation by detailed calculations ⟶ 3. Design errors due to drafting and transcribing errors in primary and secondary documents ⟵ 4. One source data bank with extensive checking on entering data

5. Assure accuracy of design before high production speed piles up scrap and rework costs ⟶ 6. Eliminate logic errors through simulation ⟶ 7. Cannot contain full design in computer storage for simulation ⟵ 8. Cut design into number of pieces, each fitting into storage ⟶ 9. Long computer run times ⟶ 10. Provide more computer power for simulation ⟵ 11. Need more manual preparation to check data used in simulation ⟶ 12. Eliminate simulation; express design in higher level language and perform autosynthesis ⟶ 13. No one language and process found suitable on all counts to permit this ⟵ 14. Do computer-aided design: permit creativity to be supplied by human, detail calculations by machine

15. Too many manufactured parts ⟶ 16. Permit standard parts to make up detail design ⟶ 17. Changes affect many parts users ⟵ 18. Develop "where-used" file
19. Nonfunctional unit-tested hardware does not work after assembly into system: find the failing part ⟶ 20. Develop diagnostic tests at circuit and wire level

21. New component: control store ⟶ 22. Develop new DA system

23. New technology: LSI ⟶ 24. Develop new interactive DA system

Fig. 1-1. Design automation evolution.

Meanwhile, manufacturing engineers considered automation further. Why should people wire panels? Why not have a computer produce punched cards with instructions that a wiring machine could follow? Couldn't the data from the engineering listings be keypunched in the proper format for the wiring machine?

The Gardner-Denver Tool Company developed a card-controlled wiring machine which came into extensive use in the late 1950s [10]. The programs that routed the wires were modified to control this machine. With its use, not only was wiring speed increased, but also fewer errors occurred. The backpanel wiring technicians could now multiply their productivity tremendously. But all could not be automated. Since the engineers were constantly improving their designs through engineering changes, manual rewiring was still necessary.

Concurrently, schemes to accumulate changing engineering information on a logic master tape were developed. This master tape could then be printed out to indicate the design status, and the printouts could be marked with colored pencil to indicate more changes. Then, only the changes would have to be keypunched to update the master tape. The engineer who provided the original sketch would have to proofread only the marked-up sections, thus saving himself a considerable amount of time.

However, the logic design had earlier depended upon draftsmen to design the packaging, do the design checking, and sometimes even add carelessly left-out details. With the draftsmen removed from the design process, the design engineer had more details to contend with. For example, a simple keypunch error could hurt the design schedule. (Fig. 1-1, item 3). To aid the designer in following details, checking programs were developed to perform a reasonableness check on the data fields of the master tape: alphabetic versus numeric formats of known functions, physical layout (so that two packages didn't occupy the same slot), and circuit fan-out (overloading). Some of these check the keypunch operators; others check the logic designers' application of the rules of design (Fig. 1-1, item 4).

One might ask if this was design automation. The answer might be that the drudgery of drawing lines, of copying numbers, and of cutting, pasting, erasing, and redrawing had been removed. The designer had more direct control over the implementation and accuracy of his design.

1.2.3. Use of Simulation

Meanwhile, an attempt was being made to reduce engineering change activity. With the high production rate of the automatic wiring machine, errors could be reproduced rapidly. Slow response could cause a lot of scrap and rework, and many dollars would be wasted. If the logic were simulated, proven once and for all to work the way it was intended to work, and checked to ensure that nothing was left out inadvertently, fewer designers would be needed, and scrap and rework would be reduced, if not eliminated (Fig. 1-1, items 5 and 6).

The engineers quickly found that in a small (4000 word model) computer they could not contain the entire design that they wanted to simulate (Fig. 1-1, item 7). But even 8000 words sometimes were not enough. Only a few small pieces of the design could be simulated in each computer run. The value of the time needed to keypunch the data and evaluate the results was often more than the value of the results themselves. But programming helped solve the problem by putting the complete design on a drum or tape storage and swapping small segments with high speed memory as required. This technique tended to use up much computer time (Fig. 1-1, items 9 and 10).

Simulation had run into trouble. The amount of data generated and printed out was unwieldy. The designer disliked having to generate data for the unnatural partitions into which his logic had to be subdivided in order to fit into the computer (Fig. 1-1, items 8 and 11). Why not validate a design before it gets into the detail level? In one sense, the true creative person in the design of a computer is the computer architect rather than the designer. If the architect could express his design in some language that could be synthesized directly into logic (Fig. 1-1, item 12), the logic then could be implemented in hardware, assigned to panels and frames, wired, and cabled automatically. It could be tested at the higher design language level by another program to prove that the architect's design, as conceived, was indeed implemented. If simulation as a means of testing a manually produced detail design did not at first succeed, maybe it should be approached differently and bypass the designer by going directly into an automatic synthesis routine. Of course, the utopia we speak of has not yet been realized. It has literally been lost in a Tower of Babel (Fig. 1-1, item 13). The languages for expressing the architect's dream are many. A discussion of higher level design languages is planned for Volume II of this series. Suffice it to say here that the search is still on for a simple, yet effective, man-machine system for computer-aided design (Fig. 1-1, item 14).

In the meantime, especially with the advent of large-scale memories, simulation was integrated into design automation systems so that the designers could make use of the already-digitized data base. Simulation thus became a practical and widely used tool.

1.2.4. The Advent of Standardization

With some success at automation behind them, manufacturing engineers introduced additional cost-saving changes by sharing fabrication techniques between models and by establishing a standard interface between design-automation-produced data and manufacturing-usable data (Fig. 1-1, items 15 and 16). In addition, bills of material, inventory control, economical ordering quantities, and long lead-time orders could be computed from the design description of a machine.

Designers of second generation equipment attempted to use, as far as possible, standard parts for many machine designs. Consequently, problems quickly developed when an engineering change to a part for one design had adverse effects on another. The establishment of the "where-used" file (Fig. 1-1, items 17 and 18) solved this problem by providing answers to such questions as: In what assemblies is a part used; which machines will be affected by a change; what plants or engineering groups are contemplating using the part; how many of the parts are being used; who has engineering control; is the part obsolete, or new, or in design [11]? The record structure, uses, and manipulation of the engineering description of the manufacturable design will be taken up in Volume II of this series.

Other manufacturing problems arose. It took too long to determine which circuit or wire was at fault if an assembled machine did not properly execute the instructions of its test program. The designers and programmers were called to help. Could they use their simulation routines for fault detection and diagnosis? Because of their simulation experience, they should be in a good position to take the information from the logic master tape and produce diagnostic tests. Through the use of logic tracing routines, they should be able to generate the stimulus signals needed to prove that circuits correctly wired together on the manufactured panels worked the way they were designed to work. If a different-than-expected result occurred, a failure would be detected. Since each test was designed to test a specific set of circuits, it was anticipated that the culprit should be easily located (Fig. 1-1, items 19 and 20).

1.2.5. Technology Changes

With the development of sufficiently fast control stores, microprogramming, which has been in the literature since 1953 [12] finally came into its own. Now the designers of simulators were faced with finding a way of simulating both microprograms and logic [13]. Further programs were needed to produce the data that could be utilized by the manufacturing organization to manufacture, test, and maintain inventories of the read-only control stores. A pioneering effort in this field, called the *controls automation system*, is described by Buckingham et al. [14] (Fig. 1-1, items 21 and 22). As with any design automation system, the controls automation system has the ability to maintain records and perform design-rules checking (such as fan-out from any one instruction or multiple use of the same facilities in one microinstruction), time checking (e.g., when a particular facility is active only during a certain time and is expected to be active at a different time), format and cross-reference checking, and automated test-pattern generating.

Since neither volume of this series covers design using microprogramming or the machine aids available or useful in this process, the interested reader is referred to S. Husson [15] for a full exposition of this topic.

With the advent of the silicon planar transistor, we entered the era of batch fabrication and integrated circuit technology, and we began to depend more and more on design automation [16]. In this technology, whose ultimate goal is referred to as *large scale integration* (LSI), layers of circuits (with possibly a thousand components) are being "grown" and printed by computer-controlled mask generators. The masks are produced from design automation data which must be thoroughly checked for electrical, logical, and physical compatibility. No repair is possible once the circuit is fabricated. Therefore, all assurances for success of the design must be computed before fabrication is attempted.

The same design automation principles that apply to discrete components also apply to LSI. The tolerances may be finer; the degree of error sensitivity higher; and mistakes costlier; but all in all, the problems have not changed. The cost in machine time per circuit for design automation may remain the same; but since the number of circuits per replaceable unit (RU) is higher, the cost per RU will naturally be higher, and the design engineers rely on more and more automation to keep the final product competitive [17]. The designers' need for fast turnaround and even more direct communication with the design automation programs becomes apparent. While the trend is not clear, interactive design automation techniques are being developed to permit design engineers immediate access to a design automation system [18] (Fig. 1-1, items 23 and 24). This change from batch to interactive design automation software will be reflected in Volume II of this series.

1.3 DESIGNING A COMPUTER SYSTEM

A computing system is composed of thousands of individually designed and interconnected components. These make up the computing element, the input/output facility, and the random access storage facility of the system. The components used in the design have undergone a steady technological improvement: from relays, to vacuum tubes, to solid state devices, to integrated circuits.

The computing element, over the generations, has become increasingly complex. More and more functions can be performed in less and less space. These transitions themselves were possible only through automating the manufacturing process. For the first generation machines, one could point to a relay that performed an AND function and hold it in one's hand. It is now almost impossible, without the aid of a microscope, to find the spot in an integrated circuit where this function is performed. Have we outfoxed ourselves? Not really. The functions once performed by Whirlwind I with its room full of equipment are now performed by a machine classified as a minicomputer. Yet all the design details that went into the one have to go into the other. All the functions that were included in Whirlwind I must also be contained in the mini. All of Whirlwind's manual wiring is now manufactured as printed circuits; the cables between frames are gone, since all the logic is contained in one frame; the power needed is less; the air-conditioning is less; but the logic, the required interconnections, the testing, and the maintenance of records... all remain. We have not cut down the sheer volume here. We have only increased the productivity of the designer by providing him with machine aids that:

1. permit him to utilize the expertise of others (such as draftsmen and layout men);
2. reduce the chance for error (by providing checking programs at every stage);

TABLE 1-1. DESIGN AUTOMATION PROGRAM USAGE

Design Step	Program Type
Subsystem building blocks	General purpose simulation
Logic diagrams	Input data checking
	Cross-reference checking
	Diagram layout and print
	File maintenance
Higher level languages	Translate to Boolean
	Simulate
Enter DA data base	Input data checking
	Cross-reference checking
	File maintenance
	Printouts
Verify design	Functional simulation
	Source-sink check
	Fan-in/fan-out check
Implement logic	Optimize
	Select part numbers
Package	Partition logic
	Assign
	Placement
	Wiring
	Cabling
Final test	Test generation
Design document maintenance	Bills of material
	Parts explosion
	Where-used
	Location charts
	Diagrams
	Tests

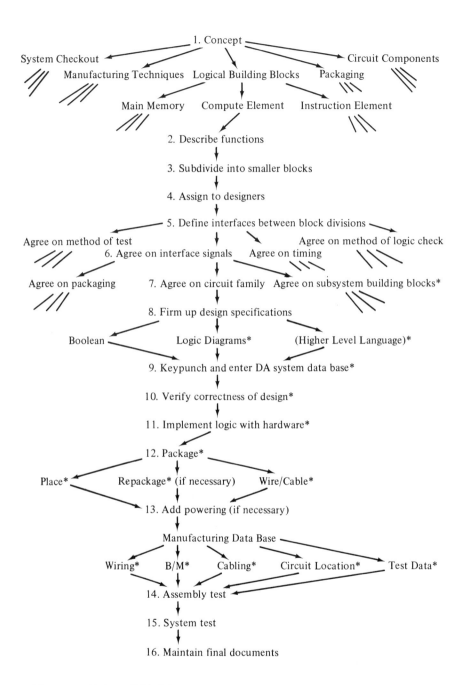

*Computer-aided; see Table 1-1.
Note: Iterations not shown.

Fig. 1-2. Computer design process.

3. reduce the information required in every part of the design (to eliminate the tedium involved in copying it by hand).

He must still end up with enough information so that someone less skilled in the art, or requiring more detailed instructions, will be able to reproduce the design from the description. This book concentrates on those aspects of the design process which have been automated successfully, or which show promise for successful automation, though they are not necessarily in wide use.

In this section we shall sketch the computer design process. We know that the process goes through a number of phases from the conceptual, to the detailed design, to the fabrication and test. Except for the one-of-a-kind machine, the need for reproducibility is immediately apparent. When a component is added to the model to correct a discrepancy, that change must be recorded and reflected throughout the engineering and manufacturing records, if the component is to appear in a subsequent machine. A designer is completely responsible, not only for the model and the proof that it works according to specifications, but also for the correctness of the design records so that all subsequently fabricated machines will work as the model. In the following exposition, we shall use Fig. 1-2 and Table 1-1 as an outline and reference.

1.3.1. From Concept to Defined Subsystem

Implementing the design of a computer main frame requires reducing the concept to building blocks (Fig. 1-2, item 1). These building blocks can be as gross as the computing element, main memory, instruction element, logic element, interconnection bus, etc. The functions of each element are then described in more and more detail, either in prose form or by breaking each building block into a set of subblocks. The subblocks may again be divided into sub-subblocks, until they can be apportioned to individual designers who are then responsible for their implementation (Fig. 1-2, items 2, 3, and 4).

The first part of the implementation process is to provide a detailed set of specifications for the connections (i.e., interfaces) between the building blocks (Fig. 1-2, items 5 and 6). Such interfaces may be as simple as "Add 1 to register 0 when line x is positive." One side of the interface has to specify under what conditions line x is made positive. The other side of the interface actually performs the function of adding 1 to register 0. By subdivision of labor in this manner, the design gradually evolves.

In addition to the functional interface description, in a synchronous machine, a timing interface or a timing convention has to be defined. Also, since the designers have to agree on a method of testing the interfaces, additional interfaces may have to be specified. During the conceptual design phase, the total system checkout and test plan concept are specified. The individual designers fill in the details, but their final detailed plan cannot depart from the overall concept.

While the conceptual design is in progress, a search for the circuit family, packaging technique, and subsystem components is made (Fig. 1-2, item 7). Before the design goes into the detailed stage, some of the salient characteristics of the subassemblies to be utilized should enter into the conceptual and interface design process. For example, if a decision is made to control the computer with a read-only control storage, undoubtedly the design will try to make maximum use of that control storage. Similarly, if a particular special component with unique features is available, and if the design group feels that this is an especially advantageous component to utilize, they will design, even conceptually, their machine around that component. To prove these concepts at this early stage, one may take advantage of simulation. The use of a register transfer simulator or a general purpose systems simulator could help the design process, especially in determining whether the use of a unique component around which the designer is slanting his design, does indeed pay off. In the past, for example, simulation has been successfully used to substantiate claims made for lookahead [19] and cache storage [20].

1.3.2. Detailing the Subsystem

When the design concepts are firm and each designer has been assigned a particular part of the total system, the detail work begins. One may use Boolean notation, logic diagrams, or a higher level language to describe, detail, and implement the design to the interface specifications (Fig. 1-2, item 8).

If a higher level language is used, two further operations may be applied to the design. The first is the simulation of the resultant partial design by means of a register transfer level simulator fed by the appropriate interface signals. These interface signals may be generated through further higher level language design. It is therefore possible to verify whether the design fits the interface conditions, as well as whether it performs the functions which the conceptual description defines. The second process that can be applied at this point is the translation of the higher level language into Boolean statements or into logic block diagrams [21].

Having arrived at the detail logic design either by manual or automatic means, the designer has to put the design data into machine-readable form (Fig. 1-2, item 9). Examples of this are given by Cray [2] and Kloomok [3]. In the process of converting from manual to machine-readable form, errors can occur. The machine-readable material is checked by various programs to discover format errors (Fig. 1-2, item 10). Errors due to carelessness are uncovered by cross-referencing. The cross-reference checking can later be used to determine whether all interfaces defined earlier are implemented, and whether a signal source is available for each signal sink. Once this checking is done, it is possible to do a logical simulation as covered in Chapter three of this volume.

After the detailed design is completed, the logic synthesis and logic minimization part of the design process, as discussed in Chapter two, is performed by the designer either before or after he has checked out his design through simulation. He simulates first so that he is reasonably sure that he is minimizing a working function. Sometimes he minimizes first to get an idea of the size of his design. If he had gone from a higher level language into the detail logic, minimization would have already been automatically included.

A note on the use of the word *minimization*. As used in conjunction with a design automation system, it is a compromise between the amount of computer time the designer is willing to spend to attain his solution and the true cost minimum, where *cost* is defined by some arbitrary function. This arbitrary function may have nothing to do with the true dollar cost of the piece of equipment. It may reflect space, usage of standard components, logic terms, or power consumption, or some other prespecified function.

1.3.3. Reduction to Hardware

Next, the designer faces the task of implementing his design (Fig. 1-2, items 11 and 12) by picking out the hardware building blocks that implement his logic. He may work with a list of standard parts and assign portions of his logic to these parts as described in Chapter four, or he may have the freedom to make up parts, in which case he resorts to partitioning and then assignment as described in Chapters four and five.

The designer must also have a very good idea of the space that his hardware design is going to occupy. At an earlier period, when the concept was discussed and the design was divided among the designers, each designer was allocated a certain amount of space for his hardware. Then too, a physical frame was drawn up, and major parts of the system were fitted into it. The memory, the control gating, the arithmetic unit, the input/output cabling, the power supplies, etc., were all assigned to some general area based on the best judgment of the conceptual designers. Now the precise placement of the parts within the frame has to be determined.

The placement problem is discussed in Chapter five. Generally, the areas originally assigned to the different designers do not change. However, errors in judgment regarding the amount of logic required by the various functions may necessitate repackaging of the design. Complete repackaging may be required if all the functions adjacent to an underestimated function use up the space allocated, so that the function cannot overflow. With an automated process available, one may postpone packaging to the very end; or one may try different placements, or redo them on occasion to control expansion space for future design changes prior to committing the design to hardware.

Once the placement has been performed, the next step is that of wiring the contents of a frame, a panel, or a card together as described in Chapter six. Depending on the kind of circuits involved, one might run into a problem not previously considered. This problem is that some circuits are not powerful enough to drive a signal over an extended length of wire. Therefore, a special driver circuit has to be placed between the source and the sink of the signal. But the drivers have to be placed somewhere, thus consuming even more of the available space (Fig. 1-2, item 13). Sometimes the designer deliberately moves a function out of a functionally packaged subcomponent, and into another subcomponent, so as to save the driver circuit. This approach is not unreasonable, but it complicates the design automation program because the placement routine now requires the ability to handle hand-placed components.

Wiring constraints may be characterized by the amount of embedding that takes place. The most complicated is that of a single-layer wiring, in which all the wires are in a single plane, with no crossovers permitted. The less complicated is two-layer wiring with "via" holes permitting the two layers to communicate with each other. Other techniques exist, such as n-layer wiring (with via holes) and discrete wiring. The last is limited, however, by what may be called wire buildup, a limitation on the number of wires crossing at a particular point.

Another form of wiring known as *cabling* occurs when a bundle of wires travels in parallel from one physical location to another. Cabling is limited by the number of wires that run in parallel, i.e., the thickness of the cable. Cabling, therefore, represents an additional problem in that a signal may not just take a point-to-point route but may also go via other points through more than one cable before reaching its destination. The cabling problem can easily turn into a traveling salesman problem.

1.3.4. Testing

Two types of tests are required to check out a fabricated design: component tests to determine whether each component meets its individual specifications; and functional tests to determine whether the designed and manufactured item performs the logic function it was designed to perform (Fig. 1-2, items 14 and 15). Just as the design is checked out in various stages, testing is performed at various stages in the manufacturing process. The individual components are tested prior to being assembled into cards; the cards are tested before they are assembled into panels; panels are tested before they are assembled into frames; and frames are tested before they are assembled into systems. While the electrical characteristics are most important in the very early part of the manufacturing process, the logic function characteristics assume major importance in some of the later stages of the testing process. Chapter seven deals with the generation of the logic tests for use at the system subassembly or assembly level.

While the process of trying to generate tests during sytem design provides a designer with valuable feedback such as design redundancies, unwanted conditions, and insufficient or redundant test points, the actual tests generated are useful only when applied to the equipment after fabrication. Since the test generation process consumes massive amounts of computer time, it becomes a design management decision as to when and how often this process should be undertaken.

1.3.5. Document Maintenance

Strictly speaking, the design process is over when the first machine comes off the assembly line. However, the records (digital data base) and blueprints that describe the machine and are, in fact, the design must be maintained until the end of life of the machine. Some modifications are always being made, either for reliability, for attachment of new devices, for implementation of additional functions, or for cost reduction. While modification and maintenance are not covered in this book specifically, these are still an engineering responsibility and the design automation records group continually answers inquiries on the status of various machines over their lifetime (Fig. 1-2, item 16).

The large number of dissimilar documents required to mass-produce a modern computer (e.g., wiring lists, panel- and circuit-location charts, edge connector and jumper lists, systems diagrams, and bills of material) is of itself a stimulus to develop accurate source data and to maintain this accuracy over the life of a design. A set of computer programs, called design mechanization or design automation programs, written to maintain this accuracy may therefore be cost-justified. While good engineering practice might dictate a parallel effort to develop programs to assist the engineer in logic design, logic minimization, translation into circuits, simulation of the resultant circuits, their placement into panels and frames, and finally their interconnection through wires and cables, the basic emphasis in any organization starting in this field must be on the creation of a good data base. The rest can follow with time.

ACKNOWLEDGMENTS

The author would like to express his sincere appreciation to Samuel B. Lee for his many helpful suggestions and his editorial guidance in the preparation of this manuscript. The author would also like to thank J.

Russell Washburne, Jr., Melvin A. Breuer, Benson H. Scheff, and Windsor B. Murley, who provided critical reviews to earlier manuscripts. Special thanks are also due to Linda M. Vouloukos who was responsible for the excellent typing.

REFERENCES

[1] R. J. Preiss, "Design Automation Survey: a Report to the AIEE Membership." AIEE Conference Paper CP61-178 (New York, 1961).

[2] S. R. Cray and R. N. Kisch, "A Progress Report on Computer Applications in Computer Design." *Proc. WJCC* (1956), pp. 82–85.

[3] M. Kloomok, P. W. Case, and H. H. Graff, "The Recording, Checking, and Printing of Logical Diagrams." *Proc. EJCC* (1958), pp. 108–118.

[4] W. L. Gordon, "Data Processing Techniques in Design Automation." *Proc. EJCC* (1960), pp. 205–209.

[5] W. A. Hannig and T. L. Mayes, "Impact of Automation on Digital Computer Design." *Proc. EJCC* (1960), pp. 211–232.

[6] R. T. Herbst, D. Leagus, and G. A. Sellers, "Machine Processing of Manufacturing Information for Digital Systems Equipment." AIEE CP61-336 (New York, 1961).

[7] Y. Kaskey, H. Lukoff, and N. S. Prywes, "Application of Computers to Circuit Design for UNIVAC LARC." *Proc. WJCC*, Vol. 19 (1961), pp. 185–205.

[8] C. W. Rosenthal, "Computing Machine Aids to a Development Project." *IRE Trans. on Electronic Computers*, Vol. EC-10 (September, 1961), pp. 400–406.

[9] A. L. Leiner, A. Weinberger, C. Coleman, and H. Loberman, "Using Digital Computers in the Design and Maintenance of New Computers." *IRE Trans. on Electronic Computers*, Vol. EC-10 (December, 1961), pp. 680–690.

[10] R. K. Grim and D. P. Brouwer, "Wiring Terminal Panels by Machine." *Control Engineering* (August, 1961), pp. 77–81.

[11] C. H. Button, H. J. Grosskamp, J. L. Kenney, M. R. Murphy, and R. L. Simek, "Parts Usage Maintenance Program (PUMP)." IBM Technical Report 00.746 (Poughkeepsie, New York, October 5, 1960).

[12] M. V. Wilkes and P. B. Springer, "Microprogramming and the Design of the Control Circuits in an Electronic Digital Computer." *Proc. Cambridge Philosophical Society*, Vol. 49, Part 2, pp. 230–238 (Cambridge, England, April, 1953).

[13] D. F. Gorman and J. P. Anderson "A Logic Design Translator." *Proc. FJCC* (1962), pp. 88–96.

References

[14] B. R. S. Buckingham, W. C. Carter, W. R. Crawford, and Y. A. Noel, "The Controls Automation System." *1965 IEEE Conference Record on Switching Theory and Logical Design* (October, 1965), pp. 279–288.

[15] S. Husson, *Microprogramming: Principles and Practice*. Englewood Cliffs, N.J., Prentice Hall, Inc., 1970.

[16] C. G. Thornton, "What Has Happened to LSI—A Supplier's View." *Proc. FJCC*, Vol. 35 (1969), pp. 369–378.

[17] C. F. O'Donnell, "Engineering for Systems Using Large Scale Integration." *Proc. FJCC*, Vol. 33 (1968), pp. 867–876.

[18] W. H. Sass and S. P. Krosner, "An 1130-Based Logic, Layout, and Evaluation System." *Proc. SHARE-ACM-IEEE 7th Design Automation* Workshop (June, 1970), pp. 64–70.

[19] J. Cocke and H. G. Kolsky, "The Virtual Memory in the STRETCH Computer." *Proc. EJCC* (1959), pp. 82–93.

[20] C. J. Conti, D. H. Gibson, and S. H. Pitkowsky, "Structural Aspects of the System/360 Model 85, Part I, General Organization." *IBM Systems Journal*, Vol. 7, No. 1 (1968), pp. 2–14.

[21] T. D. Friedman and S. C. Yang, "Methods Used in an Automatic Logic Design Generator (ALERT)." *IEEE Trans. on Computers*, Vol. C-18, No. 7 (July, 1969), pp. 593–614.

chapter two

LOGIC SYNTHESIS

MELVIN A. BREUER

Electrical Engineering Department
University of Southern California
Los Angeles, California

2.1 INTRODUCTION

In this chapter we will consider the problem of synthesizing Boolean switching functions. We will be primarily concerned with the problems of minimization, synthesis to satisfy fan-in and fan-out constraints, and converting from one family of logic gates (such as AND-OR) to another (such as NAND).

The logical structure of a digital device can be described at many different levels. At a high level, one can express the transfer of information using a register transfer language [21]. For example, consider the two statements

DECLARE F/F REGISTERS: $A[32]$, $B[32]$, $OP[4]$, $T[5]$.

$(OP = 8) \cdot (T = 3)$: $A + B \longrightarrow B, T + 1 \longrightarrow T$.

The first statement declares the existence of certain flip-flop registers, e.g., A is a 32-bit register. The second statement implies that, when the four-bit operation code register OP is in state 8 (binary 1000), and the five-bit phase counter T is in state 3 (binary 00011), then the following two operations should occur concurrently:

1. The sum of the current contents of registers A and B should replace the contents of register B;
2. Register T should be incremented by 1.

It is possible to describe the functional operation of an entire digital system in terms of a suitable high level language. This has been done, for example, with the IBM 360 series of computers, using APL as the descriptive language [26].

At a lower level of description, one can describe a system in the language of Boolean equations.† For example, the application equation for the least significant sum bit for the B register in our previous example is

$$B_{32}^{\tau+1} = [(A_{32} \cdot \bar{B}_{32} + \bar{A}_{32} \cdot B_{32}) \cdot X \cdot Y]^{\tau}$$

where

$$X = OP_1 \cdot \overline{OP}_2 \cdot \overline{OP}_3 \cdot \overline{OP}_4$$

and

$$Y = \bar{T}_1 \cdot \bar{T}_2 \cdot \bar{T}_3 \cdot T_4 \cdot T_5$$

Many logical designs of digital systems are first carried out at this level of description, sometimes with the aid of timing diagrams. The equation for the sum was obtained by knowing explicitly the Boolean switching function implied by the plus sign ($+$) in the expression $A + B$. This function, for two inputs, is shown in Table 2-1.

TABLE 2-1. Two Bit Sum (S) and Carry (C) Functions ($S = x_1 \cdot \bar{x}_2 + \bar{x}_1 \cdot x_2$; $C = x_1 \cdot x_2$)

x_1	x_2	S	C
0	0	0	0
1	0	1	0
0	1	1	0
1	1	0	1

So far the information expressed by the register transfer statements, Boolean equations and Boolean functions have indicated only the logical structure of the machine and not its gate structure. Once a given circuit family of components is selected, it is possible to describe the logical opera-

†We assume the reader has a basic knowledge of Boolean functions and expressions and their simplification.

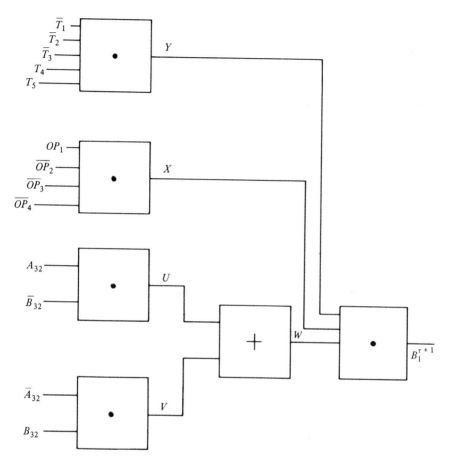

Fig. 2-1. Gate realization of $B_{32}^{\tau+1}$.

tion of a system at a gate level. For example, using AND and OR gates, we can implement our equation for $B_{32}^{\tau+1}$ as shown in Fig. 2-1. This circuit can be described in a gate level language where each gate is uniquely specified as to its inputs, its name, and a part number. For the circuit of Fig. 2-1 we may write

TYPE AND; INPUTS $\bar{T}_1, \bar{T}_2, \bar{T}_3, T_4, T_5$; OUTPUT Y

⋮

TYPE OR; INPUTS U, V; OUTPUT W

⋮

Note that there are many different circuits using AND and OR gates which will realize the function $B_{32}^{\tau+1}$. It is usually desirable to find that circuit which requires the least number of gates or inputs. Also, the circuit must satisfy all hardware circuit constraints such as fan-in, fan-out, and delay. These problems are of primary concern to logic designers.

An example of a hypothetical circuit family is shown in Table 2-2.

TABLE 2-2. A Circuit Family

Gate Type	Fan-In	Fan-Out	Normalized Cost	Delay
AND†	4	4	1	1
$NAND_1$	4	4	1	1
$NAND_2$	5	7	2	2
NOR	5	5	1	1
LATCH‡	2	7, 7	4	3

† No AND-AND gating allowed.
‡ 1-1 input not allowed.

We can now define the logic synthesis problem as follows:

1. Given a circuit family of components, including all constraints and circuit limitations associated with each circuit element type.
2. Given a logical description of a digital system in some language, such as the language of register transfers, Boolean equations or functions, or even gate equations.
3. Realize the system described in item 2 using components given in item 1, and in addition minimize the total implementation cost. This cost consists of both the circuit costs and the per-unit design costs. Usually these two costs are inversely proportional to each other.

We have described the synthesis process as implementing a given digital system in terms of elements from a given circuit family. This is analogous to the problem of machine-language translation, e.g., the compilation of a FORTRAN program into machine code. Here the FORTRAN program corresponds to a set of register transfer statements, and the instruction set of the object computer corresponds to the given circuit family. Table 2-3 further illustrates the analogy between these two problems.

Just as the aim of a compiler is to automatically translate a program from one language into another, so the aim of a synthesis system is to translate a given system (behavioral) description into an equivalent gate (structural) level description. Much work has been carried out in the area of developing algorithms for logic synthesis. Most of this work has dealt with the problem of gate reduction or simplification of Boolean switching expressions. Un-

TABLE 2-3. ANALOGY BETWEEN SYNTHESIS AND MACHINE TRANSLATION

Synthesis	Machine Translation

1. Register transfer language
   Example: $A \cdot B \rightarrow C$
	1. High level language (e.g., FORTRAN)
	   Example: $C = A.\text{AND}.B$

2. Boolean expression
   Example: AB
	2. Macro assembly language
	   Example: AND A, B

3. Synthesized expression
   Example: TYPE AND; INPUTS A, B; OUTPUT C.
	3. Object or assembly code
	   Example: CAL A
	   ANA B
	   SLW C

4. Logic simplification	4. Code optimization
5. Design time	5. Compilation time
6. Circuit costs	6. Execution time, memory requirements
7. Circuit constraints, e.g., fan-in, fan-out	7. Code restrictions, e.g., number of addresses
8. Conversion between different circuit families	8. Conversion from one object code to another

fortunately, this is only a small part of the overall synthesis problem. Very little effort has gone into developing algorithms for handling the various other problems related to synthesis. Hence, we now find that many aspects of this area appear to be still in their infancy, and more work needs to be done before a totally automated synthesis system is developed which will be capable of synthesizing circuits as efficiently as those obtained by logic designers. Currently, the use of automated synthesis systems has been typically restricted to the design of rather special functions such as instruction decoders, character recognition devices, and other applications usually dealing with decoding complex functions.

Much of the early work in the area of automated synthesis aids actually dealt with design verification and documentation. Logic verification systems checked that all fan-in, fan-out, and delay constraints for a given circuit were satisfied. Such systems could also check that all load resistors were inserted when necessary and that every signal source was properly defined. More advanced systems were capable of checking for race and logic hazards, while others were concerned with generating logic documentation.

It is interesting to note that, while almost all digital systems are sequential, most of the logical design process deals with the synthesis of combinational logic circuits. Due to this fact, in this chapter we will restrict our attention to the synthesis of combinational circuits. We will also exclude register transfer language statements as our source language because we are primarily concerned with gate level problems and not in the problem of translating register transfer statements into, say, Boolean equations. We will therefore confine our source language to be either Boolean equations of functions. For further information on the translation of register transfer statements, see references [22], [28], [29], [35], [69], [71].

In this chapter we have divided the synthesis problem into three major subproblems: (a) logical simplification or gate count reduction; (b) design to satisfy circuit constraints such as fan-in and fan-out; and (c) conversion between different circuit families. We will present algorithms to carry out single- and multiple-output simplification, as well as present factorization algorithms to synthesize fan-in and fan-out limited multiple-output NAND networks. We will also discuss techniques for converting a network from one circuit family to another. Finally, we will briefly describe a data representation for computer implementation of these algorithms.

2.2 NOTATION, DEFINITIONS, AND BASIC RESULTS

In this section we will briefly review some basic concepts of Boolean switching functions. We will use two equivalent notations: the normal form Boolean switching expression and a cubical representation. For example, the expression $x_1\bar{x}_2 + x_1x_3$ is represented in its cubical form as $\{10x, 1x1\}$. The cubical notation and associated operators to be presented are due primarily to the work of Roth [77]–[79]. This notation is used because it is concise, it lends itself to direct, easy computer implementation using arrays, and because functions represented in this notation can be efficiently manipulated by a set of operators. By means of these operators, many synthesis algorithms can be easily formulated. We will now formalize these concepts, many of which will be illustrated in Fig. 2-2 and 2-3.

Let $B = \{0, 1\}$ and B^n be the set of all $(0, 1)^n$ tuples. Then a *single-output Boolean switching function f* of n variables is a mapping from S into B, where $S \subseteq B^n$. If $S = B^n$ then f is said to be a *totally specified function;* otherwise it is a *partially specified function*. Those n-tuples which map into 1 are referred to as the *care, ON*, or *true vertices* of f, those which map into 0 are the *OFF* or *false verticies* of f, and those in $B^n - S$ are referred to as *don't care vertices* of f. Don't care conditions occur in digital design due to a number of reasons, such as when some input n-tuples cannot occur, or when the response to some input n-tuples is immaterial. Since combinational circuits define completely specified switching functions, when synthesizing a partially specified switching function each don't care vertex must be changed to either a true or false vertex. It is convenient to represent a partially specified function by the mapping $f: B^n \rightarrow \{0, 1, d\}$, where each don't care vertex maps onto d. (See Fig. 2-2a, b, and c.)

An n-tuple $c = (c_1c_2 \ldots c_n)$, where $c_i \in \{0, 1, x\}$, is said to be a *cube*.

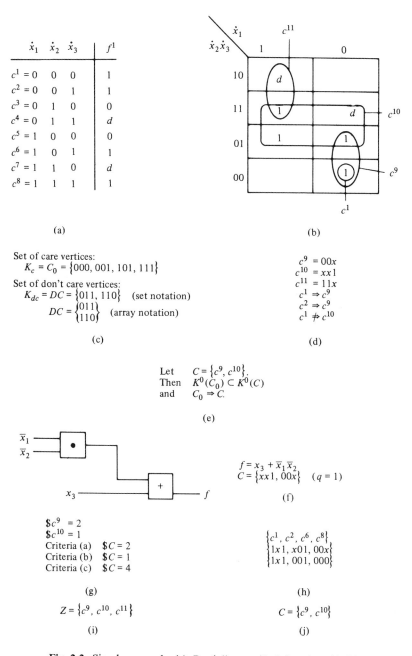

Fig. 2-2. Simple example (a) Partially specified function f^1 (b) Karnaugh chart for f^1 (c) Set of care and don't care vertices (d) Examples of subsuming (e) Example of set covering (f) Two-level AND-OR circuit (g) Circuit cost (h) Three possible initial covers for K_c (i) Set of prime implicants (j) Minimum cost cover.

As previously stated, a cube c defines a Boolean *product term* $P(c)$, and vice versa. For example, $P(10x0) = x_1 \bar{x}_2 \bar{x}_4$, where $\dot{x}_1, \dot{x}_2, \ldots, \dot{x}_n$ are the n variables of a switching function, and x_i and $\bar{x}_i$ are said to be the true and false *literals* associated with variable $\dot{x}_i$. In general, if we set $\dot{x}_i^1 = x_i$, $\dot{x}_i^0 = \bar{x}_i$, and $\dot{x}_i^x = 1$, then $P(c) = \prod_{i=1}^{n} \dot{x}_i^{c_i}$. If r elements of c are x's, we say that c is an r-cube. An r-cube is said to cover or contain 2^r 0-cubes, namely, all those 0-cubes which can be obtained from c by replacing the x's by 0's and 1's. For example, $1xx$ covers the set of cubes $\{111, 110, 101, 100\}$ or, equivalently, $x_1 = x_1 x_2 x_3 + x_1 x_2 \bar{x}_3 + x_1 \bar{x}_2 x_3 + x_1 \bar{x}_2 \bar{x}_3$. We say that a cube c *subsumes*† cube c', written $c \Rightarrow c'$, if and only if all the 0-cubes covered by c are also covered by c'. For example, $11x$ subsumes $1xx$ and, equivalently, $x_1 x_2$ subsumes x_1. (See Fig. 2-2b and d.)

Let C be a set of cubes. Then $K^0(C)$, the *0-complex* defined by C, is the set of all 0-cubes, each of which is covered by some element of C. C is said to *cover* $K^0(C)$. In general, a set of 0-cubes can have many different covers. If C and C' are two sets of cubes, then we say that C' is *covered* by C, denoted by $C' \Rightarrow C$, if and only if $K^0(C') \subseteq K^0(C)$.‡ (See Fig. 2-2e.)

A set of cubes $C = \{c^1, c^2, \ldots, c^p\}$ uniquely defines a two-level AND-OR(NAND-NAND) implementation of some function f. This implementation is obtained as follows:

1. For each c^i which is not an $(n-1)$-cube, we require an AND gate whose input is c^i and whose output is $P(c^i)$.

2. If $p > 1$ we require one OR gate whose inputs are the output of each AND gate and also each $(n-1)$-cube c^j.

The output of this OR gate is $P(C) = \sum_{i=1}^{p} P(c^i)$ which is a *normal form* Boolean expression representation for f. If c^i is a 0-cube, then $P(c^i)$ is called a *minterm*; if all $c^i \in C$ are 0-cubes, then $P(C)$ is said to be in *canonical normal form*. If there are q $(n-1)$-cubes in C, where $p > q$, then $p - q + 1$ gates are required for this circuit. (See Fig. 2-2f.)

In the algorithms to be presented, C is usually considered to be a set; however, in the actual computer implementation of an algorithm, C is represented by an array. In this chapter, these two representations are taken to be equivalent.

---

†If c subsumes c', then equivalently we can say that c *implies* c' or c is *contained in* or is *covered by* c. Also c' is said to *cover* c. If P and Q are two Boolean expressions, we say that P subsumes or implies Q if and only if $(P = 1)$ implies $(Q = 1)$.

‡Note the distinction between $C \Rightarrow C'$ and $C \subseteq C'$. The symbol, $\subseteq$, refers to set containment. Hence $(C \subseteq C')$ imples $(C \Rightarrow C')$, but not vice versa. For example, let $C = \{1\}$ and $C' = \{x\}$. Then $C \Rightarrow C'$ but $C \not\subseteq C'$. Finally, by $C - C'$ we mean set substraction, i.e., $C - C' = C - C \cap C' = \{c \mid c \in C \text{ and } c \notin C'\}$. For example, $\{11x, 101\} - \{101, x1x\} = \{11x\}$.

The cost associated with an r-cube, c, is denoted by $\$c$, where $\$c = n - r$, i.e., $\$c$ equals the number of literals in $P(c)$. Classically, one of the following three cost criteria is used for defining the cost $\$C$ associated with the two-level AND-OR realization of C:

1. Number of inputs to the AND gates,† i.e.,

$$\$C = \sum_{c \in C} \$c - q$$

2. Number of AND gates, i.e.,

$$\$C = |C| - q$$

where $|C|$ denotes the total number of cubes in C.

3. Total number of inputs to all AND gates and the OR gate, i.e., (See Fig. 2-2g.)

$$\$C = \sum_{c \in C} \$c + |C| - q$$

We can associate with a switching function f a *care complex* $K_c(f) = K_c$, which consists of the set of all true vertices of f, and a *don't care complex* $K_{dc}(f) = K_{dc}$, which consists of the set of all don't care vertices of f. Let C_0 be an initial cover of K_c, and DC an initial cover of K_{dc}, with the properties that $K_c = K^0(C_0)$ and $K_{dc} = K^0(DC)$. (In general, C_0 and DC need not consist solely of 0-cubes. See Fig. 2-2h.) Then the classical switching function simplification problem reduces to finding a minimum cost cover C of f such that

$$C_0 \Longrightarrow C \Longrightarrow C_0 \cup DC$$

or equivalently

$$K^0(C_0) \subseteq K^0(C) \subseteq K^0(C_0 \cup DC)$$

A cube c is said to be a *prime implicant* of a function f if and only if $c \Rightarrow C_0 \cup DC$, $c \not\Rightarrow DC$, and there exists no cube c' such that $c \Rightarrow c'$ and $c' \Rightarrow C_0 \cup DC$. That is, c implies the function, and no large cube containing c implies the function. Let Z be the set of all prime implicants of f. (See Fig. 2-2i.) Then for all cost functions which increase monotonically with $\$c$ and $|C|$, we have the following fundamental switching theory result.

Theorem 1: *If C is a minimum cost cover of f, then $C \subseteq Z$.* (See Fig. 2-2j.)

It is due to this result that much of the work on synthesis has actually dealt with the problem of constructing the set Z.

We have introduced the concept of cubical complexes and illustrated their

---

†The number of inputs to a circuit is a measure of the number of pins required to connect this circuit to another circuit.

relationship to Boolean functions and their implementation. We will now introduce a number of binary cube operators and relations, namely, the intersect operator, $\cap$, the subsume relation, $\Rightarrow$, the sharp product, $\#$, and the consensus operator, $\rlap{/}{c}$. In the remainder of this section we will let $a = (a_1 a_2 \ldots a_n)$ and $b = (b_1 b_2 \ldots b_n)$ be two arbitrary cubes. In Fig. 2-3 we illustrate examples of these various operations.

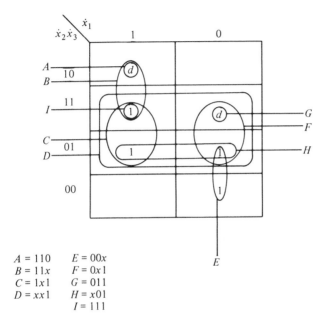

$A = 110$ $E = 00x$
$B = 11x$ $F = 0x1$
$C = 1x1$ $G = 011$
$D = xx1$ $H = x01$
             $I = 111$

Consensus	Sharp product	Intersect	Subsuming
$C \rlap{/}{c} F = D$	$F \# D = \psi$	$A \cap B = A$	$C \Rightarrow D$
$A \rlap{/}{c} C = B$	$D \# F = C$	$B \cap C = I$	$A \Rightarrow B$
$D \rlap{/}{c} E = \psi$	$D \# A = D$	$C \cap F = \psi$	$S[\{C, D, E\}] = \{D, E\}$
	$D \# B = \{H, F\}$		$S[\{A, B, \ldots, I\}] = \{B, D, E\}$
	$B \# D = A$		$S[\{B, C, H,\}] \neq \{B, H\}$

Fig. 2-3. Examples of cubical operations.

Subsuming. Recall that $a \Rightarrow b$ if and only if $K^0(a) \subseteq K^0(b)$. Hence, if C is a cover of a complex K^0, and if $a \Rightarrow b$ where $a, b \in C$, then $C - a$ is also a cover of K^0. Let C' be obtained from C by deleting all cubes $a \in C$ such that $a \Rightarrow b$ for some $b \in C$, where $a \neq b$. This operation of deriving C' from C is called *subsuming*, and we denote it by writing $C' = S[C]$. The relation, $\Rightarrow$, is reflexive and transitive, and if $a \Rightarrow b$, then $\$a \geq \$b$. Note that $a \Rightarrow b$ if and only if $a_i = b_i$ whenever $b_i \in \{0, 1\}$.

Intersection. It is often necessary to evaluate the logical conjunction of two product terms, e.g., $(x_1 x_2) \cdot (x_2 \bar{x}_3) = x_1 x_2 \bar{x}_3$ and $(x_1 x_2) \cdot (\bar{x}_2 \bar{x}_3) = 0$.

Sec. 2.2   Notation, Definitions, and Basic Results   31

In general, $P(a) \cdot P(b)$ is the largest product† which subsumes both $P(a)$ and $P(b)$. If $P(c) = P(a) \cdot P(b)$, then cube c is said to be the *intersect* of a and b, denoted by $a \cap b$, i.e., $c = a \cap b$. The intersect operator, $\cap$, is defined via the coordinate table:‡

		b_i		
$\cap$	0	1	x
0	0	Ω	0
a_i 1	Ω	1	1
x	0	1	x

Then $a \cap b = \psi$ if $a_i \cap b_i = \Omega$ for any i; otherwise $a \cap b = (a_1 \cap b_1, a_2 \cap b_2, \ldots, a_n \cap b_n)$. Note that $\cap$ is associative, commutative, and distributive. For our two previous examples, we obtain $11x \cap x10 = 110$ and $11x \cap x00 = \psi$. We can extend the intersect operator to sets by defining

$$c' \cap C = S[\bigcup_{c \in C} c' \cap c]$$

and

$$C' \cap C = S[\bigcup_{c' \in C'} (c' \cap C)]$$

Sharp-Product. If $f = x_1 \bar{x}_3 x_4$§, then by DeMorgan's theorem¶ we have that $\bar{f} = \overline{x_1 \bar{x}_3 x_4} = \bar{x}_1 + x_3 + \bar{x}_4$. Each term in this expression for $\bar{f}$ is a prime implicant of $\bar{f}$, and there are no others. Similarly, let $f = (x_1 x_2) \cdot \overline{(x_2 x_3 x_4)}$. Expanding this expression we obtain

$$f = x_1 x_2 \cdot (\bar{x}_2 + \bar{x}_3 + \bar{x}_4)$$
$$= x_1 x_2 \bar{x}_3 + x_1 x_2 \bar{x}_4$$

Again, each term in this expression is a prime implicant of f, and there are no others.

We now define the sharp product (#-product) between two cubes a and b, denoted by $a \# b$, to be the set of cubes such that $P(a \# b)$ is the set of all prime implicants of the function defined by $P(a) \overline{P(b)}$. Equivalently, $a \# b$ is the set of all prime implicants for the complex $K^0(a) - [K^0(a) \cap K^0(b)]$.

---

†Q is said to be the largest product having property P if no other product R exists which has property P and such that Q subsumes R.

‡The null cube is denoted by ψ, where $P(\psi) = 0$. The entire n-dimensional unit cube is denoted by $U_n = xx \ldots x$, where $P(U_n) = 1$. The empty set is denoted by φ. Finally, Ω is used to denote the result of an undefined operation, and we define $\{\psi\}$ to equal φ.

§Since an expression E uniquely defines a function f, we will use the notation $f = E$.

¶$\overline{A \cdot B} = \bar{A} + \bar{B}$ and $\overline{A + B} = \bar{A} \cdot \bar{B}$.

If $K^0(a) \cap K^0(b) = \varphi$, then $a \# b = a$; and if $K^0(a) \subseteq K^0(b)$, then $a \# b = \psi$.

Algebraically, $a \# b$ is defined according to the following coordinate table:

			b_i	
	#	0	1	x
	0	z	y	z
a_i	1	y	z	z
	x	1	0	z

We have that $a \# b = a$ if $a_i \# b_i = y$ for any i; $a \# b = \psi$ if $a_i \# b_i = z$ for all i; otherwise, $a \# b = \bigcup_i (a_1 a_2 \ldots a_{i-1} \alpha_i a_{i+1} \ldots a_n)$, where $a_i \# b_i = \alpha_i \in \{0, 1\}$, and the union runs over all such i.

For example, to compute $1xx \# x01$, we first note that $a_i \# b_i \neq y$ for any i. For $i = 1$, $a_i \# b_i = 1 \# x = z$; hence, no cube is formed. For $i = 2$, $a_i \# b_i = x \# 0 = 1$; hence, we form the cube $a_1 1 a_3 = 11x$. For $i = 3$, $a_i \# b_i = x \# 1 = 0$; hence, we form the cube $a_1 a_2 0 = 1x0$. For $a = 1xx$ and $b = x01$, we obtain $P(a \# b) = P(a) \overline{P(b)} = x_1 \cdot (\bar{x}_2 x_3) = x_1 x_2 + x_1 \bar{x}_3$. Note that the definition of the #-product follows directly from DeMorgan's theorem for complementation.

By definition, $C' \# b = S[\bigcup_{c' \in C'} c' \# b]$, and $a \# C = S[(\ldots (a \# c^{i_1}) \# c^{i_2}) \ldots) \# c^{i_p})]$, where $i_1, i_2, \ldots, i_p$ is any permutation of $1, 2, \ldots, p$. Also $C' \# C = S[\bigcup_{c' \in C'} c' \# C]$. The #-operation is noncommutative, nonassociative, but distributive over $\cap$ and $\cup$. Sharping can be thought of as a subtraction process because $C' \# C$ is a cover for all vertices in C' which are not in C.

Consensus. Given a set of cubes C which covers a complex K, it is sometimes desirable to generate new cubes which can be elements of a cover of K. One way of generating such cubes is by employing the consensus operation ¢ of Quine [72]. The consensus of two product terms P and Q is the largest product R such that R does not imply either P or Q, but R implies $P + Q$. The consensus of P and Q is defined only for the case where there exists exactly one i such that x_i is a literal in P and $\bar{x}_i$ is a literal in Q. In this case we can write $P = x_i \cdot P'$ and $Q = \bar{x}_i \cdot Q'$, where P' and Q' are not functions of $\dot{x}_i$, and therefore the consensus of P and Q is $P'Q'$. For example, the consensus of $\bar{x}_1 \bar{x}_2$ and $x_1 x_2$ is not defined; the consensus of $\bar{x}_1 \bar{x}_2$ and $x_1 x_3 x_4$ is $\bar{x}_2 x_3 x_4$; and the consensus of $\bar{x}_1 \bar{x}_2$ and $x_3 x_4$ is not defined. If the consensus of $P(a)$ and $P(b)$ is $P(c)$, then we write $c = a \notcent b$. The consensus operator ¢ is commutative but nonassociative, and it is defined by the coordinate table:

$\cent$	0	1	x
0	0	y	0
1	y	1	1
x	0	1	x

(with a_i on rows and b_i on columns)

If $a_i \cent b_i = y$ for exactly one i, then $a \cent b = (m(a_1 \cent b_1), m(a_2 \cent b_2), \ldots, m(a_n \cent b_n))$, where $m(0) = 0$, $m(1) = 1$, and $m(x) = m(y) = x$; otherwise, $a \cent b$ is undefined. Examples are

$$011 \cent 111 = x11$$
$$x11 \cent x00 = \Omega \quad \text{(undefined)}$$
$$x01 \cent 1x0 = 10x$$

Note that, if $a \cent b = c$, then

$$\max(\$a, \$b) - 1 \leq \$c \leq n - 1$$

We define $C \cent C$ by the equation

$$C \cent C = \{c^i \cent c^j \mid c^i, c^j \in C, i < j\}$$

To compute $C \cent C$ one needs to compute the consensus of $p(p - 1)/2$ pairs of cubes, where $|C| = p$.

2.3 COVERING A SINGLE-OUTPUTS SWITCHING FUNCTION

In this section we will consider the classical problem of finding a minimum cost cover for a single-output switching function. For a discrete component technology, one reason for seeking a minimum cost cover is to reduce the cost of the circuit. With modern LSI, MOS, and IC technologies, the cost of a single gate, transistor, or diode is not too significant. What is important is circuit area and the interconnection between gates and modules. The number of interconnections is important in these technologies since, to a large extent, it dictates the final size of a circuit module. Pin connections also contribute to a large percentage of malfunctions and manufacturing costs. Most modules turn out to be pin-limited rather than gate-limited. Also, due to the many steps involved in the manufacturing process, it is important to decrease circuit area in order to increase the production yield. Hence, logic simplification is important.

We have previously defined the classical switching function simplification problem, and we have defined the total synthesis cost as consisting of two components, the design cost and the circuit cost C. The design cost is related to the algorithm employed and the details of how it is implemented on a computer. It is possible to obtain algorithms which execute faster and/or require less memory if we do not require a minimum cost cover, but only an irredundant cover. A cover C of f is said to be *minimal* or *irredundant* if each cube in C is a prime implicant of f, and if no proper subset of C is a cover of f. Hence, one can form a computational hierarchy of procedures for simplifying switching functions, where some of the parameters differentiating the various algorithms are computation time, memory requirements, and optimality criteria used such as minimum vs irredundant. When developing an algorithm to simplify a function of a large number of variables, such considerations are very important because the computational complexity of these problems often grows exponentially with the number of variables.†

In this section we will discuss a few such algorithms for solving this simplification problem. We have divided the discussion of this problem into three topics, namely, the generation of the set Z of prime implicants, the selection of a minimum cost cover C^*, and the construction of an irredundant cover.

We will assume that the function f for which a cover is to be found is defined by two sets of cubes, C_0 and DC, which represent an initial cover for the care and don't care complexes, respectively.

2.3.1. Generation of the Set Z of All Prime Implicants

Let $U_n = xx \ldots x$ be the n-cube which covers the set of all 2^n 0-cubes. Then $U_n \# C$ [25], [77], is the set of all prime implicants of the complement of the function defined by C, and therefore $U_n \# (U_n \# C)$ is the set Z of all prime implicants of the function defined by C‡. (See problems 5 and 6.)

Algorithm 1: Generation of Z via $\#$-product
Step 1: $Z' = U_n \# (U_n \# (C_0 \cup DC))$.
Step 2: $Z = \{c \mid c \in Z' \text{ and } c \cap C_0 \neq \psi\}$.

---

†See references [57]–[59] for statistical properties of the number of prime implicants and minterms of Boolean switching functions.

‡It is sufficient to construct a cover of the complement of the function, not all prime implicants. To reduce the computation and memory requirements, a modified $\#$-product ($\#'$) [25] can be used for computing a prime implicant cover, but not necessarily all prime implicants. Hence, we can write $Z = U_n \# (U_n \#' C)$. This algorithm is given in Appendix A. In addition $Z = U_n \# F_0$, where F_0 is an initial cover for the false vertices of f.

Step 2 is necessary since a cube which is totally contained in K_{dc} is not considered a prime implicant of f. Note that the result will be correct even if $K^0(C_0) \cap K^0(DC) \neq \varphi$ and if we assume each cube in this intersect is in the care complex K_c.

As an example of this procedure, let $C_0 = \{10, 01\}$ and $DC = \{11\}$. Then $U_n \# (C_0 \cup DC)$ is computed as follows

$$(((xx \# 10) \# 01) \# 11) = (\{0x, x1\} \# 01) \# 11 = \{00, 11\} \# 11 = \{00\}$$

Therefore, $Z = xx \# 00 = \{1x, x1\}$.

In many cases, the set of prime implicants $U_n \# (C_0 \cup DC)$ (or $U_n \#' (C_0 \cup DC)$) of the complement function is quite large and it is advantageous to avoid its enumeration. This can be accomplished by employing the consensus operator to generate Z directly.

Recall that if C is a cover for f, then so is $C \cup (C \not\subset C)$.

Algorithm 2: Generation of Z via iterative consensus

Step 1: Set $C' = S[C \cup (C \not\subset C)]$.

Step 2: If sets C and C' are identical, go to step 3; otherwise replace C by C' and return to step 1.

Step 3: Set $Z = \{c \mid c \in C' \text{ and } c \cap C_0 \neq \psi\}$.

As the set C becomes large, $C \not\subset C$ requires an increasing amount of computation time and storage. The following procedure generates the prime implicants in the order of increasing dimension [25], [60].

In this algorithm we will evaluate Z according to the identity $Z = \bigcup_{r=0}^{n} Z^r$, where Z^r is the set of all prime implicants of f which are r-cubes. Let $A_0 = S[C_0 \cup DC]$. If 0-cube $a \in A_0$ is not a prime implicant, then there must exist and $a' \in A_0$ such that $a \not\subset a'$ is a 1-cube. We then have that Z^0 is the set of 0-cubes $a \in A_0$ such that $a \cap C_0 \neq \psi$, and no element in the set $a \not\subset A_0$ is a 1-cube. Note that, since a is a 0-cube, $c = a \not\subset b$ is at most a 1-cube. Therefore, if there are any 0-cubes which are prime implicants, they must be in A_0 and they cannot be contained in any higher order cube also contained in the function.

We now iterate by setting

$$A_1 = S[A_0 \cup (A_0 \not\subset A_0)] - Z^0$$

from which Z^1 is determined.

The general algorithm is as follows:

Algorithm 3: Generalized Quine procedure for generating Z

Step 1: $A_0 = S[C_0 \cup DC]$.

Step 2: For $r = 0, 1, \ldots$, we have:
  (a) Z^r is the set of all r-cubes $a \in A_r$ such that $a \cap C_0 \neq \psi$ and no element in the set $a \not{c} A_r$ is an $(r+1)$-cube.
  (b) $A_{r+1} = S[A_r \cup (A_r \not{c} A_r)] - Z^r$
  (c) If $A_{r+1} = \varphi$, the process can be terminated since $Z^s = \varphi$ for all $r + 1 \leq s \leq n$.

If A_0 consists of only 0-cubes, then this algorithm is analogous to the well-known Quine–McCluskey procedure [51], [52]. Its main advantage is that one can compute Z without enumerating the minterms of the function. This is important because a function of n variables can contain up to 2^n minterms.

To illustrate this algorithm, let $C_0 = \{110, 0x1\}$ and $DC = \{101\}$. Then $A_0 = \{110, 0x1, 101\}$ and

$$A_0 \not{c} A_0 = \begin{Bmatrix} 110 \not{c} 0x1 \\ 110 \not{c} 101 \\ 0x1 \not{c} 101 \end{Bmatrix} = \begin{Bmatrix} \psi \\ \psi \\ x01 \end{Bmatrix}$$

Now $110 \not{c} A_0$ does not contain any 1-cubes, while $101 \not{c} 0x1 = x01$; hence, $Z^0 = \{110\}$. Now

$$A_1 = S[\{110, 0x1, 101, x01\}] - \{110\} = \{0x1, x01\}$$

Since $A_1 \not{c} A_1 = \varphi$, $Z^1 = A_1$, $A_2 = \varphi$, and we have

$$Z = Z^0 \cup Z^1 = \begin{Bmatrix} 110 \\ 0x1 \\ x01 \end{Bmatrix}$$

In the Quine-McCluskey procedure for generating Z, the consensus operation is again employed, starting with $C = K_c \cup K_{dc}$. Now, however, only cubes of the same size are compared, i.e., we form the consensus of c and c' only if $\$c = \$c'$. Note that if $\$c = \$c'$ and if $c \not{c} c' \neq \psi$, then the number of one components in c and c' must differ by exactly one. Therefore, some procedures sort the c on the number of one component in c prior to applying the consensus operation.

A particular efficient implementation of this technique has been reported by Morreale [61]–[63]. He employs a procedure referred to as a *partitioned list* which we will now illustrate on a four-variable problem.

For this technique we use an operator $\not{c}_i$. This operator is similar to $\not{c}$ except that only position i is inspected for the true (x_i) and false ($\bar{x}_i$) forms

Sec. 2.3 Covering a Single-Outputs Switching Function 37

of the variable $\dot{x}_i$. That is, $c \not\phi_i c' = c \phi c'$ if $P(c) = x_i R$ and $P(c') = \bar{x}_i S$; otherwise, $c \not\phi_i c' = \psi$. Figure 2-4 illustrates the order of applying the consensus operators ϕ_i. The notation $C^k \xrightarrow{\phi_i} C^j$ implies that the set C^j is obtained from set C^k by applying ϕ_i, i.e., $C^j = \{c \not\phi_i c' \mid c, c' \in C^k\}$.

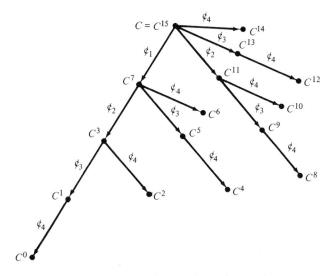

Fig. 2-4. Order for applying ϕ_i for $n = 4$.

As an example, consider the initial set

$$C = C^{15} = \begin{Bmatrix} 0 & 0 & 0 & 1 \\ 0 & 0 & 1 & 0 \\ 0 & 0 & 1 & 1 \\ 0 & 1 & 1 & 1 \\ 1 & 0 & 0 & 0 \\ 1 & 0 & 1 & 0 \\ 1 & 0 & 1 & 1 \\ 1 & 1 & 0 & 0 \\ 1 & 1 & 0 & 1 \\ 1 & 0 & 0 & 1 \end{Bmatrix} \begin{matrix} - \\ - \\ - \\ - \\ - \\ - \\ - \\ - \\ - \\ - \end{matrix}$$

Applying ϕ_4 to C^{15} produces the set

$$C^{14} = \begin{Bmatrix} 0 & 0 & 1 & x \\ 1 & 0 & 0 & x \\ 1 & 0 & 1 & x \\ 1 & 1 & 0 & x \end{Bmatrix} \begin{matrix} \checkmark \\ \checkmark \\ \checkmark \\ \checkmark \end{matrix}$$

Whenever $c \not\mathrel{\phi}_i c'$ exists, we mark (—) c and c'. Continuing, we obtain

$$C^{13} = \begin{Bmatrix} 0 & 0 & x & 1 \\ 1 & 0 & x & 0 \\ 1 & 0 & x & 1 \end{Bmatrix} \begin{matrix} \checkmark \\ — \\ — \end{matrix}$$

and $C^{12} = \{1 \ 0 \ x \ x\}$. Finally, we obtain

$$C^{11} = \begin{Bmatrix} 0 & x & 1 & 1 \\ 1 & x & 0 & 0 \\ 1 & x & 0 & 1 \end{Bmatrix} \begin{matrix} \\ — \\ — \end{matrix} \quad C^{10} = \{1 \ x \ 0 \ x\} \quad C^9 = C^8 = \varphi$$

$$C^7 = \begin{Bmatrix} x & 0 & 0 & 1 \\ x & 0 & 1 & 0 \\ x & 0 & 1 & 1 \end{Bmatrix} \begin{matrix} — \\ \\ — \end{matrix} \quad C^6 = \{x \ 0 \ 1 \ x\} \quad C^5 = \{x \ 0 \ x \ 1\}$$

and $C^4 = C^3 = C^2 = C^1 = C^0 = \varphi$.

Now, returning to C^{14}, we apply $\not\phi_3$, $\not\phi_2$, and $\not\phi_1$ and check ($\checkmark$) each unmarked cube that has a consensus with some other cube in C^{14}. Note, for example, that $001x \not\mathrel{\phi}_1 101x = x01x$; hence, we check $001x$ and $101x$. Similarly, for C^{13} we apply $\not\phi_1$, $\not\phi_2$, and $\not\phi_4$. In general, for C^k we apply all $\not\phi_i$ not yet applied in constructing C^k. All cubes remaining in the sets C^k, $k = 0, 1, \ldots, 15$, which are not marked (—) nor checked ($\checkmark$) are prime implicants.

The main attributes of this technique are that each cube generated is unique and at most n sets (arrays or lists) must be stored at any one time. Table 2-4 indicates some typical computational results obtained by this technique.

TABLE 2-4. Time to Generate Z

Number of minterms	100	200	400
t (seconds)	5	30	80

Source: Morreale [62]
Machine: IBM 7040 (about four times slower than the 7090)
Language: Algol
Comment: Ten variable problems randomly selected

2.3.2. Selection of a Minimum Cost Cover

There are two general procedures for determining a minimum cost cover: one is an extraction technique and the other is a matrix method. Both techniques are quite similar and utilize the concepts of extremals and dominance. The extraction technique also requires a branch and bound with backtracking procedure. The matrix approach can also use this procedure or employ an integer linear programming algorithm. The main difference between these two approaches is that the matrix approach requires either a partial or total enumeration of the set of care vertices K_c.

Extraction Algorithm. In the extraction algorithm [25], [38], [56], [60], [78], [79], a minimum cost cover C^* is found via an iterative process. Prime implicants are removed from Z and are either inserted into a set C, which represents a partial cover of f, or discarded. Each time a cube c is inserted into C, we replace C_0 by $C_0 \mathbin{\#} c$, and we denote this operation by $C_0 \mathbin{\#} c \rightarrow C_0$. We also delete c from Z, denoted by $Z - c \rightarrow Z$.

At a specific step in the process, $c \in Z$ is said to be an *extremal* with respect to C and Z if and only if c covers some true vertex v of f not yet covered by C and not covered by any other cube in Z. Hence, if c is an extremal, it must be made an element of C.

The concept of extremal is a generalization of the concept of an essential prime implicant, which is a prime implicant which covers a care vertex not covered by any other prime implicant.

Given a partial cover C, it is possible that some of the vertices of f covered by $c \in Z$ are already covered by C. If this is the case, then it is possible to delete some of the cubes $c \in Z$ from further consideration.

We illustrate these concepts in Fig. 2-5. Here, the cubes are indicated by lines, and we assume that they all are of the same cost. For this example, we let $C_0 = Z = \{c^1, c^2, \ldots, c^{12}\}$, as shown in Fig. 2-5a. Clearly, c^1 and c^5 are extremals and hence must be elements of our final cover. Removing these elements and the vertices they cover produces the result shown in Fig. 2-5b. Now c^3 covers the same remaining uncovered vertices as $c^2(c^4)$ covers, as well as one additional one. Since $\$c^3 = \$c^2(\$c^4)$, it is clear that a minimum cost solution can be found which does not include $c^2(c^4)$. We say that c^3 *dominates* $c^2(c^4)$. Deleting c^2 and c^4, we find that now c^3 is an extremal and hence can be deleted from Z and inserted into our partial cover C (Fig. 2-5b, c, and d).

Formally, we say that c *dominates* c' if $\$c \leq \$c'$ and if

$$c' \cap C_0 \Longrightarrow c \cap C_0$$

If c dominates c' and c' dominates c, then we can arbitrarily delete one or the other from further consideration.

In some cases, after deleting all cubes in Z dominated by another, one reaches a situation where there are no extremals. At this point an impasse has been reached, and it is necessary to employ a new technique, called *branching*, in order to continue. It is this situation which makes the covering problem computationally nontrivial.

This situation is illustrated in Fig. 2-5d. Consider cube c^7; either it is an

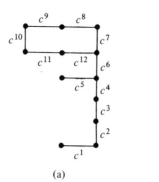
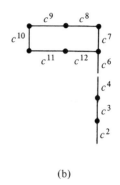

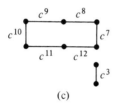
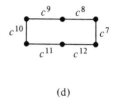

Fig. 2-5. Function to be covered.

element in some minimum cost cover or it is not. Assume c^7 is an element in a minimal cost cover. Then, placing c^7 in C, we find that c^9 dominates c^8 and c^{11} dominates c^{12}; hence, $Z = \{c^9, c^{10}, c^{11}\}$ and $C_0 \# c^7 \rightarrow C_0$. Now c^9 and c^{11} are extremals; hence, we place c^9 and c^{11} into C and set $Z = \{c^{10}\}$ and $(C_0 \# c^9) \# c^{11} \rightarrow C_0$. Since $C_0 = \varphi$, we have obtained a cover $C^{(1)}$. We must now backtrack and attempt to find a cover $C^{(2)}$ not containing c^7. In other words, deleting c^7 from the set $Z = \{c^7, c^8, c^9, c^{10}, c^{11}, c^{12}\}$ produces the extremals c^8 and c^{12}. We then find that c^{10} dominates c^9 and c^{11}; therefore, the second cover is

$$C^{(2)} = \{c^1, c^3, c^5, c^8, c^{10}, c^{12}\}$$

In this case both covers are of equal cost. In general, this is not the case, and the cover of lowest cost represents the minimum cost cover C^*. In large problems it may become necessary to branch within branches. Each time a branch occurs, the number of possible covers doubles. To reduce the total amount of computations, the computation along any branch path can be terminated when the estimated minimum cost of the corresponding cover exceeds the cost of a previously determined cover or a predefined upper bound on the cost. This technique is known as *branch and bound*.

One major disadvantage of this branching procedure is that, once computation along a path has terminated, it is necessary to backtrack to the last branch point and try and alternate path. Hence, it is necessary to store the status of the partial cover at each branch point. If there are many branches, this procedure can require a great deal of memory and computation time.

One can obtain a cover, which may not be minimum cost, by never backtracking. This approximate solution technique is very fast, greatly reduces the memory requirements, and generally produces solutions which are not far from optimal.

The complete algorithm will now be summarized.

Algorithm 4: Single-output minimum cost cover via extraction

Step 1: (Initialization) Determine C_0, DC, and an initial cost bound B. Set $C = \varphi$ and $i = 0$.

Step 2: Form the set of prime implicants Z.

Step 3: Generate the set E of extremals, where

$$E = \{c \in Z \mid [c \# (Z - c)] \cap C_0 \neq \varphi\}$$

Step 4: $C \cup E \to C, Z - E \to Z, C_0 \# E \to C_0, \varphi \to E$ where C is the new partial solution.

Step 5: (Check for bound) If $\$C > B$, go to step 10; otherwise continue.

Step 6: If $C_0 = \varphi$, save $C(C \to C^*)$ as the best solution obtained so far, set $B = \$C$, and go to step 10. If $C_0 \neq \varphi$, continue.

Step 7: If no extremals were found in step 3 and if $i \neq 0$, go to step 9; otherwise continue.

Step 8: (Dominance) Delete from Z all cubes c' in Z dominated by some other cube c in Z i.e., $\$c' \geq \$c$ and $c' \cap C_0 \Rightarrow c \cap C_0$. If at least one cube is deleted from Z, then go to step 3; otherwise continue.

Step 9: (Branching) Pick some cube $c \in Z$, say that cube of least cost. Save the current status of the solution, i.e., C, C_0, Z, c. For example, set $i = i + 1$, $C \to \alpha^i$, $C_0 \to \beta^i$, $Z \to \gamma^i$, and $c \to \delta^i$. Set $C \cup c \to C, Z - c \to Z$, and $C_0 \# c \to C_0$. Go to step 8.

Step 10: (Backtracking—retrieve last solution saved) If $i = 0$, we are done, and C^* is the minimum cost solution; otherwise, set $\alpha^i \to C$, $\beta^i \to C_0$, $\gamma^i \to Z$, $\delta^i \to c$, and $i - 1 \to i$. Set $Z - c \to Z$ and go to step 3.

There are many other variations to the procedure just described for finding a minimum cost cover. Some of these [25], [60], [92], use a technique called the *local extraction algorithm*, which does not require the initial generation of the total set of prime implicants Z but rather generates Z and C^* as the computation proceeds. For this technique, one first generates the prime implicants in the "periphery" of a prime implicant, z, and then determines whether z is an extremal or is dominated by other cubes. This procedure can sometimes significantly reduce the total storage requirements.

Matrix Algorithm. Given the set $Z = \{c^1, c^2, \ldots, c^n\}$ of prime implicants and the set $K_c = \{v^1, v^2, \ldots, v^m\}$ of care vertices of a function f, a minimum cost cover C^* can be found by solving the following integer programming form of a covering problem.
Minimize

$$\sum_{j=1}^{n} c_j x_j \tag{2-1}$$

subject to

$$\sum_{j=1}^{n} a_{ij} x_j \geq 1 \quad i = 1, 2, \ldots, m \tag{2-2}$$

where the x_j's are 0, 1 variables, and $a_{ij} = 1$ if $v^i \Rightarrow c^j$; otherwise $a_{ij} = 0$. The constant c_j is a cost assigned to x_j. Typically, c_j equals either 1, $\$c^j$, or $\$c^j + 1$. If in solving this problem we obtain $x_j = 1$, then prime implicant c^j is an element of C^*; otherwise it is not. We define matrix $A = [a_{ij}]$; A is sometimes referred to as a prime implicant table or core matrix.

In Fig. 2-6 we illustrate the A matrix for the same problem shown in Fig. 2-5a, where now every care vertex corresponds to a row in A, and every prime implicant corresponds to a column. For simplicity, the 0 elements are not shown. Since in row v^1 there is one 1 element, c^1 is an extremal. By selecting c^1 to be an element of our minimum cost cover, we also cover v^2. Hence rows v^1 and v^2 can be deleted from further consideration. Similarly c^5 is an extremal. The reduced matrix is shown in Fig. 2-6b. Note that c^3 covers all rows covered by c^2 and c^4; thus c^2 and c^4 can be deleted, in which case c^3 becomes an extremal. Also c^6 can be deleted. The reduced matrix is shown in Fig. 2-6c. This final matrix form of our problem can be solved by either a branch and bound algorithm with backtracking, as previously described, or by means of an integer programming algorithm [3], [11], [12], [30], [33]. Table 2-5 indicates some computer times required to solve this covering problem. Finally, Petrick's method [68] can be employed; this is an enumerative procedure for generating all irredundant covers of A. To illustrate this approach, note that for the matrix in Fig. 2-6c, only c^7 and c^{12} cover v^7; c^7 and c^8 cover v^8; etc. Therefore, letting c^i be a Boolean variable, a condition for covering this matrix is that

$$(c^7 + c^{12})(c^7 + c^8) \ldots (c^{11} + c^{12}) = 1$$

Sec. 2.3   Covering a Single-Outputs Switching Function   43

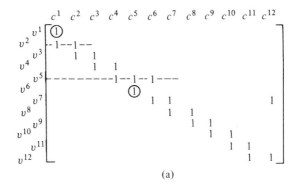

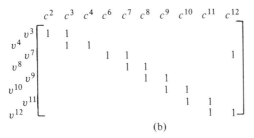

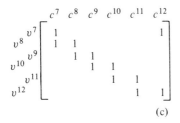

Fig. 2-6. Reduction of core matrix A.

TABLE 2-5. TIME TO SOLVE MATRIX COVERING PROBLEM

Size of Matrix	t (second)
20×20	2.3
45×45	31.2

Source: Geoffrion [30]
Machine: IBM 7090
Language: Fortran

or

$$c^7 c^9 c^{11} + c^8 c^{10} c^{12} = 1$$

Each term in this expression represents a cover.

One further technique is quite useful for reducing the A matrix. Consider the two rows shown below.

$$a^i \quad 1\ 1\ 1\ 0\ 0\ 0\ \ldots\ 0$$
$$a^j \quad 1\ 1\ 1\ 1\ 1\ 0\ \ldots\ 0$$

Note that if v^i is covered by some prime implicant, then so must v^j be covered. Thus we need not concern ourselves with row a^j, and it can be deleted from A. We will now summarize the three rules used for reducing the A matrix.

Rule 1. (Extremals) If row a^i has a single 1 element in, say, column j, then c^j is an extremal and all rows covered by c^j can be deleted from A.

Rule 2. (Row dominance) If there exists two rows a^i and a^k such that $(a_{ij} = 1)$ implies $(a_{kj} = 1)$ for all j, then row a^k can be deleted.

Rule 3. (Column dominance) If there exist two columns j and k such that $(a_{ji} = 1)$ implies $(a_{ki} = 1)$ for all i, and if $c_k \leq c_j$, then column j can be deleted.

For a more detailed discussion of the technique for simplifying the system defined by (2-1) and (2-2), see references [8], [31], [34], [42], [47], [54].

2.3.3. Irredundant Covers

Recall that a cover C is said to be an irredundant cover of f if $C \subseteq Z$ and if, for each $c \in C$, $(C - c)$ is not a cover of f. Referring to Fig. 2-7,

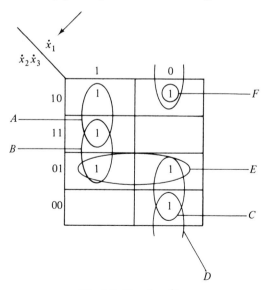

Fig. 2-7. Function f^2.

we see that $\{A, B, C, D\}$ is an irredundant nonminimum cost cover. To reduce computation time and storage, one can seek an irredundant cover, sometimes called a minimal cost cover, rather than a minimum cost cover. In this section we will outline an algorithm which produces an irredundant cover from an initial cover. The procedure consists of a cube (clause) deletion operation and an element (literal) deletion operation.

Given the cover $\{A, B, C, D, E\}$ for f^2, cube B can be deleted since all 0-cubes in the care complex covered by B are covered by the remaining cubes, i.e., B is not an extremal with respect to $\{A, B, C, D, E\}$ and f^2.

Consider the cover $\{A, C, E, F\}$, where $F = \bar{x}_1 x_2 \bar{x}_3$. Can the clause F in this cover be replaced by $F' = \bar{x}_1 \bar{x}_3$? Note that $\bar{x}_1 \bar{x}_3 = \bar{x}_1 x_2 \bar{x}_3 + \bar{x}_1 \bar{x}_2 \bar{x}_3$, and therefore F can be replaced by F' if and only if $\{A, C, E, F, G\}$ is also a cover for f^2, where $G = \bar{x}_1 \bar{x}_2 \bar{x}_3$.

We will now formalize these concepts.

Let C_0 be an initial cover of the care complex K_c, let DC be an initial cover of the don't care complex K_{dc}, and let $C = C_0 \cup DC$. Let $c = (c_1 c_2 \ldots c_n) \in C$, where $c_i = \alpha \in \{0, 1\}$. Set $\bar{c}(i) = (c_1 c_2 \ldots c_{i-1} \bar{\alpha} c_{i+1} \ldots c_n)$ and $c' = (c_1 c_2 \ldots c_{i-1} x c_{i+1} \ldots c_n)$, where $\bar{\alpha} = 1 - \alpha$. Obviously $c \Rightarrow c'$ and $\$c > \$c'$. Let $C' = (C - c) \cup c'$.

The *element deletion operation* states that c can be replaced by c' if and only if $K(C') = K(C)$. To test for this condition we use the following result.

Theorem 2: $c'(i) \mathbin{\#} (C - c) = \varphi$ if and only if $K(C') = K(C)$

Whenever $\bar{c}(i) \mathbin{\#} (C - c) = \varphi$, we replace c by c' in C, and we repeat this procedure for another value of i, using cube c' in place of c. Let $\Delta_1(C, c, i) = c'$ if $\bar{c}(i) \mathbin{\#} (C - c) = \varphi$; otherwise $\Delta_1(C, c, i) = c$.

By the definition of prime implicant, we obtain the next theorem.

Theorem 3: Cube c is a prime implicant if and only if for all i such that $c_i \in \{0, 1\}$, $\bar{c}(i) \mathbin{\#} (C - c) \neq \varphi$

Hence, after applying the element deletion operation on each non-x position, we obtain a prime implicant. By applying this technique to each $c \in C$, we obtain a new cover for f which consists of no more cubes than in the original cover C.

The *cube deletion* operation is based upon the fact that $C - c$ is a cover for f if and only if each true vertex covered by c is also covered by some other cube in $C - c$. A cube c is of this type if and only if

$$c \mathbin{\#} (C - c) \mathbin{\#} DC = \varphi$$

or equivalently

$$(c \mathbin{\#} (C - c)) \cap C_0 = \varphi$$

If c has this property, then c can be deleted from C. Let $\Delta_2(C, DC, c) = C - c$ if $c \mathbin{\#} (C - c) \mathbin{\#} DC = \varphi$; otherwise $\Delta_2(C, DC, c) = C$. The concept of cube deletion is a generalization of subsuming, i.e., rather than c subsumes c' we have that c subsumes $(C - c)$.

The final simplification algorithm, given initial cover $C = C_0 \cup DC$, is as follows:

Algorithm 5: Single-output irredundant cover

Step 1: For each $c \in C$, replace C by $\Delta_2(C, \varphi, c)$. (Delete superfluous cubes.)

Step 2: For each $c \in C$ in turn: (a) delete c if $c \Rightarrow c'$ for some $c' \in C$; otherwise (b) replace c by $\Delta_1(C, c, i)$. (Generate prime implicants.) Do this operation for each i such that $c_i \in \{0, 1\}$.

Step 3: For each $c \in C$, replace C by $\Delta_2(C, DC, c)$.

As an example, consider the function f^3 shown in Fig. 2-8, where

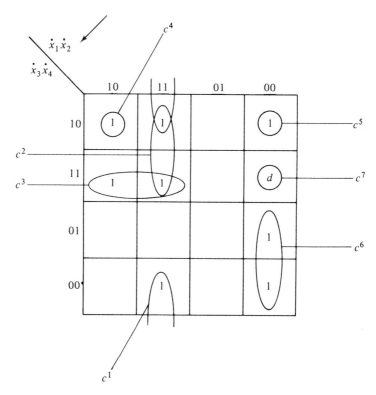

Fig. 2-8. Function f^3 and its initial cover.

Sec. 2.3   Covering a Single-Outputs Switching Function   47

$$C_0 = \begin{cases} 1 & 1 & x & 0 \\ 1 & 1 & 1 & x \\ 1 & x & 1 & 1 \\ 1 & 0 & 1 & 0 \\ 0 & 0 & 1 & 0 \\ 0 & 0 & 0 & x \end{cases} \begin{matrix} c^1 \\ c^2 \\ c^3 \\ c^4 \\ c^5 \\ c^6 \end{matrix}$$

$$DC = \{0 \quad 0 \quad 1 \quad 1\} \quad c^7$$

and $C = C_0 \cup DC$.

Step 1: Now $\Delta_2(C, DC, c^2) = C - c^2$ since

$$c^2 \mathrel{\#} \{c^1, c^3\} = (111x \mathrel{\#} 11x0) \mathrel{\#} 1x11$$
$$= 1111 \mathrel{\#} 1x11 = \varphi$$

Therefore, $c^2 \mathrel{\#} (C - c^2) \mathrel{\#} DC = \varphi$. $\Delta_2(C, DC, c^i) = C$ for all $i \neq 2$; hence we have

$$C = \{c^1, c^3, c^4, c^5, c^6, c^7\}$$

Step 2: (a) c^1 does not subsume any cubes, and $\Delta_1(C, c^1, i) = c^1$ for $i = 1, 2, 4$.

(b) c^3 does not subsume any cubes, and $\Delta_1(C, c^3, i) = c^3$ for $i = 1, 3$.

However, $\Delta_1(C, c^3, 4) = 1x1x$, since for $i = 4$ we have $\bar{c}^3(4) = 1x10$ and $\bar{c}^3(4) \mathrel{\#} (C - c^4) = \varphi$ as shown by the following calculation

$$\bar{c}^3(4) \mathrel{\#} \{c^1, c^4\} = (1x10 \mathrel{\#} 11x0) \mathrel{\#} 1010$$
$$= 1010 \mathrel{\#} 1010 = \varphi$$

We therefore set $c^3 = 1x1x$.

(c) Since $c^4 = 1010 \subset c^3 = 1x1x$, c^4 can be deleted.

(d) c^5 does not subsume any cubes. However, $\Delta_1(C, c^5, 1) = x010$ and $\Delta_1(C, c^5, 4) = x01x$. Note that, starting with the original c^5, we have $\Delta_1(C, c^5, 3) = 00x0$ and $\Delta_1(C, c^5, 4) = 00x0$. Hence different results can be obtained depending on the order in which elements are tested for deletion.

(e) $\Delta_1(C, c^6, 3) = 00xx$.

(f) $c^7 \Rightarrow c^6$.

We now have

$$C = \begin{cases} 1 & 1 & x & 0 \\ 1 & x & 1 & x \\ x & 0 & 1 & x \\ 0 & 0 & x & x \end{cases} \begin{matrix} c^1 \\ c^3 \\ c^5 \\ c^6 \end{matrix}$$

48 Logic Synthesis   Chap. 2

Step 1: (Repeated) $\Delta_2(C, DC, c^i) = C$ for $i = 1, 3, 6$, while $\Delta_2(C, DC, c^5) = C - c^5$ since

$$x01x \not\# \{11x0, 1x1x, 00xx\} \not\# 0011$$
$$(((x01x \not\# 11x0) \not\# 1x1x) \not\# 00xx) \not\# 0011 =$$
$$((x01x \not\# 1x1x) \not\# 00xx) \not\# 0011 =$$
$$(001x \not\# 00xx) \not\# 0011 =$$
$$\psi \not\# 0011 = \varphi$$

The final solution is

$$C = \begin{Bmatrix} 1 & 1 & x & 0 \\ 1 & x & 1 & x \\ 0 & 0 & x & x \end{Bmatrix} \begin{matrix} c^1 \\ c^3 \\ c^6 \end{matrix}$$

In this case, we have obtained a minimum cost cover, though in general this need not be the case. Also, different solutions can be obtained depending on the order in which the cubes are processed.

A technique for carrying out the operations of this algorithm without using the $\#$-operator is described by Breuer [7], and some computational results are given in Table 2-6.

TABLE 2-6. IRREDUNDANT COVER
  (a) Reduction of 2^n minterms to U_n as a function of n.
  (b) Reduction of an initial set of cubes C as a function of $|C|$.
  (c) Reduction of an initial set C as a function of n.

	n	6	7	8
(a)	t (seconds)	3.5	16	90
	$\|C\|$	16	24	32
(b) ($n = 7$)	t (seconds)	1	1.7	2.3
	n	7	8	9
(c) ($\|C\| = 32$)	t (seconds)	2.3	2.7	3

Source: Breuer (unpublished)
Machine: NCR 315 RMC
Language: Assembly

By applying the cube deletion criteria to the set of prime implicants shown in Fig. 2-5a, we see that c^2, c^4, c^6, c^7, c^9, and c^{11} can be deleted. Hence, in one pass through a prime implicant cover, we can obtain an irredundant cover.

In summary, this simplification procedure has the following properties:

1. Minterms are not generated.
2. The number of cubes in C monotonically decreases as the algorithm proceeds; hence, the storage requirements are quite minimal.
3. The algorithm can be terminated at any time, and the present C represents a cover for f. This is a very important property since in many algorithms, if the computation is not carried to completion, no solution is obtained.
4. Large problems can be partitioned and solved separately. For example, let $C = \bigcup_{i=1}^{r} C_i$. Then each C_i can be simplified to C'_i using this algorithm, and then $C' = \bigcup_{i=1}^{r} C'_i$ can be simplified as a final step.
5. Computation time tends to grow linearly with the size of the initial cover.

Summary. As one can see from the references, there are numerous papers dealing with the problem of finding a cover for a Boolean switching function. Here we have divided the problem into three subproblems, namely, the generation of the set of prime implicants Z, the selection of a minimum cost cover C^* which is a subset of Z, and the construction of an irredundant cover which also is a subset of Z. For large problems one usually must seek a nonminimum cost cover in order to reduce computation time. Some techniques attempt to generate extremals and continually reduce the complex to be covered. Thus the generation of prime implicants and covering are accomplished simultaneously. The main inefficiency in seeking a minimum cost cover is due to the fact that, as n increases, the value of the ratio $E(|Z(n)|)/E(|C^*(n)|)$ also increases, where $E(|Z(n)|)$ is the expected number of prime implicants in an n-variable function f, and $E(|C^*(n)|)$ is the expected number of cubes in the minimal cost cover for f. Hence a great deal of effort is wasted in generating prime implicants which will not appear in C^*. Note that $|C^*(n)| \leq 2^{n-1}$, and for $n > 10$, $E(|Z(n)|) > 2^n$.

Though many algorithms exist for this problem, very little if any comparative analytic analysis has been made concerning the relative efficiency of these various approaches. Also, very little computational experience has been reported comparing these techniques. Those computer times reported in the literature are difficult to interpret because each refers to a different computer working on a different set of problems. A few results have been included here only to give the reader a general indication for some of the computation times involved for these procedures.

2.4 COVERING A MULTIPLE-OUTPUT SWITCHING FUNCTION

In Fig. 2-9a and b we illustrate two functions which are to be synthesized. If we implement the functions in the form $f^1 = x_2 x_3 + \bar{x}_1 x_2$ and $f^2 = x_1 \bar{x}_3 + x_1 x_2$, then four AND gates would be required. However, the circuit in Fig. 2-9c indicates a realization requiring only three AND gates. Note that the term $x_1 x_2 x_3$ is not a prime implicant of either f^1 or f^2 but is a prime implicant of the function $(f^1 \cdot f^2)$ whose care complex equals the intersection of the care complex of f^1 and f^2. (See Fig. 2-9d.) The main distinction between synthesizing single- and multiple-output switching functions is that we must first broaden our definition of a prime implicant and also note that a cube can cover elements in the complex of more than one single-output function. In Fig. 2-9e we illustrate a cubical representation $\hat{C}$, referred to as a connection cover, for the circuit shown in Fig. 2-9c. Here, for example, the cube $01x1x$ can be partitioned into two parts. The first part is $c = (c_1 c_2 c_3) = (01x)$, and it represents the product $\bar{x}_1 x_2$. The second part is $e = (e_1 e_2) = (1x)$, which indicates that c is one element in the circuit for $f^1 (e_1 = 1)$, and c is not an element in the circuit for $f^2 (e_2 = x)$. In Fig. 2-9f we illustrate a functional array $\tilde{C}$ for this function. Its interpretation should be self-explanatory. Finally, in Fig. 2-9g we illustrate a tagged array T for f. Here, for example, the cube $01xx0$ can be partitioned into two parts. The first part is $c = (c_1 c_2 c_3) = (01x)$, and it again represents the product $\bar{x}_1 x_2$. The second part is $t = (t_1 t_2) = (x0)$, which indicates that $c \Rightarrow f^1 (t_1 = x)$. The distinction between $\hat{C}$ and T will be made after they have been normally defined, which we will now do.

Let $f: \{(0, 1)^n\} \to \{(0, 1, d)^m\}$ be a multiple-output switching function. One can consider that f consists of an ordered m-tuple of single-output functions, denoted by $f = (f^1, f^2, \ldots, f^m)$. We assume that each function f^i is defined by an initial care cover C_0^i and an initial don't care cover DC^i.

We will now define and state four properties related to tagged cubes.

1. Let $\tau = ct$ be a *tagged cube*, where $c = (c_1 c_2 \ldots c_n)$ is the input cube portion of τ, $t = (t_1 t_2 \ldots t_m)$ is the tag portion of τ, $c_j \in \{0, 1, x\}$, and $t_i \in \{0, x\}$. A tagged cube τ is said to be *consistent* with respect to f if $t_i = x$ if and only if c implies f^i, i.e., $c \Rightarrow C_0^i \cup DC^i$.
2. Let $T = \{\tau^1, \tau^2, \ldots, \tau^p\}$ and $B^i = \{c \mid \tau = ct \in T \text{ and } t_i = x\}$. Then T is said to be a *tagged cover* of f if, for each $i = 1, 2, \ldots, m$, B^i covers f^i, i.e., $C_0^i \Rightarrow B^i \Rightarrow C_0^i \cup DC^i$.
3. A tagged cube is said to be a *multiple-output prime implicant* of f if:
   (a) c implies f^i if and only if $t_i = x$;
   (b) no larger cube c' containing c has this property.
   Note that if f is a prime implicant, then it is consistent.

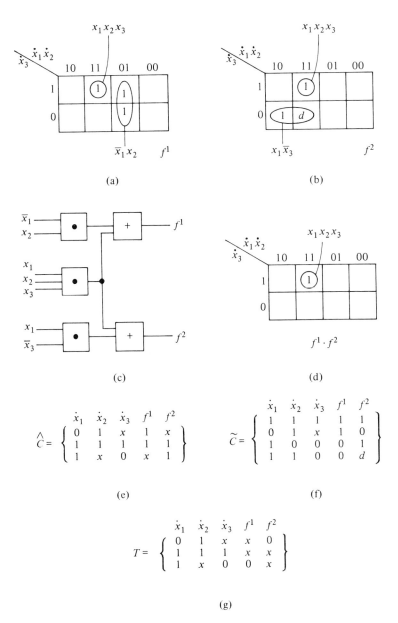

Fig. 2-9. Synthesis of a multiple-output function $f = (f^1, f^2)$.

Let τ be a consistent cube. Then, by the preceding definition, it follows that τ is a prime implicant of f if and only if c is a prime implicant of the single-output function $f(\tau)$ having care complex $K_c(\tau)$ and don't care complex $K_{dc}(\tau)$ constructed as follows:

$$K_c(\tau) = \bigcap_{\forall i \ni t_i = x} K^0(C_0^i)$$

$$K_{dc}(\tau) = \bigcap_{\forall i \ni t_i = x} K^0(DC^i)$$

4. Let $\$\tau = \$c$. If T contains p cubes, we can define the cost of T as some monotonically increasing function of $\$\tau$ and/or p, such as

$$\sum_{\tau \in T} \$\tau \quad \text{or} \quad \sum_{\tau \in T} \$\tau + p$$

Most procedures for finding minimum cost tagged covers for f are based upon the following theorem.

Theorem 4: *If Z is the set of all multiple-output prime implicants of f, then there exists a minimum cost tagged cover T^* of f such that $T^* \subseteq Z$.*

Unlike the result for single-output functions, not all minimum cost tagged covers need consist of a subset of Z. This follows since we have not included the cost associated with the tag cube t in our definition of $\$\tau$.

We will now consider the problem of construction Z. One procedure, due to Muller [65], consists of transforming an n-imput m-output function f into an $(n + m)$-input single-output function F. From a minimum cost cover for F, an irredundant cover for f can be obtained after proper decoding of the cover for F and the elimination of possible redundancies. Unfortunately, this transformation produces a function F having a great number of prime implicants. For this reason this technique is often not feasible for large problems.

A second procedure for generating Z, referred to as the tag method [6], [51], consists of first forming all intersection complexes from the C_0^i, i.e., we set $C(I) = \underset{i \in I}{\in} C_0^i$ for each subset I of $\{1, 2, \ldots, m\}$ containing two or more elements. There are $2^m - (m + 1)$ such subsets. Note that $C(I)$ is a cover for the true vertices in the function $\prod_{i \in I} f^i$. We now add m coordinates, referred to as tags, to each cube $c = (c_1 c_2 \ldots c_n)$ in $C(I)$, forming a new cube $\tau = ct = (c_1 c_2 \ldots c_n t_1 t_2 \ldots t_m)$, where $t_i = x$ if $i \in I$; otherwise $t_i = 0$. Let the set of all tagged cubes define the set C_0. In similar fashion we can form the set of tagged cubes DC associated with the don't care vertices of the f^i.

Now C_0 and DC define a complex $K^0(C_0 \cup DC)$, and let Z' be the set of prime implicants for this complex obtained by any of the techniques described in the previous section. Then Z is the subset of Z' obtained by deleting all cubes τ from Z' having all 0's in the tag coordinates. Note that there will never be any 1's in these positions.

Sec. 2.4 Covering a Multiple-Output Switching Function

A minimum cost tagged cover T^* for f can now be found using a procedure analogous to the one used for the single-output case. All that is required is to properly define the concepts of extremals and dominance for this multiple-output case.

Let $C^i = \{c \mid \tau = ct \in Z \text{ and } t_i = x\}$, that is, C^i is the set of all input cubes of Z which can be used to cover f^i. Consider a cube $\tau = ct \in Z$, where $t_i = x$. Then τ is said to be a K_i-extremal if c is an extremal of C^i. It is clear that, if c is an extremal of C^i, then τ can be included as one cube in our minimum cost tagged cover.

Let $\tau = ct$ and $\tau' = c't'$. Then τ dominates τ' if:

1. $t' \Rightarrow t$;
2. $\$c \leq \$c'$;
3. $c' \,\#\, C_0^i \Rightarrow c \,\#\, C_0^i$ for all i such that $t_i' = x$.

The branching technique is the same as that described earlier. With these definitions, the extraction procedure for finding a minimum cost cover for a single-output function can be used to find a minimum cost tagged cover T^* for a multiple-output function.

Let $\hat{c} = ce$ be a *connection cube*, where $c = (c_1 c_2 \ldots c_n)$, $e = (e_1 e_2 \ldots e_m)$, and $e_i = (x, 1)$. Given a tagged cube $\tau = ct$, we can construct a corresponding connection cube $\hat{c}$ by setting $e_i = 1$ if and only if $t_i = x$. For example $\tau = c0xx$ becomes $\hat{c} = cx11$. By this transformation, we can convert a tagged cover T into a connection cover $\hat{C}$.

A connecting cover $\hat{C} = \{\hat{c}^1, \hat{c}^2, \ldots, \hat{c}^p\}$ for a multiple function defines a two-level AND-OR (NAND-NAND) implementation consisting of p second-level AND (NAND) gates and m first-level OR (NAND) gates. The input to the kth second-level AND gate is c^k, and its output is $P(c^k)$. This AND gate can be eliminated if $\$c = 1$. The output of the kth second-level gate is an input to the ith first-level gate if $e_i^k = 1$. If $e_i^k = 1$ for only one value of k, then this OR gate can be eliminated.

In general, this circuit can contain some redundancy, that is, some of the inputs to the OR gates can be eliminated. For example, consider the functions defined in Fig. 2-10a, where $C_0^1 = \{1x1, 00x\}$ and $C_0^2 = \{x01\}$. Then $C_0^1 \cap C_0^2 = x01$

$$C_0 = \begin{Bmatrix} 1 & x & 1 & x & 0 \\ 0 & 0 & x & x & 0 \\ x & 0 & 1 & 0 & x \\ x & 0 & 1 & x & x \end{Bmatrix}$$

$$T^* = Z = \begin{Bmatrix} 1 & x & 1 & x & 0 \\ 0 & 0 & x & x & 0 \\ x & 0 & 1 & x & x \end{Bmatrix} \text{ and } \hat{C} = \begin{Bmatrix} 1 & x & 1 & 1 & x \\ 0 & 0 & x & 1 & x \\ x & 0 & 1 & 1 & 1 \end{Bmatrix}$$

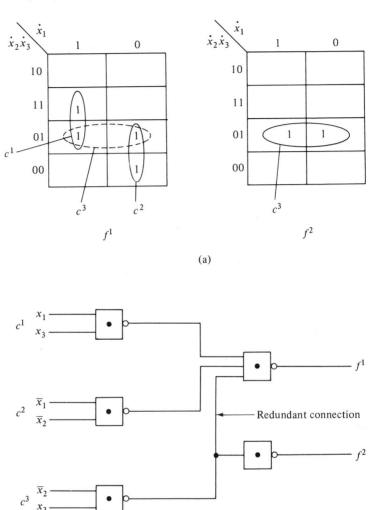

Fig. 2-10. (a) Multiple-output function $\{f^1, f^2\}$ and (b) NAND implementation.

Since each cube in Z is an extremal, we have that $Z = T^*$. The circuit corresponding to $\hat{C}$ is shown in Fig. 5-10b, along with a redundant connection. The cube c^3 in the cover of f^1 is redundant since $c^3 \Rightarrow \{c^1, c^2\}$, i.e., $c^3 \# \{c^1, c^2\} = \varphi$. This redundancy has occurred because we have not included the cost of the tag cube t in defining the cost of τ.

We will now consider the problem of eliminating the redundancies implied by the coordinates of the output cube e.

Let $\hat{c} = ce$ be a cube in $\hat{C}$, where $e_i = 1$. From $\hat{c}$ we can form cube $\hat{c}(i) = c\hat{e}$, where $\hat{e}$ and e are identical except in coordinate i where $\hat{e}_i = x$. Now $\hat{c}$ in connection cover $\hat{C}$ of f is said to be *redundant* if $(\hat{C} - \hat{c}) \cup \hat{c}(i)$ is also a logically correct connection cover of f. If we now let $\$\hat{c} = \$c + \$e$, then we have that $\$\hat{c}(i) = \$\hat{c} - 1$. Hence a minimal cost connection cover is one in which all cubes are irredundant. One procedure for removing these redundancies is the following [94].

Algorithm 6: Reduction of connection cover
Step 1: Set $A^i = \{c \mid ce \in \hat{C}, e_i = 1\}$.
Step 2: $\hat{c} = ce$ in $\hat{C}$, where $e_i = 1$, can be replaced by $\hat{c}(i)$ if and only if

$$(c \cap C_0^i) \# (A^i - c) = \varphi$$

The term $c \cap C_0^i$ represents all the true vertices of f^i contained in c. If no don't care terms exist in f^i, then $c \cap C_0^i$ can be replaced by just c. The $\#$-product determines whether or not all these true vertices are covered by the other numbers of A^i.

A minimal cost connection cover $\hat{C}$ is obtained when no further redundancies exists. If a cube is obtained where $e_i = x$ for all i, then $\hat{c} = ce$ can be deleted from $\hat{C}$, since such a cube applies to none of the output variables.

Applying this procedure to our previous example, we obtain

$$\hat{C} = \left\{ \begin{matrix} 1 & x & 1 & 1 & x \\ 0 & 0 & x & 1 & x \\ x & 0 & 1 & x & 1 \end{matrix} \right\}$$

Letting $\$\hat{C} = \sum_{c \in C} \$\hat{c}$, we have $\$\hat{C} = 9$ for this example.

One difficulty with the tag algorithm is that it requires the generation of the complexes $C(I) = \bigcap_{i \in I} C_0^i$. This can require a great amount of computation which grows exponentially with both n and m.

In a recent paper by Roth and Wagner [84], the extraction algorithm is formally generalized to the case of multiple-output functions. They have also generalized the $\#$-product to operate over cubes $\hat{c}$. (See Appendix B.) Their procedure will find a minimum cost cover over a very broad class of cost functions.

The main difficulty with the simplification procedures discussed so far is that they require the generation of the set Z which can be quite large.

We will now describe a technique for rapidly obtaining an initial connection cover $\hat{C}_0$. Once $\hat{C}_0$ is obtained, redundancies in it can be eliminated, and an irredundant connection cover obtained. In this procedure, we will simplify each function f^i individually; hence, the amount of computation grows only linearly with m. To each single-output prime implicant c obtained, we will append an appropriate e-cube, thus forming $\hat{c}$.

Algorithm 7: Generation of initial connection cover $\hat{C}_0$

Step 1: Generate a minimal cost cover C^i for f^i from some initial covers C_0^i and DC^i.

Step 2: For each $c \in \bigcup_{i=1}^{m} C^i$, we obtain an element ce in $\hat{C}_0$, where $e_i = 1$ if and only if $c \not\sharp C^i = \varphi$.

We will now illustrate this algorithm and the reduction of the resulting connection cover on the multiple-output function $f = \{f^1, f^2, f^3, f^4\}$ [94] shown in Fig. 2-11. Using any of the techniques described in the previous section, we can obtain a minimal cost cover C^i for each single-output function f^i. For example, $C^1 = \{c^1, c^2, c^3, c^4\}$. Let $C = C^1 \cup C^2 \cup C^3 \cup C^4 = \{c^1, c^2, \ldots, c^{15}\}$. The initial connection cover for f is $\hat{C}_0 = \{\hat{c}^1, \hat{c}^2, \ldots, \hat{c}^{15}\}$, where

$$\hat{c}^1 = c^1 \quad 1xxx$$
$$\hat{c}^2 = c^2 \quad 1xxx$$
$$\hat{c}^3 = c^3 \quad 1xxx$$
$$\hat{c}^4 = c^4 \quad 11xx$$
$$\hat{c}^5 = c^5 \quad x11x$$
$$\hat{c}^6 = c^6 \quad 11x1$$
$$\hat{c}^7 = c^7 \quad xx11$$
$$\hat{c}^8 = c^8 \quad xx1x$$
$$\hat{c}^9 = c^9 \quad xx1x$$
$$\hat{c}^{10} = c^{10} \quad 1x11$$
$$\hat{c}^{11} = c^{11} \quad 1111$$
$$\hat{c}^{12} = c^{12} \quad 11x1$$
$$\hat{c}^{13} = c^{13} \quad xx11$$
$$\hat{c}^{14} = c^{14} \quad xxx1$$
$$\hat{c}^{15} = c^{15} \quad xx1x$$

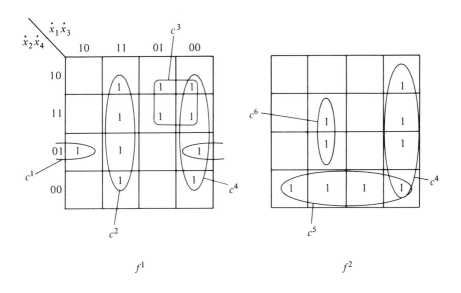
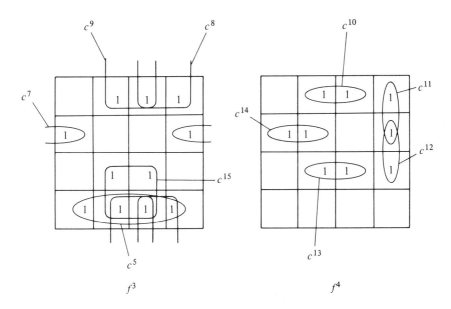

Fig. 2-11. Multiple-output function $f = \{f^1, f^2, f^3, f^4\}$.

The output cube e^4, where $\hat{c}^4 = c^4 e^4$, is constructed as follows:
1. $c^4 \,\#\, C^i = \varphi$; hence, $e_i^4 = 1$ for $i = 1, 2$;
2. $c^4 \,\#\, C^i \neq \varphi$; hence, $e_i^4 = x$ for $i = 3, 4$.

Therefore, $e^4 = 11xx$. The remaining e-cubes are constructed in the same manner.

Recall that if all components of e are x, then $P(c)$ need not be formed at the output of a gate. Hence we divide the reduction of $\hat{C}$ into two steps, namely, the elimination of entire cubes $\hat{c}$, followed by the elimination of redundancies in the components of e. Now $\hat{c}$ can be deleted if, for *each* i such that $e_i = 1$, we have that

$$(c \cap C_0^i) \,\#\, (A^i - c) = \varphi$$

That is, c can be deleted if in each function every true vertex covered by c is also covered by other cubes also contained in that function.

Using this intuitive definition, we see that $\hat{c}^8$ can be deleted, since $c^8 = 0xx0$ contains 0-cubes $0000 \Rightarrow c^5$, $0010 \Rightarrow c^5$, $0100 \Rightarrow c^{11}$, and $0110 \Rightarrow c^9$. Similarly $\hat{c}^9$, $\hat{c}^{14}$, and $\hat{c}^{15}$ can be deleted.

Eliminating redundancies in the e_i components, we obtain

$$\hat{c}^6 = c^6 \quad x1x1$$
$$\hat{c}^{10} = c^{10} \quad xx11$$
$$\hat{c}^{11} = c^{11} \quad xx11$$
$$\hat{c}^{12} = c^{12} \quad xxx1$$

No other reductions are possible. For this final solution we have $\$\hat{C} = 47$ vs 45 in the minimum cost solution.

An alternate approach for achieving an initial connection cover, which operates on all functions f^i simultaneously, is discussed by Su [94].

Summary. In this section we have outlined the extension of the techniques for finding covers for single-output functions to the case for multiple-output functions. The resulting covering problem for large problems usually turns out to be quite complex, though both extraction [84] and matrix techniques [34] have been proposed for its solution. Though this problem is of much greater practical importance than the single-output problem, little work has been done in developing fast procedures for constructing good irredundant covers. Finally, we should point out that, if we restrict ourselves to just two-level circuits which are not necessarily of the form AND-OR, then minimum cost covers need not represent minimum cost circuit realizations [100].

2.5 SYNTHESIS OF CONSTRAINED NAND NETWORKS VIA FACTORIZATION

In general, a two-level circuit is not the most economical circuit realization of a function. Unfortunately, the synthesis of multilevel minimum cost networks is an exceedingly difficult problem. So far in our discussion we have not considered any engineering or circuit constraints on what constitutes an acceptable network. In this section we will restrict our attention to a subclass of these two problems. We will consider the problem of synthesizing multiple-output switching functions in terms of multiple-level NAND networks, where fan-in and fan-out constraints are specified. The technique will not be based upon a true decomposition of the functions; rather, it will involve the simpler concept of factorization.

Consider the problem of implementing a cover in terms of elements from a specific circuit family of available gates. Associated with a specific type of gate are two integer constants FI and FO, where FI is the fan-in limit and equals the maximum number of inputs to this gate, and FO is the fan-out limit and equals the maximum number of gates which can be connected to the output of this gate. For simplicity, we will assume that all gate types have the same FI and FO values. To implement a cube c with a single gate g implies that g has $\$c$ inputs. If $\$c > FI$, we say that g exceeds the fan-in constraint. The fan-out of a literal x_i with respect to a set of cubes C, denoted by $h(x_i)$, is given by the equation $h(x_i) = |\{c \,|\, c_i = 1, c \in C\}|$. If $h(x_i) > FO$, we say that x_i exceeds the fan-out constraint. $h(\bar{x}_i)$ is defined analogously.

By employing factorization techniques, one can convert a given circuit description to a logically equivalent one in such a way that all gates and literals in the circuit implementation implied by this new description satisfy the fan-in and fan-out constraints. These factorization techniques will be the subject of this section.

To illustrate a few of the concepts involved, consider the following two expressions and their associated numbers of gate inputs

$F_1 = x_1 x_2 x_3 x_4 + x_1 x_2 \bar{x}_3 \bar{x}_4$ (AND-OR logic; 10 inputs, 3 gates)

$F_2 = x_1 x_2 x_3 x_4 + x_1 x_2 \bar{x}_3 \bar{x}_4 + x_1 x_2 x_5 + x_1 x_2 \bar{x}_6$

(AND-OR logic; 18 inputs, 5 gates)

We can rewrite these expressions in factored form as

$F'_1 = x_1 x_2 (x_3 x_4 + \bar{x}_3 \bar{x}_4)$ (AND-OR-AND logic; 9 inputs, 4 gates)

$F'_2 = x_1 x_2 (x_3 x_4 + \bar{x}_3 \bar{x}_4 + x_5 + \bar{x}_6)$ (AND-OR-AND logic;

11 inputs, 4 gates)

where $x_1 x_2$ is a factor.

In general, implementing a factored form of an expression leads to the following three results:
1. an increase in the logic levels; hence delay;
2. an increase or decrease in the cost of the circuit;
3. a decrease in both fan-in and fan-out of some of the gates and literals.

If each product term in the resulting factored form consists of more than one literal, and if the cost of a circuit is equal to the number of gates in the circuit, then factored expressions always lead to more expensive circuits. Note, however, that the fan-out of x_1 in expression F_2 is 4, whereas in F'_2 it is only 1. Also, the average fan-in of the four AND gates in F_2 is $\frac{7}{2}$, while the average fan-in of the three AND gates in F'_2 is $\frac{7}{3}$.

Sets of expressions can be factored simultaneously. For example

and
$$F_1 = x_1 x_2 x_3 x_4 + x_1 x_2 \bar{x}_3 \bar{x}_4 + x_2 \bar{x}_3 x_5$$

$$F_2 = x_1 x_2 \bar{x}_4 \bar{x}_5 + x_1 \bar{x}_3$$

can be rewritten as

$$F_1 = x(x_3 x_4 + \bar{x}_3 \bar{x}_4) + x_2 \bar{x}_3 x_5$$
$$F_2 = x \bar{x}_4 \bar{x}_5 + x_1 \bar{x}_3$$

where factor $x = x_1 x_2$ is fanned-out to both F_1 and F_2.

We will now formalize the factorization process. Given two cubes $\gamma = (\gamma_1 \gamma_2 \ldots \gamma_n)$ and $c = (c_1 c_2 \ldots c_n)$, γ is said to be a *factor* of c, or equivalently c is a *parent* of γ, if and only if $c \Rightarrow \gamma$. If γ is a factor of c, then the *residue* of c under factorization by γ, denoted by c/γ, equals $(\beta_1 \beta_2 \ldots \beta_n)$ where β_i is defined below.

		γ_i	
/	0	1	x
c_i 0	x	Ω	0
1	Ω	x	1
x	Ω	Ω	x

A Ω component indicates that γ is not a factor of c. If $c \not\Rightarrow \gamma$, then let $c/\gamma = \psi$.

In the following examples we will use the product notation rather than the cubical notation. We have that $x_1 \bar{x}_2$ is a factor of $x_1 \bar{x}_2 \bar{x}_4$, and $x_1 \bar{x}_2 \bar{x}_4 / x_1 \bar{x}_2 = \bar{x}_4$, while $x_1 \bar{x}_2$ is not a factor of $x_1 x_2 \bar{x}_4$.

Given a set of cubes C, let $C/\gamma = \{c/\gamma \mid c \in C, c/\gamma \neq \psi\}$. Then the *height* of γ, denoted by $h(\gamma)$, is given by $h(\gamma) = |C/\gamma|$. The *width* $w(\gamma)$ of γ is given by $w(\gamma) = \$\gamma$. The *figure of merit* $FM(\gamma)$ of γ is given by the equation $FM(\gamma) =$

$h(\gamma) \cdot w(\gamma)$, or simply by $FM = h \cdot w$ [18], [20]. Let $C = \sum_{c \in C} \$c$ be a measure of the cost of implementing C. Given C and γ, the reduced form of C under factorization by γ is denoted by $C//\gamma$, where $C//\gamma = \{c \in C \mid c/\gamma = \psi\} \cup C/\gamma$. Since $\$(C//\gamma) = \$C - FM(\gamma)$, it is seen that, to minimize the value of $\$(C//\gamma)$, one should seek a γ having the largest figure of merit.

2.5.1. Construction of Factors

Given $C = \{c^1, c^2, \ldots, c^q\}$, we say that γ is a *common factor* of C if γ is a factor of every cube $c \in C$. To find such a γ of maximum width, we can employ the factor operator $\mathscr{F}$ defined below.

	$\mathscr{F}$	0	1	x
	0	0	x	x
c_i^k	1	x	1	x
	x	x	x	x

with column header c_i^j.

By applying $\mathscr{F}$ to C we obtain

$$\gamma = (\ldots (c^1 \mathscr{F} c^2) \mathscr{F} c^3) \ldots) \mathscr{F} c^q) \ldots)$$

The cube $\gamma = U_n$ implies that no common factor exists. As an example, we have $((x1001 \mathscr{F} 110x1) \mathscr{F} 11000) = (x10x1 \mathscr{F} 11000) = x10xx$.

We now consider the problem of finding a factor γ of large figure of merit, where γ need not be a common factor of C. An enumerative procedure for finding that factor of maximum figure of merit was suggested by Dietmeyer and Schneider [20] and was later refined by Dietmeyer and Su [18] in their Algorithm 1. These procedures are efficient for small and intermediate size problems. A faster procedure has also been suggested, referred to as Algorithm 2 [18], which does not guarantee finding a factor having a maximum figure of merit. A modified version of this procedure is now given.

Algorithm 8: State splitting

Step 1: Let $M = 1$ and $\gamma = U_n$. For $i = 1, 2, \ldots, n$, perform the following step.

Step 2: Let $X = \{c^j \mid c_i^j = 0\}$. Apply the factor operator $\mathscr{F}$ to X, obtaining γ', and calculate $FM(\gamma')$. If $X = \varphi$ set $\gamma' = U_n$, and if $X = \{c\}$ set $\gamma' = c$. If $FM(\gamma') \succ M$, set $M = FM(\gamma')$ and $\gamma = \gamma'$. Repeat for $X = \{c^j \mid c_i^j = 1\}$.

The last γ formed is the desired factor. The process of forming the set X is referred to as splitting on the value of c_i^j. For large covers C, this procedure usually results in finding a factor γ of very small width. To illustrate this algorithm, consider the set

$$C = \begin{Bmatrix} 1 & 0 & 1 & x \\ 1 & 0 & 0 & 1 \\ 0 & 1 & x & 0 \\ x & 1 & 1 & 0 \end{Bmatrix}$$

For $i = 1$ by splitting on 0 we obtain $X = \{01x0\}$, $\gamma = \gamma' = 01x0$, and $M = 3$. By splitting on 1 we obtain $X = \{101x, 1001\}$, $\gamma = \gamma' = 10xx$, and $M = 4$. Continuing for $i = 2, 3, 4$, we find no factor of larger figure of merit; hence $\gamma = 10xx$.

We will now describe an algorithm, referred to as the Grow Factor Algorithm [9], which produces a sequence of factors $\gamma^1, \gamma^2, \ldots$, each of which is one unit wider than the previous, i.e., $w(\gamma^{j+1}) = w(\gamma^j) + 1$; hence the name "grow factor." Only factors of height two or more are sought. Factors of height one consist of a subset of C and are trivial to find. Their figure of merit is simply $\$c$. The basic premise behind this technique is that if, for all factors of width w, the height of γ is a maximum, then there exists a factor γ' such that, for all factors of width $w + 1$, $h(\gamma')$ is a maximum and $\gamma' \Rightarrow \gamma$. Though this assumption is not always true, the procedure leads to quite good results and is relatively fast.

Algorithm 9: Grow factor

Step 1: Set $\gamma^0 = U_n$, $k = 0$, and $h^0(l) = 2$ for all literals l in $C^0 = C$.

Step 2: $k + 1 \rightarrow k$.

Step 3: Let $C^k = \{c \,|\, c/\gamma^{k-1} \neq \psi, c \in C^{k-1}\}$. (By construction $C^1 = C^0$.)

For all i such that $\gamma_i^{k-1} = x$, determine, with respect to C^k only:

(a) $h^k(x_i)$ if $h^{k-1}(x_i) \geq 2$;
(b) $h^k(\bar{x}_i)$ if $h^{k-1}(\bar{x}_i) \geq 2$.

Let l be that literal such that $h^k(l)$ is a maximum and $h^k(l) \geq 2$. If no such literal exists, we are finished. Otherwise, set $\gamma^{k-1} \rightarrow \gamma = (\gamma_1 \gamma_2 \ldots \gamma_n)$; set $\gamma_i = 1$ if $l = x_i$, or $\gamma_i = 0$ if $l = \bar{x}_i$. Let $\gamma \rightarrow \gamma^k$, and $FM(\gamma^k) = h^k(l) \cdot k$. Return to step 2.

Figure 2-12 illustrates the use of this algorithm. Given the initial array $C^0 = C^1$, we calculate the vectors $h^1(1)$ and $h^1(0)$ which are the heights for the literals x_i and $\bar{x}_i$, $i = 1, 2, \ldots, n$. The largest height value is 4, and we select

$$\quad 1\ 2\ 3\ 4\ 5\ 6\ 7\ 8\ 9\ 10 \qquad C^1\ C^2\ C^3\ C^4\ C^5$$

$$C^0 = \left\{\begin{matrix} x & 0 & 0 & 0 & 1 & x & 1 & 1 & x & 0 \\ x & 0 & 1 & 0 & 1 & x & 1 & 0 & x & 0 \\ 0 & x & 1 & 0 & 0 & 1 & x & x & x & x \\ 1 & x & 1 & 1 & 1 & x & x & x & x & x \\ x & 1 & x & x & x & x & 0 & 1 & 0 & x \\ x & 1 & 1 & 0 & 0 & 1 & 1 & x & x & 1 \end{matrix}\right\} \qquad \begin{matrix} \checkmark & \checkmark & \checkmark & & \\ \checkmark & \checkmark & \checkmark & \checkmark & \checkmark \\ \checkmark & \checkmark & \checkmark & \checkmark & \checkmark \\ \checkmark & \checkmark & & & \\ \checkmark & & & & \\ \checkmark & \checkmark & \checkmark & \checkmark & \checkmark \end{matrix}$$

(a)

$FM(\gamma)$

γ^0	x	x	x	x	x	x	x	x	x	x	0
$h^1(1)$	1	2	4*	1	3	2	3	2	0	1	
$h^1(0)$	1	2	1	4	2	0	1	1	1	2	
γ^1	x	x	1	x	x	x	x	x	x	x	4
$h^2(1)$	-	1	-	-	2	2	2	0	-	-	
$h^2(0)$	-	1	-	3*	2	-	-	-	-	1	
γ^2	x	x	1	0	x	x	x	x	x	x	6
$h^3(1)$	-	-	-	-	1	2	2	-	-	-	
$h^3(0)$	-	-	-	-	2*	-	-	-	-	-	
γ^3	x	x	1	0	0	x	x	x	x	x	6
$h^4(1)$	-	-	-	-	-	2*	1	-	-	-	
$h^4(0)$	-	-	-	-	-	-	-	-	-	-	
γ^4	x	x	1	0	0	1	x	x	x	x	8

(b)

$$\begin{matrix} x & 0 & 0 & 0 & 1 & x & 1 & 1 & x & 0 \\ x & 0 & 1 & 0 & 1 & x & 1 & 0 & x & 0 \\ 0 & x & x & x & x & x & x & x & x & x \\ 1 & x & 1 & 1 & 1 & x & x & x & x & x \\ x & 1 & x & x & x & 0 & 1 & 0 & x \\ x & 1 & x & x & x & 1 & x & x & 1 \end{matrix}$$

(c)

Fig. 2-12. (a) Original cover C^0 (b) Calculations (c) $C^0//\gamma^4$.

$\gamma^1 = x_3$ since x_3 is the last such literal encountered (scanning right to left) which possesses a maximum height value. This method of breaking ties is of course arbitrary. We next calculate $C^2 = C^1/\gamma^1$. In general, the cubes in C^k are identified by those rows having a check ($\sqrt{}$) mark in column C^k of Fig. 2-12a. Now $h(C^2) = 4$ and $FM(\gamma^1) = 4$. Continuing, we calculate $h^2(1)$ and $h^2(0)$ for only those literals which satisfy the conditions of step 3. Those literals which do not satisfy these conditions have a dash (–) entry. The last factor found is γ^4, where $FM(\gamma^4) = 8$. In this instance, the last factor found has the largest figure of merit among the factors we have found so far; but this is not always the case. Figure 2-12c indicates $C^0//\gamma^4$, obtained from C^0 by replacing c^3 and c^6 by c^3/γ^4 and c^6/γ^4.

By a simple modification to this algorithm, we can generate more factors. In fact, all factors can be generated if desired. This modification is achieved by selecting not just one literal having the largest height, but also, in turn, those K literals having the largest height values.

2.5.2. Synthesis of Single-Output Fan-In Limited NAND Networks

We now investigate the problem of implementing the cover C using only NAND gates, where at present we will consider only the fan-in constraint. If c is applied to the input of a NAND gate, the output is $U_n \# c$. If $U_n \# c$ is applied to the input of a NAND gate, the output is $U_n \# (U_n \# c) = c$ or, equivalently, $P(c)$ as illustrated in Fig. 2-13. This pair of NAND gates is said to form an *AND function*.

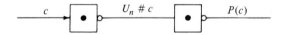

Fig. 2-13. The AND function.

In Fig. 2-14 we illustrate how the cube $c = (1111111)$ is implemented using two AND functions, where $\$c = 7$ and $FI = 4$. Here we have arbitrarily removed the four literals x_4, x_5, x_6, and x_7 from the input to a gate and have replaced them by one input equal in value to their product. It therefore requires three NAND gates to implement the logical equivalent of a seven-input NAND gate where $FI = 4$. In general, for NAND gates having a fan-in limit of FI, $\mathscr{G}(\xi)$ gates are required to implement the logical equivalent of a ξ-input gate, where

$$\mathscr{G}(\xi) = \begin{cases} 0 & \text{if } \xi = 1 \\ 1 & \text{if } 1 < \xi \leq FI \\ 1 + 2\left[\dfrac{\xi - FI}{FI - 1}\right] & \text{if } \xi > FI \end{cases}$$

Sec. 2.5 Synthesis of Constrained NAND Networks via Factorization 65

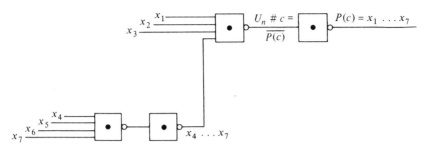

Fig. 2-14. Implementation of $c = (1111111)$ using two AND functions, where $FI = 4$.

and $[x]$ denotes the smallest integer greater than or equal to x. In our previous example, $\xi = \$c = 7$. We set $\mathscr{G}(1) = 0$ because we assume that the complement of each literal is always available.

Using AND functions to implement gates which exceed the fan-in limit, we can transform any NAND circuit into a logically equivalent one where all fan-in constraints are satisfied. Unfortunately, this procedure usually requires many more gates than is necessary. We will now illustrate a more efficient procedure for satisfying the fan-in constraint.

The expression $F = x_1 x_2 x_3 x_4 + x_1 x_2 \bar{x}_3 \bar{x}_4$, which can be written in factored form as $x_1 x_2 (x_3 x_4 + \bar{x}_3 \bar{x}_4)$, can be implemented by a number of different NAND networks, three of which are shown in Fig. 2-15.

We generalize these results by defining three types of circuits as shown in Fig. 2-16. Set $C = C' \cup (C - C')$; hence, $\bar{C} = \bar{C}' \cap \overline{C - C'}$ and $C = U_n \# (\bar{C}' \cap \overline{C - C'})$. Assume $C' = \{c^1, c^2, \ldots, c^q\}$ has a common factor γ; hence, $C'/\gamma = D = \{d^1, d^2, \ldots, d^q\}$ where $d^i = c^i/\gamma$. Let $\Delta = \{\delta^1, \delta^2, \ldots, \delta^s\}$ be an irredundant cover for the complex $\bar{D} = U_n \# D$. Given C, C', Δ, and γ, we can evaluate the cost of implementing C' using each circuit type. This cost is an estimate of the number of gates required to implement the corresponding circuit.

For type 1, this cost is denoted by $\$_1(C')$ and is given by the equation

$$\$_1(C') = \sum_{i=1}^{q} \mathscr{G}(\$c^i) + \mathscr{G}(q + 1)$$

The term $\mathscr{G}(\$c^i)$ is the cost of forming $\overline{P(c^i)}$, and $\mathscr{G}(q + 1)$ is the cost of implementing a $(q + 1)$-input gate. In general, the term $\mathscr{G}(q + 1)$ is just an estimate of the complexity of the equivalent of this first-level gate, since it is not known how many inputs will correspond to the term $\overline{C - C'}$. We have counted only one input in our equation. Note that this gate may already have had other inputs assigned to it due to the iterative nature of the factorization algorithm.

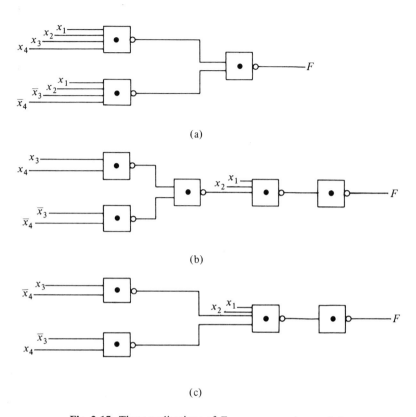

Fig. 2-15. Three realizations of $F = x_1 x_2 x_3 x_4 + x_1 x_2 \bar{x}_3 \bar{x}_4$.

In a similar manner, we obtain the cost associated with the other circuits, namely

$$\$_2(C') = \sum_{i=1}^{q} \mathscr{G}(\$d^i) + \mathscr{G}(q) + \mathscr{G}(\$\gamma + 1) + 1$$

$$\$_3(C') = \sum_{i=1}^{s} \mathscr{G}(\$\delta^i) + \mathscr{G}(\$\gamma + s) + 1$$

Referring to our previous example, we have that $C = C' = \{1111, 1100\}$ and $\gamma = 11xx$. Now $D = \{xx11, xx00\}$ and $\Delta = \{xx10, xx01\}$. For $FI = 3$, we have $\$_1(C') = 7$, $\$_2(C') = 5$, and $\$_3(C') = 6$; therefore a type 2 circuit should be used to implement F.

Note that if $\$c = 1$, then $P(c)$ consists of a single literal, say $x_i(\bar{x}_i)$, and $\bar{x}_i(x_i)$ can be applied directly to the input of the first-level gate.

We will now outline an algorithm for multilevel NAND synthesis [18]. Let C be a cover for f. In general, C is either an irredundant or a minimum cost cover, but this condition is not necessary.

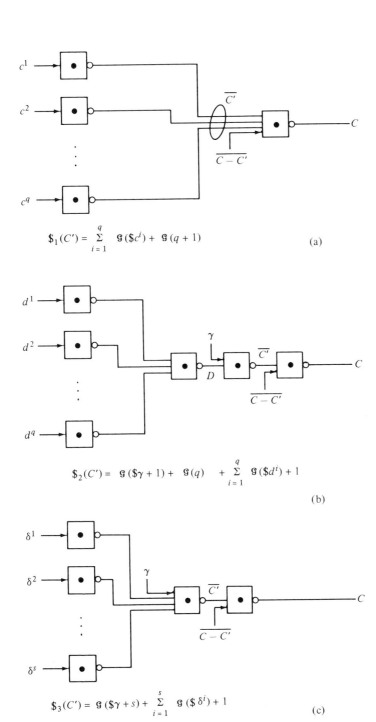

Fig. 2-16. (a) Type 1 (unfactored) circuit (b) Type 2 (post-factored) circuit (c) Type 3 (complement-factored) circuit.

Algorithm 10: Fan-in limited NAND synthesis of a single output function

Step 1: Form $C' = \{c \in C \mid \$c = 1\}$. Set $C - C' \to C$. For each $c \in C'$, set $\overline{P(c)}$ as an input literal to the first-level gate.

Step 2: Find a factor γ of C having a large (preferably maximum) figure of merit. Let $C' = \{c \in C \mid c/\gamma \neq \psi\}$; hence C' contains the common factor γ.

Determine D and Δ and calculate $\$_1(C')$, $\$_2(C')$, and $\$_3(C')$. Implement C' using a type k circuit, where $\$_k(C') = \min\{\$_1(C'), \$_2(C'), \$_3(C')\}$. Set $C - C' \to C$. If $C \neq \varphi$, return to the beginning of step 2; otherwise continue.

Step 3: If the number of inputs to any gate exceeds FI, reduce this number using AND functions.

To illustrate this algorithm, let $FI = 3$ and

$$C = \begin{Bmatrix} 1100xx \\ 110x0x \\ 11x0x0 \\ 11xx00 \\ 0x011x \\ x1100x \\ 1xx001 \end{Bmatrix}$$

Now $\gamma = 11xxxx$ and

$$C/\gamma = D = \begin{Bmatrix} xx00xx \\ xx0x0x \\ xxx0x0 \\ xxxx00 \end{Bmatrix}$$

$$\Delta = \begin{Bmatrix} xxx11x \\ xx1xx1 \end{Bmatrix}$$

and

$$C' = D \cap \gamma = \begin{Bmatrix} 1100xx \\ 110x0x \\ 11x0x0 \\ 11xx00 \end{Bmatrix}$$

Now $\$_1(C') = 15$, $\$_2(C') = 9$, and $\$_3(C') = 6$; therefore, a type 3 circuit is selected. In Fig. 2-17, $\bar{C}'$ is implemented by gates 2, 3, and 4.

Sec. 2.5 Synthesis of Constrained NAND Networks via Factorization   69

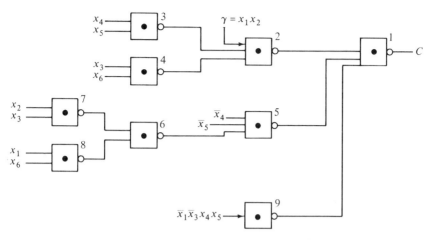

Fig. 2-17. Fan-in limited NAND synthesis.

The new cover $(C - C' \to C)$ is

$$C = \begin{Bmatrix} 0x011x \\ x1100x \\ 1xx001 \end{Bmatrix}$$

Now $\gamma = xxx00x$ and

$$C/\gamma = D = \begin{Bmatrix} x11xxx \\ 1xxxx1 \end{Bmatrix}$$

A type 2 circuit is now the least expensive circuit for implementing C', and this implementation is indicated by gates 5, 6, 7, and 8. Finally we obtain

$$C = \{0x011x\}$$

which is implemented by gate 9. Since the number of inputs to gates 2 and 9 exceeds 3, two AND functions are required, resulting in a network having thirteen gates. If the original cover C were implemented using only AND functions, twenty-six gates would be required.

A number of modifications can be added to this synthesis algorithm. For example, note that D and Δ are themselves sets of cubes and can be factored in the same manner that C was factored. That is, we can make the algorithm recursive by setting $D \to C$ and returning to step 1.

Now that the iterative nature of the algorithm is clear, the estimate of the cost of the first-level gate which implements C can be made more accurate in the following manner. Assume that, in evaluating $\$_1(C')$, the first-level gate already has been assigned τ inputs. Then, in the expression for $\$_1(C')$, the term $\mathscr{G}(q+1)$ should be replaced by $\mathscr{G}(\tau+q+1)$. In the expressions for $\$_2(C')$ and $\$_3(C')$, the term 1 should be replaced by $\mathscr{G}(\tau+2)$.

Another extension to this algorithm would be to include the use of dot or wired AND functions obtained by tying together the outputs of two or more NAND gates. Also, note that it may be cheaper to implement C by first implementing $\bar{C}$ directly and then inverting this result.

The main attributes of this procedure are that it is effective in the sense that the final network is physically realizable, it leads to network costs which are substantially less than those for other procedures (e.g., using only AND functions), and it is computationally efficient for moderate size problems.

In the next section we will extend this factorization procedure to the case of multiple-output functions.

2.5.3. Synthesis of Multiple-Output Fan-In Limited NAND Networks

Let $f = (f^1, f^2, \ldots, f^m)$ be a multiple-output function. A cube $\tilde{c} = ca$ is said to be a *functional multiple-output cube*, where c has the same meaning as defined previously, and $a = (a_1 a_2 \ldots a_m)$ has the following properties: $a_i \in \{0, 1, d\}$, and $a_i = 1$ if $c \Rightarrow K_c(f^i)$; $a_i = d$ if $c \Rightarrow K_{dc}(f^i)$; otherwise $a_i = 0$. Then $\tilde{C} = \{\tilde{c}\}$ is a functional cover or array for f if and only if for each $v \in K_c(f^i)$ there exists a cube $\tilde{c} \in \tilde{C}$ such that $v \Rightarrow c$, and for each $v \in K_{dc}(f^i)$ there exists a cube $\tilde{c} \in \tilde{C}$ such that $v \Rightarrow c$, where $i = 1, 2, \ldots, m$.

For example, one possible functional array for the function shown in Fig. 2-9 is

$$\tilde{C} = \begin{Bmatrix} \dot{x}_1 & \dot{x}_2 & \dot{x}_3 & f^1 & f^2 \\ 1 & 1 & 1 & 1 & 1 \\ 0 & 1 & x & 1 & 0 \\ 1 & x & 0 & 0 & 1 \end{Bmatrix}$$

Let $\tilde{C}$ be an initial functional array of a multiple-output switching function f, and let $\tilde{C}'$ be a subarray of $\tilde{C}$, where $\tilde{C}' = C'A'$ and C' contains a common factor γ. For example, let

$$\tilde{C}' = \begin{Bmatrix} \dot{x}_1 & \dot{x}_2 & \dot{x}_3 & \dot{x}_4 & f^1 & f^2 & f^3 & f^4 & \\ 1 & 0 & 0 & 0 & 1 & 1 & 1 & d & \tilde{c}^1 \\ 1 & 1 & 1 & x & d & 1 & 0 & d & \tilde{c}^2 \\ 1 & 0 & 1 & 0 & d & d & 1 & d & \tilde{c}^3 \\ 1 & 1 & 0 & 1 & 1 & 1 & 0 & d & \tilde{c}^4 \end{Bmatrix}$$
$$\underbrace{}_{C'} \quad \underbrace{}_{A'}$$

If the d entry a_1^2 is assigned the value 1, then columns one and two of A'

Sec. 2.5 Synthesis of Constrained NAND Networks via Factorization

are identical and contain no 0 entries. We disregard column 4 of A' since all its entries are d. It follows that the cubes $c^j, j = 1, 2, 3, 4$, can be shared between the functions f^1 and f^2. Let $C_0 = \{c^1, c^2, c^4\}$ and $DC = \{c^3\}$ be initial covers for a single-output function. Then an irredundant cover for this function is

$$C^* = \begin{Bmatrix} 1 & 1 & 1 & x \\ 1 & 1 & x & 1 \\ 1 & 0 & x & 0 \end{Bmatrix}$$

We can now implement C^* using the techniques described in Section 2.5.2 and fan-out the result to f^1 and f^2. Now $\gamma = 1xxx$ is a common factor of C^*, and

$$C^*/\gamma = D = \begin{Bmatrix} x & 1 & 1 & x \\ x & 1 & x & 1 \\ x & 0 & x & 0 \end{Bmatrix}$$

and

$$\Delta = \bar{D} = \begin{Bmatrix} x & 1 & 0 & 0 \\ x & 0 & x & 1 \end{Bmatrix}$$

In Fig. 2-18 we show the implementation of this portion of $\tilde{C}$ using a type 3 circuit, where $FI = 3$. Since the vertices covered by the original cubes

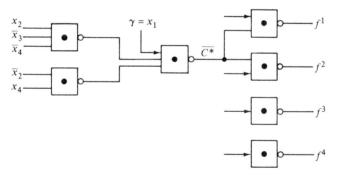

Fig. 2-18. Partial implementation of f.

$c^j, j = 1, 2, 3, 4$, in f^1 and f^2 have now been covered, we can change all 1's in columns one and two of A' to d's, and obtain

$$\tilde{C}' = \begin{Bmatrix} 1 & 0 & 0 & 0 & d & d & 1 & d \\ 1 & 1 & 1 & x & d & d & 0 & d \\ 1 & 0 & 1 & 0 & d & d & 1 & d \\ 1 & 1 & 0 & 1 & d & d & 0 & d \end{Bmatrix} \begin{matrix} \tilde{c}^1 \\ \tilde{c}^2 \\ \tilde{c}^3 \\ \tilde{c}^4 \end{matrix}$$

Now $\tilde{c}^2$ and $\tilde{c}^4$ can be either deleted from further consideration, since a^2 and a^4 contain no 1 elements, or else put into a special don't care array used only in finding a minimal cover C^*. For simplicity, we will choose the former of these two options. With these modifications to $\tilde{C}$, we now select another subarray of $\tilde{C}$ and repeat this same implementation process.

We now give one criterion for selecting the subarray $\tilde{C}' = C'A'$ from $\tilde{C} = CA$. Assume subarray $\tilde{C}'$ of $\tilde{C}$ has the property that C' has a common factor γ. The figure of merit of this factor is $FM(C', \gamma) = \$\gamma \cdot |C'|$. Assume that A' contains exactly w columns with no 0 elements and at least one 1 element. Let these columns be given by the index set $\lambda \subseteq \{1, 2, \ldots, m\}$, where $|\lambda| = w$. Let $FM(A', \lambda)$ equal the total number of 1 elements in the columns λ of A'. In our example $\lambda = \{1, 2\}$, $w = 2$, and $FM(A', \lambda) = 5$. Note that it is only possible to share cubes of C' among two or more outputs if $w \geq 2$. We can now define a figure of merit for $\tilde{C}'$ to be

$$FM(\tilde{C}, \gamma, \lambda) = FM(C', \gamma) + FM(A', \lambda)$$

We select the $\tilde{C}'$ such that $w \geq 2$ and $FM(C', \gamma, \lambda)$ is a maximum. Other criteria can be formulated for selecting $\tilde{C}'$.

We now present a modified version of an algorithm of Su [93] for implementing a multiple-output function f employing fan-in limited NAND gates, where f is defined by some initial array $\tilde{C}$.

Algorithm 11: Fan-in limited NAND synthesis of multiple-output functions

Step 1: Delete cubes $ca \in \tilde{C}$ if $a_i \neq 1$ for at least one value of $i = 1, 2, \ldots, m$.

Step 2: Find $\tilde{C}'$, a subarray of $\tilde{C}$, with the largest figure of merit and such that (a) C' has a common factor γ and (b) $w \geq 2$. If no such $\tilde{C}'$ exists, select $\tilde{C}'$ neglecting condition a. If still no $\tilde{C}'$ exists, neglect conditions a and b. (This case corresponds to implementing the functions separately using the single-output NAND synthesis algorithm.)

Step 3: Let $C_0 = \{c \mid ca \in \tilde{C}'$ and $a_i = 1$ for some $i \in \lambda\}$, and $DC = \{c \mid ca \in \tilde{C}'$ and $c \notin C_0\}$, and find a minimal cover C^*.

Step 4: Synthesize $\bar{C}^*$ using a type k circuit, where $\$_k(C^*) = $ min $[\$_1(C^*), \$_2(C^*), \$_3(C^*)]$. Fan-out the result $\bar{C}^*$ to functions $f^i, i \in \lambda$.

Step 5: In array A, for each $a \in A'$ such that $a_i = 1$ where $i \in \lambda$, change a_i to d. Return to step 1.

Note that in this synthesis procedure it was never necessary to simplify a multiple-output function.

2.5.4. Synthesis of Multiple-Output Fan-Out Limited NAND Networks

In the circuit implementatin of C shown in Fig. 2-17, the literal x_1 is an input to gate 3 only; therefore, the fan-out of x_1, denoted by $h(x_1)$, is one. If this same cover C were implemented using only AND functions, then the fan-out of x_1 would be four, i.e., $h(x_1) = |\{c \in C|\ c_1 = 1\}|$. Thus the fan-in synthesis algorithm has reduced the fan-out of x_1 by three. The reduction in the fan-out of all literals for this example is summarized below. In general, the fan-in factorization tends to reduce the fan-out of the literals.

x_1	3	x_3	0	x_5	0
$\bar{x}_1$	0	$\bar{x}_3$	1	$\bar{x}_5$	0
x_2	0	x_4	1	x_6	0
$\bar{x}_2$	3	$\bar{x}_4$	0	$\bar{x}_6$	0

If the fan-out $h(l)$ of a literal l in a given circuit is greater than FO, then this circuit is not yet in realizable form. This problem can be handled by either directly fanning-out this literal using additional gates, or by employing a factorization process.

To illustrate these techniques, consider the circuit shown in Fig. 2-19a. Here the fan-out of x_1 and x_2 is four. Assume $FO = 2$. Then these literals can be directly fanned-out as shown in Fig. 2-19b, where now the output of each gate is fanned-out to at most two gates. Some gates have inputs which are encircled. Here, either the solid input line or the dashed input line should be chosen. In general, the dashed inputs represent fewer input leads but greater signal delay.

In general, to directly fan-out a literal l to $h(l)$ gates requires $\mathscr{H}(l)$ gates, where

$$\mathscr{H}(l) = \begin{cases} 0 & h(l) \leq FO \\ \left\lceil \dfrac{h(l) + 1}{FO} \right\rceil & FO < h(l) < FO(FO + 1) \end{cases}$$

Note that the direct fan-out of a literal is analogous to using AND functions in the fan-in synthesis procedure.

In Fig. 2-19c we illustrate the concept of first implementing the product $x_1 x_2$ and then directly fanning-out this term. Note that this circuit requires three less gates than the previous one.

We will now formalize the procedure for synthesizing fan-out limited NAND networks. Let $C = \{c^1, c^2, \ldots, c^p\}$ be a set of cubes, each of which represents an input to a gate. Let $h(l)$ represent the number of appearances of literal l among the products $P(c)$, $c \in C$, and set $S = \{l\,|\,h(l) > FO\}$. S is

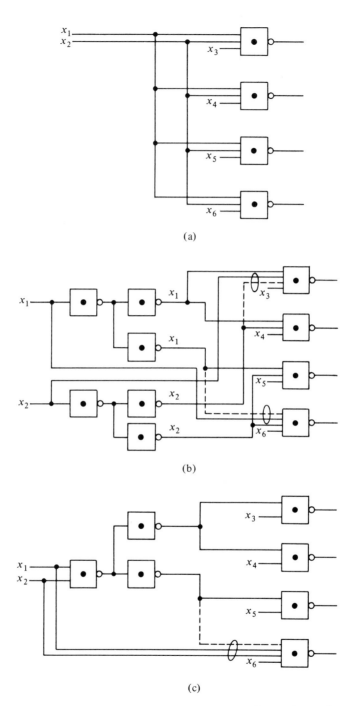

Fig. 2-19. (a) Original circuit ($FO = 4$) (b) Direct fan-out ($FO = 2$) (c) Factorization ($FO = 2$).

Sec. 2.5 Synthesis of Constrained NAND Networks via Factorization

the set of literals which do not satisfy the fan-out constraint. Now γ is said to be a *proper factor* if γ is a factor of C and if each literal in $P(\gamma)$ is an element of S. If c^i is a parent of γ, then upon factorization the input to its associated gate becomes the pair $(c^i/\gamma, \gamma)$.

For example, for the circuit shown in Fig. 2-19a, we have

$$C = \begin{cases} 1 & 1 & 1 & x & x & x & & c^1 \\ 1 & 1 & x & 1 & x & x & & c^2 \\ 1 & 1 & x & x & 1 & x & & c^3 \\ 1 & 1 & x & x & x & 1 & & c^4 \end{cases}$$

For $FO = 2$ we obtain $S = \{x_1, x_2\}$ and $\gamma = (1\ 1\ x\ x\ x\ x)$. To form the factor γ requires the use of the literals x_1 and x_2. We reflect this fact by adding this cube to the set of cubes which must be implemented. We therefore obtain, upon factoring C by γ, the set of cubes

$$C = \begin{cases} x & x & 1 & x & x & x & & c^1/\gamma \\ x & x & x & 1 & x & x & & c^2/\gamma \\ x & x & x & x & 1 & x & & c^3/\gamma \\ x & x & x & x & x & 1 & & c^4/\gamma \\ 1 & 1 & x & x & x & x & & \gamma \end{cases}$$

Since S is now null, no additional factoring is required. Note that, for each literal l in the term $P(\gamma)$, $h(l)$ has been reduced by the amount $h(\gamma) - 1$.

In general, it is not necessary to factor each cube c^i which is a parent of γ. For example, for the set of cubes

$$C = \begin{cases} x & x & 1 & x & x & x & & c^1/\gamma \\ x & x & x & 1 & x & x & & c^2/\gamma \\ x & x & x & x & 1 & x & & c^3/\gamma \\ 1 & 1 & x & x & x & 1 & & c^4 \\ 1 & 1 & x & x & x & x & & \gamma \end{cases}$$

we again have $S = \varphi$. The conditions under which the input to a gate c^i is replaced by $(c^i/\gamma, \gamma)$ are given in the following algorithm. Note that, by placing γ in the set C, we allow factors to be factored.

Algorithm 12: Fan-out factorization

Step 1: Given C, determine $h(l)$ for all literals.
Step 2: Set $S = \{l \mid h(l) > FO\}$. If $S = \varphi$, exit; otherwise continue.
Step 3: Delete c^i if the intersect of no literal in $P(c^i)$ is an element of S.

Step 4: Determine a proper factor γ of C having height of at least 2. If no proper factor exists, go to step 6; otherwise continue.

Step 5: Let $M = \max_l [h(l) - (h(\gamma) - 1)]$ where l is a literal in $P(\gamma)$.

(a) If $M \geq FO$, replace each parent c^i or γ by c^i/γ, and set $h(l) - (h(\gamma) - 1) \rightarrow h(l)$ for each literal l appearing in $P(\gamma)$.

(b) If $M < FO$, replace only $h(\gamma) - (FO - M)$ parents c^i of γ by c^i/γ, preferably those of maximum width, and set $h(l) - [h(\gamma) - (FO - M) - 1] \rightarrow h(l)$ for each literal l appearing in $P(\gamma)$. Note that additional parents of γ may be factored as long as no new gates are required to form γ.

(c) Insert γ as a new cube in C, directly fan-out γ to each cube factored, and return to step 2.

Step 6: Directly fan-out each $l \in S$ to degree $h(l)$.

In our example, $M = 1 < FO = 2$; hence, only $h(l) - (FO - M) = 4 - 1 = 3$ parents of γ need to be factored.

Summary. The complete synthesis procedure for multiple-output fan-in, fan-out limited NAND networks is summarized below.

1. Construct and initial functional cover for the function to be synthesized, and determine the parameters FI and FO.
2. Carry out the multiple-output synthesis algorithm for fan-in limited networks. This algorithm employs a subroutine to find an irredundant or minimum cost cover for a single-output function.
3. Carry out the fan-out factorization algorithm to obtain the final solution.

In this section we have outlined a few techniques for factorization and synthesis of fan-in and fan-out limited NAND networks. Here our building block, namely, the NAND gate, was a fairly elementary element. Currently there is much interest in trying to develop synthesis techniques where the basic elements are more complicated than just single gates such as AND, OR, and NAND. Much more work needs to be done in this area before an efficient, useful synthesis system will exist for this more general problem.

Significant contributions to this problem of synthesizing multiple-output switching functions in terms of multiple-level networks, not discussed in this section, have been made by many researchers. At present, some of these techniques appear to be practical only for small problems. We will now briefly mention a few of these contributions:

1. Dietmeyer and Schneider [19] have outlined a synthesis procedure where the individual logic elements can be complex logic modules.
2. Davidson [14] has formulated the branch-and-bound technique with back-

Sec. 2.6 Circuit Modification Techniques 77

tracking so that minimum cost NAND networks (multilevel with fan-in constraints) can be synthesized.

3. Logan [45], [46], has proposed and implemented a synthesis system based upon his concept of "logical determinism."
4. Roth [80], [82], has extended his original work on minimization to cover the problem of multilevel synthesis.

Other work dealing with multilevel and/or multiple-output switching function synthesis is reported in references [4], [10], [32], [41], [50], [96], [101].

2.6 CIRCUIT MODIFICATION TECHNIQUES

In the previous sections we presented techniques for synthesizing circuits starting from a functional representation of the problem. In this section† we will assume we are given a circuit, and it is desired to modify the type of logical elements used in it without changing its overall logical transfer func-

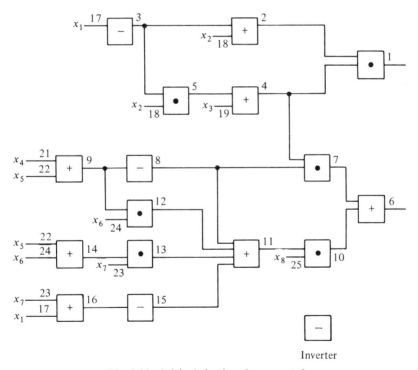

Fig. 2-20. Original circuit to be converted.

†Many of the techniques discussed in this section are based upon the IBM System CIMPL.

tion. For example, an engineer may have designed a network using AND and OR gates and must now convert it into NAND-NOR technology.

We will assume all gates (signals) are indexed by a unique integer. Input signals to the circuit are called primary inputs (pi), and those signals which represent the functions to be implemented are called primary outputs (po). Let $U^{-1}(i)$ be the set of inputs to gate i, and $U(i)$ the set of gates having signal i as an input. For example, in Fig. 2-20, we have $U^{-1}(4) = \{5, 19\}$, $U(4) = \{1, 7\}$, the set of primary outputs is $\{1, 6\}$, and the set of primary inputs is $\{17, 18, 19, 20, 21, 22, 23, 24, 25, 26\}$. Since each gate is assumed to have only one output, signal and gate indices are equivalent, e.g., the output of gate 4 is signal 4.

2.6.1. Gate Type Conversion

There are a number of different approaches to the proplem of converting a logic network from one family of gate types to another [2], [56]. In this section we will outline one procedure for converting from a family of AND, OR, INVERT gate types to NAND, NOR, INVERT.

The first part of this procedure consists of tracing a path from each output terminal backward until either an input terminal, another net output, or a previously encountered logic gate is reached. As each gate is encountered, a level of either even (E) or odd (O) is assigned to it. The level of every output is defined to be odd. The inputs to all even (odd) level gates must be from odd (even) level gates. If, when tracing backwards, we find that this condition cannot be satisfied, then an inverter must be inserted between the two gates. It is necessary to think of primary inputs being associated with an odd level; therefore, some inverters may be required on primary input lines. Once a level has been assigned to each gate, a simple procedure can be followed for replacing the present gate types by the new gate types.

The algorithm for tracing back through the network, assigning levels, and noting the positions where inverters are to be inserted is given next.

Algorithm 13: Level assignment and inverter insertion

Step 1: Place the index number of each primary output gate onto stack† S, and set the level of these gates to odd (O).

Step 2: If S is empty we are done; otherwise, pop S and label the value obtained i.

†A stack is a particular storage structure. Data are placed on the stack by a *push* operation and are removed by a *pop* operation. The stack has the property that a pop operation removes the data in the opposite order in which it was pushed. That is, the first piece of data pushed is the last to be retrieved, and the last piece of data pushed is the first to be retrieved.

Step 3: For each $j \in U^{-1}(i)$, do steps (a) and (b).

(a) If j has a level assigned to it, and if this level is opposite to that of i, do nothing; otherwise, mark the line from j to gate i (an inverter must be inserted here).

(b) If j is a gate and has no level yet assigned to it, set its level opposite to that of i, and push j onto S.

Step 4: Return to step 2.

Applying this algorithm to the circuit in Fig. 2-20 gives the results shown in Fig. 2-21. The contents of the stack S after each push and pop operation are as follows

Stack S

1	φ	10	9, 10
2	1, 6	11	10
3	2, 4, 6	12	11
4	3, 4, 6	13	12, 13, 15
5	4, 6	14	13, 15
6	5, 6	15	14, 15
7	6	16	15
8	7, 10	17	16
9	8, 10	18	φ

We first place the indices 1 and 6 on the stack and assign these two gates a level value of odd. We then pop the top of the stack, obtain $i = 1$, and push $U^{-1}(1) = \{2, 4\}$. Eventually we obtain $i = 3$ and assign gate 3 a level of odd. Since gate 3 has a primary input, we mark this input line to indicate that an inverter is to be inserted at this point. When we later obtain $i = 5$ from the stack and assign it a level of odd, we see that one of its inputs, namely gate 3, has already been assigned a logic level of odd; hence we must mark the line connecting these two gates.

The next part of the algorithm is to carry out the gate conversions according to the following rules:

1. Replace each odd level AND(OR) gate by a NOR(NAND) gate.
2. Replace each even level AND(OR) gate by a NAND(NOR) gate.
3. Delete all inverters in the original circuit, and insert inverters at the points marked by the previous algorithm.

The result of this operation, when applied to the circuit in Fig. 2-21, is shown in Fig. 2-22. This conversion could have been carried out simultaneously with the previous algorithm, but it has been separated here for clarity of presentation.

80   Logic Synthesis   Chap. 2

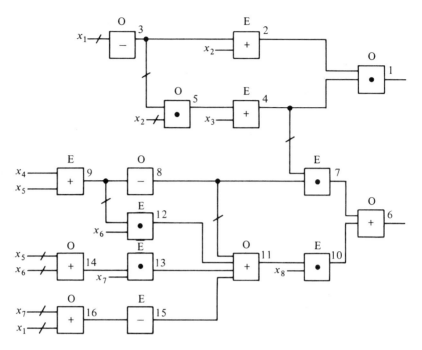

Fig. 2-21. Determination of levels and points (indicated by a slash) where inverters must be inserted.

As a final step, we can eliminate those inverters producing more than one signal inversion in a path, i.e., we do not need two or more inverters in a sequence. In our example, signal x_1 can be applied directly to gate 5, and gate 27 can be deleted.

The gate count of this result can be reduced if we take advantage of the equivalence between certain structures, as shown in Fig. 2-23. Applying these transformations to the circuit of Fig. 2-22 produces the result shown in Fig. 2-24.

In general, if a NAND or NOR gate has p inputs emitting from inverters, each with a fan-out of one, and q inputs from noninverters, then these transformations can be employed at a net increase in gates of $q + 1 - p$. Hence, if $p > q + 1$, the gate count can be decreased. Gates 14 and 16 are the only gates in the circuit whose transformation will lead to a gate count reduction.

Note that solutions other than the one shown here can also be obtained by this conversion procedure if the elements are placed on the stack in a different order. For example, if the stack contents were

Sec. 2.6   *Circuit Modification Techniques*   81

```
         Stack S
      1   φ
      2   6, 1
      3   7, 9, 1
      4   4, 8, 9, 1
          etc.
```

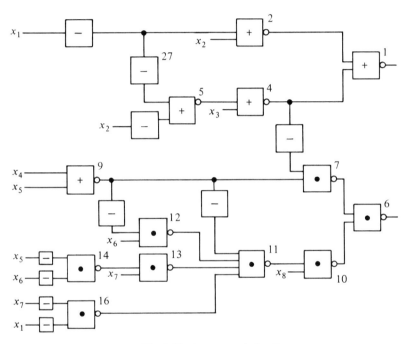

Fig. 2-22. A converted circuit.

then gate 4 would be assigned an odd level, there would be no inverter inserted between gates 4 and 7, but there would be one inserted between gates 4 and 1.

If only NAND gates are available, then each NOR gate can be replaced by a NAND gate and an inverter on each input and output line, where an inverter would consist of a one input NAND gate.

2.6.2. Functional Packages

In many circumstances one is given a circuit element which is more complex than just a single gate. One such element, say E_1, is shown in Fig.

82   Logic Synthesis   Chap. 2

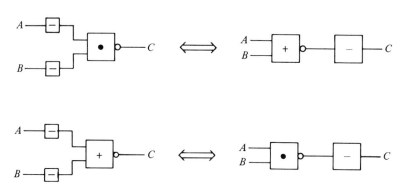

Fig. 2-23. Some equivalent circuits.

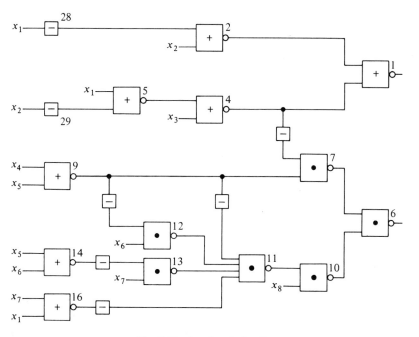

Fig. 2-24. A reduced circuit.

2-25a. Assume it is desired to synthesize a network given a family of four circuit element types, namely, a NAND gate, a NOR gate, an inverter, and E_1, where each of these circuit elements has a cost of one unit. In Fig. 2-25 we illustrate three circuit elements, E_1, E_2, E_3, all of which are logically equivalent. Since E_1 is the cheapest of the three, if a subcircuit configuration corresponding to, say, type E_2 is found in a large circuit, then this subcircuit

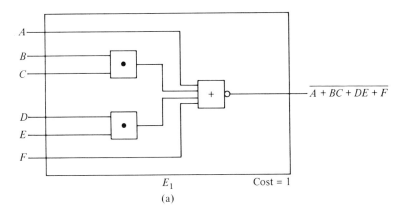

(a)

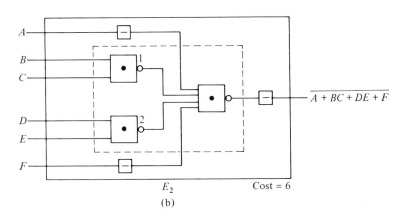

(b)

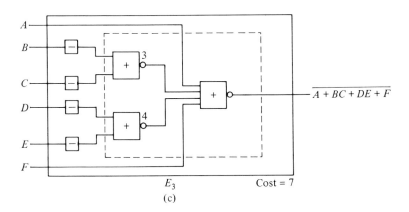

(c)

Fig. 2-25. Three equivalent circuit elements.

can be replaced by circuit element E_1 at a net savings in cost of five units. If this same subcircuit differed from E_2 in that it did not have an inverter in the input line corresponding to signal A, then this subcircuit could again be replaced by an E_1 circuit element if an inverter is inserted in the line corresponding to signal A. The net savings in cost is now only three units.

In general, the problem is to transform a given circuit into a form where the given circuit element types can be used more efficiently, i.e., when a reduction in circuit cost is realized. In applying these transformations, certain fan-out constraints must be observed, namely, gates 1, 2, 3, and 4 in Fig. 2-25 must have a fan-out of 1.

We will now trace back through our circuit (Fig. 2-24) and carry out this transformation process. We first encounter NOR gate 1 having an input from NOR gate 2. It also has an input from NOR gate 4; but since this gate has a fan-out greater than 1, it cannot be used at this time. Using the equivalence between E_3 and E_1, we obtain the result shown in Fig. 2-26. Note that an inverter was added on line x_2 and one was deleted from line x_1. In similar fashion, we can replace gates 4, 5, and 29 using the equivalence between

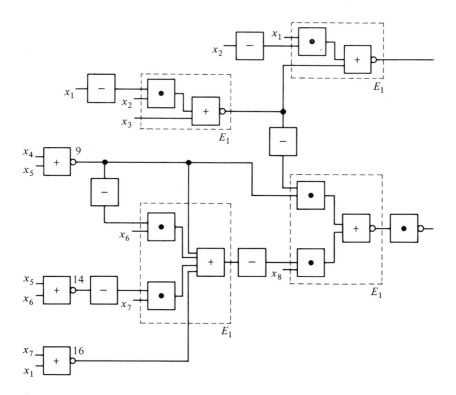

Fig. 2-26. Implementation using standard circuit element E_1.

E_3 and E_1; gates 6, 7, and 10 using the equivalence between E_2 and E_1; and again for gates 11, 12, and 13.

In general, one is given a circuit described in terms of one family of circuit elements, and it is necessary to generate a logically equivalent description in terms of a second family of circuit elements. For this case, many different transformations may exist. No general algorithm exists which will efficiently carry out this transformation. All we have shown is that, for specific cases, procedures can be developed for converting between circuit families, and that these procedures can be easily implemented on a computer.

2.7 DATA STRUCTURES

There are many different data structures which can be employed when implementing logic synthesis algorithms. One factor which influences the structure to be used is the computer on which the program is to execute. For example, the IBM 1620 is a byte-oriented, variable word-length, decimal machine which is best utilized by a data structure different from that used on a word-oriented binary machine.

Equivalent operations on cubes can be carried out in any number system, such as decimal, octal, ternary, or binary.

Two main considerations in selecting how to represent a cube c are the amount of memory required and the computation time required to manipulate c. To reduce memory requirements, the components of c can be packed into a word(s) rather than stored one component per word. A second advantage of this technique is that it allows the processing of many components of c to be carried out simultaneously.

To briefly illustrate these concepts, consider a work-oriented binary machine having n data bits (excluding sign) per word.

Let $c^i = (c_1^i c_2^i \ldots c_n^i)$ where $c_j^i \in \{0, 1, x\}$. Since c_j^i can take on three different values, at least two bits are required to represent c_j^i. We will therefore use two words, V_i and T_i, to describe c^i, as described by House [36] and Breuer [7]. We have $V_i = (v_{i1} v_{i2} \ldots v_{in})$ and $T_i = (t_{i1} t_{i2} \ldots t_{in})$ where: v_{ij} and t_{ij} are the jth bit positions of V_i and T_i; where $v_{ij} = 1$ if $c_j^i \neq x$, otherwise $v_{ij} = 0$; and where $t_{ij} = 1$ if $c_j^i = 1$, otherwise $t_{ij} = 0$. Thus V is the characteristic function for the variables, and T is the "true" indication. For example, corresponding to $c^1 = (10x1x \ldots x)$ we have $V_1 = (11010 \ldots 0)$ and $T_1 = (10010 \ldots 0)$.

We now illustrate a few examples of how all elements of c^i can be processed in parallel.

1. $c^i \cap c^j = \varphi$ if and only if

$$P_1(i, j) \doteq V_i \cdot V_j \cdot (T_i \oplus T_j) \neq \underline{0}$$

where $\underline{0} \doteq (0\,0\,\cdots\,0)$.

Examples: (a) Let $c^i = (10x)$ and $c^j = (00x)$. Then

$$P_1(i,j) = (11x) \cdot (11x) \cdot ((100) \oplus (000))$$
$$= (11x) \cdot (100) = (100) \neq \underline{0}$$

and $c^i \cap c^j = \varphi$.

(b) Let $c^i = (10x)$ and $c^j = (1x0)$. Then

$$P_1(i,j) = (110) \cdot (101) \cdot ((100) \oplus (100))$$
$$= (100) \cdot (000) = \underline{0}$$

and $c^i \cap c^j \neq \varphi$.

2. $c^i \Rightarrow c^j$ if and only if

$$P_2(i,j) \doteq V_i \cdot V_j = V_j$$

and

$$P_3(i,j) \doteq V_j \cdot (T_i \oplus T_j) = \underline{0}$$

Examples: (a) Let $c^i = (111)$ and $c^j = (1xx)$. Then

$$P_2(i,j) = (111) \cdot (100) = (100) = V_j$$

and

$$P_3(i,j) = (100) \cdot ((111) \oplus (100))$$
$$= (100) \cdot (011) = \underline{0}$$

(b) Let $c^i = (1xx)$ and $c^j = (xx0)$. Then

$$P_2(i,j) = \underline{0} \neq V_j$$

(c) Let $c^i = (100)$ and $c^j = (000)$. Then

$$P_3 = (111) \cdot (100) = (100) \neq \underline{0}$$

3. If $c^i \mathcal{F} c^j = c^k$, then we have that

$$V_k = V_i \cdot V_j \cdot \overline{(T_i \oplus T_j)}$$

and

$$T_k = V_k \cdot T_i = V_k \cdot T_j$$

Example: Let $c^i = (100)$ and $c^j = (101)$. Then

$$V_k = (111) \cdot (111) \cdot (\overline{(100) \oplus (101)})$$
$$= (111) \cdot (\overline{001})$$
$$= (111) \cdot (110) = 110$$

and

$$T_k = 110 \cdot 100 = 100$$

Therefore $c^k = (10x)$,

4. Let $c^i \not\subset c^j = c^k$; then we have that $c^k = \psi$ if

$$\|P_1(i,j)\| \neq 1$$

otherwise

$$V_k = (V_i + V_j) \cdot \overline{P_1(i,j)}$$

and

$$T_k = (T +_i T_j) \cdot \overline{P_1(i,j)}$$

The symbol $\|W\|$ indicates the number of *one*-bits in word W. Note that $\|W\| = 1$ if and only if $W \neq \underline{0}$ and $W \cdot (W - 1) = \underline{0}$, where $W - 1$ represents normal binary subtraction, and $1 = (00 \cdots 0\overline{1})$. For example, let $W = (101)$. Then $W \cdot (W - 1) = (101) \cdot (100) = (100) \neq \underline{0}$. If $W = (100)$, then $W \cdot (W - 1) = (100) \cdot (011) = \underline{0}$.

5. For the sharp operator we have
   (a) $c^i \# c^j = c^i$ if and only if $P_1(i,j) \neq \underline{0}$.
   (b) $c^i \# c^j = \psi$ if and only if

$$P_2(i,j) = V_j$$

and

$$P_3(i,j) = \underline{0}$$

(c) Otherwise

$$c^i \# c^j = \cup \ a^k$$

where each a^k is represented by the pair $(V(k), T(k))$ determined as follows:

Step 1: $0 \longrightarrow k$, $1 \longrightarrow m$; $(00 \cdots 01) \longrightarrow P$.

Step 2: If $P \cdot \bar{V}_i \cdot V_j \neq \underline{0}$, then set $k + 1 \longrightarrow k$ and

$$V(k) = V_i + P; \quad T(k) = T_i + P \cdot \bar{T}_j$$

Step 3: $P \times 2 \longrightarrow P$ (shift P left one position)
$m + 1 \longrightarrow m$.

Step 4: If $m > n$, stop; otherwise go to step 2.

It is seen that, with very few computer instructions, most of the cubical operators can be implemented. Hence the cubical representation is advantageous because it leads to both condensed data representation as well as efficient computational procedures.

2.8 MISCELLANEOUS TOPICS

In this section we will briefly discuss two additional areas where programs have been developed for carrying out parts of the synthesis procedure, namely, the areas of functional decomposition and sequential machine. Other areas of activity include the synthesis of threshold logic networks, synthesis using universal elements, and synthesis using cellular arrays.

2.8.1. Decomposition

A function f is said to possess a simple decomposition if there exist functions h and g such that $f = h(g(A), B))$, where A and B are subsets of $X = \{\dot{x}_1, \dot{x}_2, \ldots, \dot{x}_n\}$. If $A \cap B = \varphi$, then h and g define a simple disjoint decomposition, as illustrated in Fig. 2-27. Decomposition reduces a problem

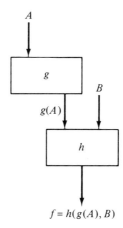

Fig. 2-27. Simple decomposition of f.

$f = h(g(A), B)$

of synthesizing an n-variable function into two problems, namely, synthesizing the $|A|$ variable function g and the $|B| + 1$ variable function h.

A general decomposition of f has the form

$$f = h(g_1(A_1), g_2(A_2), \ldots, g_r(A_r), A_{r+1})$$

where

$$\bigcup_{i=1}^{r+1} A_i = X \quad \text{and} \quad |A_i| \geq 2 \quad \text{for} \quad i = 1, 2, \ldots, r.$$

For example, the odd-parity function, f, on four variables can be decomposed as follows

$$f(\dot{x}_1, \dot{x}_2, \dot{x}_3, \dot{x}_4) = h(g_1(\dot{x}_1, \dot{x}_2), g_2(\dot{x}_3, \dot{x}_4)) = g_1 \oplus g_2$$

where

$$g_1(\dot{x}_1, \dot{x}_2) = x_1 \oplus x_2 \quad \text{and} \quad g_2(\dot{x}_3, \dot{x}_4) = x_3 \oplus x_4$$

Synthesis procedures based upon decomposition are reported in references [1], [4], [13], [14], [39]. At present, only problems in a few variables can be efficiently synthesized by this approach.

2.8.2. Sequential Machine Synthesis

A number of algorithms have been programmed for carrying out the synthesis of sequential machines [89], [90]. Typically the input represents the state table of the machine, and the program carries out the following functions:
1. state reduction;
2. state assignment;
3. synthesis of logic.

Both synchronous and asynchronous circuits can be handled, though computation time typically limits the size of problems to circuits having less than fifty states and four inputs.

ACKNOWLEDGMENTS

I would like to thank Dr. Y-H Su for his many helpful suggestions.

PROBLEMS

1. Show geometrically that $U_3 \# 01x = \{1xx, x0x\}$.
2. Show that the $\#$-product has the following properties:
   (a) $u \# v \neq v \# u$ noncommutative;
   (b) $(u \# v) \# w \neq u \# (v \# w)$ nonassociative;
   (c) $\{u, v\} \# w = \{(u \# w), (v \# w)\}$
      $(u \cap v) \# w = (u \# w) \cap (v \# w)$ } distributive;
   (d) $u \# v = \varphi$ if $u \Rightarrow v$;
   (e) $u \# v \Rightarrow u$ and $u \# v = u$ if $u \cap v = \varphi$;
   (f) $(u \# v) \# w = (u \# w) \# v$.
3. Show that $U_n \# (u \cap (U_n \# v)) = U_n \# (u \# v)$ and $U_n \# (u \cap (U_n \# v)) = \{v, (U_n \# u)\}$.
4. If A and B are two sets of cubes, show that $A - B \neq A \# B$ unless $(A - B) \cap B = \varphi$.
5. Prove that $a \# b$ is the set of all prime implicants of the complex $K^0(a) - K^0(a) \cap K^0(b)$ and, therefore, $P(u \# v) = P(u) \overline{P(v)}$.
6. Let C be the set of all prime implicants of a complex K. Prove that $C \# b$ is the set of all prime implicants of the complex $K^0(C) - K^0(C) \cap K^0(b)$. Algorithm 1 follows from this result.
7. Prove that $c = a \cap b$ is the largest cube which subsumes a and b.
8. Given the definition of the consensus operator via its coordinate table, prove that $c = a \mathbin{\tilde{\cancel{c}}} b$ is the largest cube such that $P(c) \not\Rightarrow P(a)$, $P(c) \not\Rightarrow P(b)$, and $P(c) \Rightarrow P(a) + P(b)$.

9. Prove that the process of iterative consensus (Algorithm 2) produces the set of all prime implicants Z.
10. Prove that, if C_0 is an initial cover for f, and if 0-cube c is a prime implicant of f, then $c \in C_0$.
11. Prove Theorem 1.
12. Prove Theorem 2.
13. Prove Theorem 3.
14. Let C_0 and C_1 be two sets of cubes. Prove the following:

    Theorem: $K(C_0) \subseteq K(C_1)$ if and only if, for all $c \in C_0$, $c \mathrel{\#} C_1 = \varphi$.

    Corollary: $K(C_0) = K(C_1)$ if and only if, for all $c \in C_\delta$, $c \mathrel{\#} C_{1-\delta} = \varphi$ for $\delta \in \{0, 1\}$.

    This result provides one with a convenient way for testing the equivalence of two covers.

15.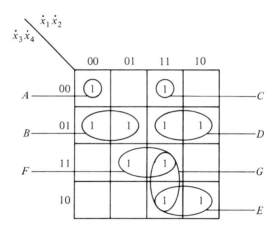

    Fill in the missing spaces according to the above figure.

    $A = 0000$ $D =$ $G =$
    $B =$ $E =$
    $C =$ $F =$
    $A \mathrel{\not\subset} B =$ $B \mathrel{\#} A =$ $S[\{E, F, G\}] =$
    $A \mathrel{\not\subset} C =$ $G \mathrel{\#} F =$ $S[\{A, B\}] =$
    $B \mathrel{\not\subset} F =$ $G \mathrel{\#} \{E, F\} =$

16. Given the initial covers for a function f

    $$C_0 = \{000x, 0111, 1110, 1011, 100x, 11x1\}$$

    and

    $$DC = \{1100, 0101\}$$

(a) Determine Z according to:
   (i) algorithm 1;
   (ii) algorithm 2;
   (iii) algorithm 3;
   (iv) partitioned list procedure.
(b) Find a minimum cost cover via algorithm 4.
(c) Construct the corresponding core matrix A for this function and simplify using the three rules of reduction.
(d) Find an irredundant cover for f via algorithm 5.
(e) Synthesize this function using NAND gates, where $FO = FI = 2$.

17. Remove all redundancies from the connection cover $\hat{C}$.

$$\hat{C} = \left\{ \begin{array}{cccc|ccc} \multicolumn{4}{c}{\overbrace{\hspace{4em}}^{c}} & \multicolumn{3}{c}{\overbrace{\hspace{3em}}^{e}} \\ x & 1 & 1 & 1 & 1 & 1 & 1 \\ 1 & 1 & 1 & x & 1 & x & 1 \\ 1 & x & x & 1 & 1 & x & 1 \\ 0 & x & 1 & x & x & 1 & 1 \\ 0 & 0 & x & x & x & 1 & 1 \\ 1 & x & 1 & x & 1 & x & x \\ x & 0 & 0 & x & x & x & 1 \end{array} \right\}$$

18. Given a set of N literals with K variables appearing in both the complemented and uncomplemented form, show that the number of unique product terms of width w which can be formed from these N literals is given by the expression

$$\sum_{i=0}^{w} 2^i \binom{K}{i} \binom{N-2K}{w-i}$$

19. Prove that, if $w = \$\gamma$ is the width of common factor γ of any subset of C, where C is a minimal cover of f, then $w \leq n - 2$.

20. Prove that an irredundant prime implicant cover of a Boolean function of n variables consists of 2^{n-1} or fewer cubes. (Therefore, for large n, it is usually very inefficient to generate Z since, for most functions, $|Z| \gg 2^{n-1}$ for $n > 10$.)

21. Prove that, if C' is a subset of C, where $h(C') = q$ and C' has a common factor of width w, then $q \leq 2^{n-w-1}$.

22. Let $C^* = \{c^i | c^i \in C, \$c^i > w\}$ and $h(C^*) = p^*$. Prove that the figure of merit of any common factor of width w of any subset of C is less than or equal to $w \cdot \min(p^*, 2^{n-w-1})$.

23. Find a factor of the set

$$C = \left\{ \begin{array}{ccccccccc} x & 0 & 0 & 0 & 1 & x & 1 & 1 & x & 0 \\ x & 0 & 1 & 1 & 1 & x & 1 & 1 & x & 0 \\ x & 0 & 1 & 0 & 1 & x & 1 & 0 & x & 0 \\ x & 0 & 0 & 1 & 0 & x & 1 & 0 & x & 0 \end{array} \right\}$$

using the $\mathscr{F}$-operator, the grow-factor algorithm, and the state-splitting algorithm.

24. Synthesize the following cover using algorithm 10

$$C = \begin{Bmatrix} x & 0 & 0 & 0 & 1 & x & 1 & 1 & x & 0 \\ x & 0 & 1 & 1 & 1 & x & 1 & 1 & x & 0 \\ x & 0 & 1 & 0 & 1 & x & 1 & 0 & x & 0 \\ x & 0 & 0 & 1 & 0 & x & 1 & 0 & x & 0 \\ 0 & x & 1 & 0 & 0 & 1 & x & x & x & x \\ x & 1 & x & x & x & x & 0 & 1 & 0 & x \\ 1 & x & 1 & 1 & 1 & x & x & x & x & x \\ 1 & x & 0 & x & 1 & x & x & 0 & x & x \\ 1 & x & 0 & 0 & 0 & 1 & x & x & x & x \\ x & 1 & 1 & 0 & 0 & 1 & 1 & x & x & 1 \end{Bmatrix}$$

(a) for $FI = 4$;
(b) for $FI = 6$.

25. Apply algorithm 12 to the result of problem 24, using $FO = 2$. How many gates in this final result have a fan-in of 4?

26. Consider the multiple-output function defined by the 0-cubes shown below.

$\dot{x}_1$	$\dot{x}_2$	$\dot{x}_3$	$\dot{x}_4$	f^1	f^2	f^3	f^4
0	0	0	0	0	1	1	0
0	0	0	1	0	1	d	d
0	0	1	0	1	d	1	d
0	0	1	1	1	0	0	1
0	1	0	0	0	0	1	1
0	1	0	1	1	d	d	d
0	1	1	0	d	0	1	1
0	1	1	1	d	1	1	d
1	0	0	0	d	1	1	d
1	0	0	1	0	0	0	0
1	0	1	0	1	1	1	d
1	0	1	1	0	0	d	1
1	1	0	0	0	0	0	0
1	1	0	1	0	0	0	0
1	1	1	0	1	1	0	1
1	1	1	1	1	1	1	1

(a) Find a minimum cost cover.
(b) Find an irredundant cover.
(c) Synthesize using NAND gates, where $FO = FI = 2$.

27. Convert the following circuit using NAND-NOR gates.

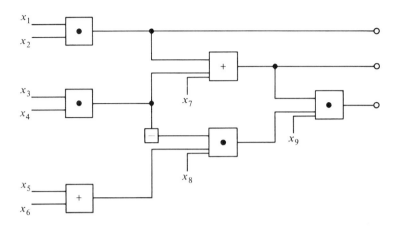

Research Problems

28. Determine an analytic measure of the computational complexity of algorithms 1, 2, and 3.
29. Investigate how delay constraints can be included in the NAND synthesis algorithm.

REFERENCES

[1] R. L. Ashenhurst, "The Decomposition of Switching Functions." *Annals of the Harvard Computational Laboratory*, Vol. 29 (1959), pp. 74–116.

[2] S. P. Asija, "Instant Logic Conversion." *IEEE Spectrum*, Vol. 5 (December, 1968), pp. 77–80.

[3] E. Balas, "An Additive Algorithm For Solving Linear Programs With Zero-One Variables." *J. Oper. Res.*, Vol. 13 (July–August, 1965), pp. 517–546.

[4] D. F. Barnard and D. F. Holman, "The Use of Roth's Decomposition Algorithm in Multilevel Design of Circuits." *The Computer Journal*, Vol. 11 (November, 1968), pp. 269–276.

[5] T. C. Bartee, "The Automatic Design of Logical Networks." *Proc. WJCC* (March, 1959), pp. 103–107.

[6] T. C. Bartee, "Computer Design of Multiple-Output Logical Networks." *IEEE Trans. on Electronic Computers*, Vol. EC-10 (March, 1960), pp. 21–30.

[7] M. A. Breuer, "Heuristic Switching Expression Simplification." *Proc. 23rd Nat. Conf. of the ACM* (1968), pp. 241–250.

[8] M. A. Breuer, "Simplification of the Covering Problem With Application to Boolean Expressions." *J. ACM*, Vol. 17, No. 1 (January, 1970), pp. 166–181.

[9] M. A. Breuer, "Generation of Optimal Code For Expressions Via Factorization." *Comm. ACM*, Vol. 12, No. 6 (June, 1969), pp. 333–340.

[10] K. K. Chakrabarti, A. K. Choudhury, and H. S. Basu, "Complementary Function Approach to the Synthesis of Three-Level NAND Networks." *IEEE Trans. on Computers*, Vol. C-19, (June, 1970), pp. 509–514.

[11] A. Cobham, R. Fridshal, and J. H. North, "A Statistical Study of the Minimization of Boolean Functions Using Integer Programming." *IBM Research Report RC-756*, June, 1962.

[12] A. Cobham, R. Fridshal, and J. H. North, "An Application of Linear Programming to the Minimization of Boolean Functions." *AIEE Symposium on Switching Circuit Theory and Logical Design*, pp. 3–9, 1961; and *IBM Research Report RC-472*, June 9, 1961.

[13] H. A. Curtis, *A New Approach to the Design of Switching Circuits*. Princeton, N. J., D. Van Nostrand Company, Inc., 1962.

[14] E. S. Davidson, "An Algorithm for NAND Decomposition Under Network Constraints." *IEEE Trans. on Computers*, Vol. C-18 (December, 1969), pp. 1098–1109.

[15] E. S. Davidson and G. Metze, "Module Complexity and NAND Network Design Algorithms." *Proc. 6th Ann. Allerton Conf. on Circuit and System Theory*, October 2–4, 1968.

[16] E. S. Davidson and G. Metze, "Comments on 'An Algorithm For Synthesis of Multiple-Output Combinational Logic'." *IEEE Trans. on Computers (Short Notes)*, Vol. C-17 (November, 1968), pp. 1091–1092.

[17] H. F. De Francesco and T. R. LaCrosse, "Automated Logical Design." *1963 IEEE International Conv. Rec.*, Vol. 11, March 27, 1963.

[18] D. L. Dietmeyer and Y. H. Su, "Logic Design Automation of Fan-In Limited NAND Networks." *IEEE Trans. on Computers*, Vol. C-18 (January, 1969), pp. 11–22.

[19] D. L. Dietmeyer and P. R. Schneider, "Identification of Symmetry, Redundancy, and Equivalence of Boolean Functions." *IEEE Trans. on Electronic Computers*, Vol. EC-16 (December, 1967), pp. 804–817.

[20] D. L. Dietmeyer and P. R. Schneider, "A Computer-Oriented Factoring Algorithm For NOR Logic Design." *IEEE Trans. on Electronic Computers*, Vol. EC-14 (December, 1965), pp. 868–875.

[21] J. R. Duley and D. L. Dietmeyer, "A Digital System Design Language (DDL)." *IEEE Trans. on Computers*, Vol. C-17 (September, 1968), pp. 850–860.

[22] J. R. Duley and D. L. Dietmeyer, "Translation of a DDL Digital System Specification to Boolean Equations." *IEEE Trans. on Computers*, Vol. C-18 (April, 1969), pp. 305–313.

[23] S. N. Einhorn, "The Use of the Simplex Algorithm in the Mechanization of Boolean Switching Functions by Means of Magnetic Cores." *IRE Trans. on Electronic Computers*, Vol. EC-10 (December, 1961), pp. 615–622.

[24] L. D. Enochson and J. P. Malbrain, "Circuit Loading Calculations from Boolean Equations in an Automated Computer Design System." *Proc. 15th Nat. Conf. ACM*, 1960.

[25] A. C. Ewing, J. P. Roth, and E. G. Wagner, "Algorithms for Logical Design." *AIEE Trans. on Comm. Electr.* (September, 1961), pp. 450–458.

[26] A. D. Falkoff, K. E. Iverson, and E. H. Sussenguth, "A Formal Description of System/360." *IBM System Journal*, Vol. 3, 1964.

[27] D. C. Forslund and R. Waxman, "The Universal Logic Block (ULB) and Its Application to Logic Design." *Proc. Seventh Annual Symposium on Switching and Automata Theory* (1966), pp. 236–250.

[28] T. D. Friedman and S. C. Yang, "Methods Used in an Automated Logic Design Generator (ALERT)." *IEEE Trans. on Computers*, Vol. C-18 (July, 1968), pp. 593–613.

[29] T. D. Friedman and S. C. Yang, "Quality of Design From an Automatic Logic Generator (ALERT)." *Proc. 1970 Design Automation Conference (SHARE-ACM-IEEE)*, pp. 71–89.

[30] A. M. Geoffrion, "An Improved Implicit Enumeration Approach for Integer Programming." *J. Oper. Res.*, Vol. 17 (May–June, 1969), pp. 437–454.

[31] J. F. Gimpel, "A Reduction Technique For Prime Implicant Tables." *IEEE Trans. on Electronic Computers*, Vol. EC-14 (1965), pp. 535–541.

[32] J. F. Gimpel, "The Minimization of TANT Network." *IEEE Trans. on Electronic Computers*, Vol. EC-16 (February, 1967), pp. 18–38.

[33] R. E. Gomory, "Outline of an Algorithm For Integer Solutions to Linear Programs." *Bull. Amer. Math. Soc.*, Vol. 64 (September, 1958), pp. 275–278.

[34] B. B. Gordon, R. W. House, A. P. Lechler, L. D. Nelson, and T. Rado, "Simplification of the Covering Problem For Multiple Output Logical Networks." *IEEE Trans. on Electronic Computers*, Vol. EC-15 (December, 1966), pp. 891–897.

[35] D. F. Gorman and J. P. Anderson, "A Logic Design Translation." *Proc. 1962 Fall Joint Computer Conference*, pp. 86–96.

[36] R. W. House and T. Rado, "On a Computer Program For Obtaining Irreducible Representations For Two-Level Multiple Input-Output Logical Systems." *J. ACM*, Vol. 10 (January, 1963), pp. 48–77.

[37] M. Karnaugh, "The Map Method For Synthesis of Combinational Logic Circuits." HIEE trans. in *Comm. and Electr.* (November, 1953), pp. 593–599.

[38] R. M. Karp, F. E. McFarlin, J. P. Roth, and J. R. Wilts, "A Computer Program For the Synthesis of Combinational Switching Circuits." *Proc. Second Annual Symposium on Switching Circuit Theory and Logical Design* (September, 1961), pp. 182–194.

[39] R. M. Karp, "Functional Decomposition of Switching Circuit Design." *J. SIAM*, Vol. 11 (June, 1965), pp. 291–335.

[40] V. D. Kazakov, "The Minimization of Logical Functions of a Large Number of Variables." *Autom. Remote Contr.*, Vol. 23 (March, 1963), pp. 1161–1165.

[41] E. L. Lawler, "An Approach to Multilevel Boolean Minimization." *J. ACM*, Vol. 11 (July, 1964), pp. 283–296.

[42] E. L. Lawler, "Covering Problems: Duality Relations and a New Method of Solution." *J. SIAM*, Vol. 14 (September, 1963), pp. 1115–1132.

[43] C. Y. Lee, "Switching Functions on an N-Dimensional Cube." *AIEE Trans. on Comm. Electr.*, No. 14 (September, 1954), pp. 289–291.

[44] D. W. Lewin and M. C. Waters, "Computer Aids to Logic System Design." *Computer Bulletin*, Vol. 13 (November, 1969), pp. 382–388.

[45] J. R. Logan, P. Lucas, and C. Wilkening, "Deterministic Network Synthesis and Large Scale Integration." *AFAL-TR-68-13*, Vols. 1 and 2 (Wright-Patterson Air Force Base, February, 1969).

[46] J. R. Logan, "Logical Determinism in Network Synthesis." *AFAL-TR-65-72* (Wright-Patterson Air Force Base, 1965).

[47] F. Luccio, "A Method for the Selection of Prime Implicants." *IEEE Trans. on Electronic Computers*, Vol. EC-15 (April, 1966), pp. 205–211.

[48] J. M. Mage, "Application of Iterative Consensus to Multiple-Output Functions." *IEEE Trans. on Computers (Correspondence)*, Vol. C-19, No. 4 (April, 1970), p. 359.

[49] Y. K. Maki, J. A. Tracey, and R. J. Smith II, "Generation of Design Equations in Asynchronous Sequential Circuits." *IEEE Trans. on Computers*, Vol. C-18 (May, 1969), pp. 467–472.

[50] M. A. Marin, *Investigation of the Field of Problems for the Boolean Analyzer*. Ph. D. dissertation, UCLA, Engineering Report No. 68–28, June, 1968.

[51] E. J. McCluskey, *Introduction to the Theory of Switching Circuits*. New York, McGraw-Hill, Inc., 1965.

[52] E. J. McCluskey, "Minimization of Boolean Functions." *Bell Sys. Tech. J.*, Vol. 35 (1956), pp. 1417–1444.

[53] E. J. McCluskey and H. Schorr, "Essential Multiple-Output Prime Implicants." *Proc. Symposium on Mathematical Theory of Automata*, Vol. 12, pp. 437–458 (Polytechnic Press, April 1962).

[54] E. J. McCluskey and I. B. Pyne, "An Essay on Prime Implicant Tables." *J. SIAM*, Vol. 9 (December, 1961), pp. 604–631.

[55] R. McNaughton and B. Mitchell, "The Minimality of Rectifier Nets With Multiple Incompletely Specified Outputs." *J. Franklin Inst.*, Vol. 26 (December, 1957), pp. 257–280.

[56] R. E. Merwin and J. L. Sanborn, "Digital Computers For Logical Designs," in M. Klerer and G. Korn, eds., *Digital Computer User's Handbook* (New York, McGraw-Hill Book Company, 1967), Part 4, pp. 167–192.

[57] F. Mileto and G. Putzolu, "Statistical Complexity of Algorithms For Boolean Function Minimization." *J. ACM*, Vol. 12 (July, 1965), pp. 356–364.

[58] F. Mileto and G. Putzolu, "Average Values of Quantities Appearing in Boolean Function Minimization." *IEEE Trans. on Electronic Computers*, Vol. EC-13 (April, 1964), pp. 87–92.

[59] F. Mileto and G. Putzolu, "Average Values of Quantities Appearing in Multiple Output Boolean Minimization." *IEEE Trans. on Electronic Computers*, Vol. EC-14 (August, 1965), pp. 542–552.

[60] R. E. Miller, *Switching Theory, Vol. 1: Combinational Circuits.* New York, John Wiley & Sons, Inc., 1965.

[61] E. Morreale, "Partitioned List Techniques for the Synthesis by Computer of Two-Level AND-OR Networks." *Proc. 23rd Nat. Conf. ACM* (1968), pp. 355–364.

[62] E. Morreale, "Partitioned List Algorithms For Prime Implicant Determination From Canonical Forms." *IEEE Trans. on Electronic Computers*, Vol. EC-16 (October, 1967), pp. 611–620.

[63] E. Morreale, "Computational Complexity of Partitioned List Algorithms." *IEEE Trans. on Computers*, Vol. C-19 (May, 1970), pp. 421–429.

[64] E. Morreale, "Recursive Operators For Prime Implicant and Irredundant Normal Form Determination." *IEEE Trans. on Computers*, Vol. C-19 (June, 1970), pp. 504–509.

[65] D. E. Muller, "Application of Boolean Algebra to Switching Circuit Design and to Error Detection." *IRE Trans. on Electronic Computers*, Vol. EC-3 (September, 1954), pp. 6–12.

[66] E. F. Norris and T. E. Wohr, "Automatic Implementation of Computer Logic." *Comm. ACM*, Vol. 1 (May, 1958), pp. 14–20.

[67] Y. N. Patt, "Complex Logic Module For Synthesis of Combinational Switching Circuits." *Proc. 1967 SJCC*, Vol. 30, pp. 699–705.

[68] S. R. Petrick, "A Direct Determination of the Irredundant Forms of a Boolean Function from the Set of Prime Implicants." *AFCRC-TR-56-110* (Air Force Cambridge Research Center, April 1956).

[69] H. Potash, "A Digital Control Design Language." Ph.D. dissertation UCLA Technical Report No. 69–21, May, 1969.

[70] R. Prather, "Computational Aids For Determining the Minimal Form of a Truth Function." *J. ACM*, Vol. 7 (October, 1960), pp. 299–310.

[71] R. M. Proctor, "A Logic Design Translator Experiment Demonstrating Relationships of Language to Systems and Logic Design." *IEEE Trans. on Electronic Computers*, Vol. EC-13 (August, 1964), pp. 422–430.

[72] W. V. Quine, "A Way to Simplify Truth Functions." *Am. Math. Monthly*, Vol. 62 (1955), pp. 627–631.

[73] W. V. Quine, "The Problem of Simplifying Truth Functions." *Am. Math. Monthly*, Vol. 59 (1952), pp. 521–531.

[74] S. U. Robinson III and R. W. House, "Gimpel's Reduction Technique Extended to the Covering Problem With Costs." *IEEE Trans. on Electronic Computers*, Vol. EC-16 (August, 1967), pp. 509–514.

[75] J. G. Root, "An Application of Symbolic Logic to a Selection Problem." *J. Oper. Res.*, Vol. 12 (July–August, 1964), pp. 519–529.

[76] R. Roth, "Computer Solutions to Minimal Cover Problems." *J. Oper. Res.* Vol. 17 (May, 1969), pp. 455–465.

[77] J. P. Roth, "Algebraic Topological Methods in Synthesis." The proceedings of an international symposium on the theory of switching in April, 1957, in *Annals of Computation Laboratory of Harvard University*, Vol. 29 (1959), pp. 57–73.

[78] J. P. Roth, "Algebraic Topological Methods for the Synthesis of Switching Systems. I." *Trans. Amer. Math. Soc.*, Vol. 88 (1958), pp. 301–326.

[79] J. P. Roth, "Algebraic Topological Methods for the Synthesis of Switching Systems. II." *Ann. Comp. Lab. Harvard Univ.*, Vol. 29 (1959), pp. 57–73.

[80] J. P. Roth, "Minimization Over Boolean Trees." *IBM Journal of Research and Development*, Vol. 5 (November, 1960), pp. 543–558.

[81] J. P. Roth, E. G. Wagner, and A. C. Ewing, "Algorithms For Logical Design." *AIEE Trans. on Comm. Electr.*, No. 56 (September, 1961), pp. 450–458.

[82] J. P. Roth and R. M. Karp, "Minimization Over Boolean Graphs." *IBM Journal of Research and Development*, Vol. 6 (April, 1962), pp. 227–238.

[83] J. P. Roth, "Systematic Design of Automata." *Proc. FJCC* (1965), pp. 1093–1100.

[84] J. P. Roth and E. G. Wagner, "A Calculus and an Algorithm for a Logic Minimization Problem Together With an Algorithmic Notation." *IBM Research Report RC-2280*, November, 1968.

[85] J. P. Roth and M. Perlman, "Space Applications of a Minimization Algorithm." *IEEE Trans. on Aerospace and Electronics Systems*, (September, 1969), pp. 703–711.

[86] H. Schorr, "Computer-Aided Digital System Design and Analysis Using a Register Transfer Language." *IEEE Trans. on Electronic Computers*, Vol. EC-13 (December, 1964), pp. 730–737.

[87] V. Y. Shen and A. C. McKellar, "An Algorithm for the Disjunctive Decomposition of Switching Functions." *IEEE Trans. on Computers*, Vol. C-19 (March, 1970), pp. 239–248.

[88] J. R. Slagle, C. L. Chang, and R. C. T. Lee, "A New Algorithm For Generating Prime Implicants." *IEEE Trans. on Computers*, Vol. C-19, No. 4 (April, 1970), pp. 304–310.

[89] R. J. Smith, J. H. Tracy, W. L. Schreffel, and G. K. Maki, "Automation in the Design of Asynchronous Sequential Circuits." *Proc. 1968 SJCC*, Vol. 32, pp. 53–60.

[90] R. J. Smith and J. H. Tracy, "A Simplification Heuristic For Large Flow Tables." *Proc. 1970 Design Automation Workshop (SHARE-ACM-IEEE)*, pp. 47–53.

[91] Y. H. Su, "Computer-Oriented Algorithms For Synthesizing Multiple-Output Combinational and finite Memory Sequential Circuits." Ph.D.

dissertation, Department of Electrical Engineering, University of Wisconsin, Madison, 1967.

[92] Y. H. Su, "Automated Logic Design Via a New Local Extraction Algorithm." *Proc. NEC* (December, 1969), pp. 651–656.

[93] Y. H. Su, "Synthesis of Multiple-Output Fan-In Limited NAND Logic Networks." Unpublished notes.

[94] Y. H. Su and D. L. Dietmeyer, "Computer Reduction of Two-Level Multiple-Output Switching Circuits." *IEEE Trans. on Computers*, Vol. C-18 (January, 1969), pp. 58–63.

[95] Y. H. Su and R. A. Weingarten, "Computer-Aided Design of Multiple-Output Logic Networks." Unpublished notes.

[96] A. Svoboda, "Ordering of Implicants." *IEEE Trans. on Electronic Computers*, Vol. EC-16 (February, 1967), pp. 100–105.

[97] R. H. Urbano and R. K. Mueller, "A Topological Method for the Determination of the Minimal Forms of a Boolean Function." *IRE Trans. on Electronic Computers*, Vol. EC-5 (1956), pp. 126–132.

[98] G. C. Vandling, "The Simplification of Multiple-Output Switching Networks Composed of Unilateral Devices." *IRE Trans. on Electronic Computers*, Vol. EC-9, No. 4 (December, 1960), pp. 477–486.

[99] R. Waxman, M. T. McMahon, B. J. Crawford, and A. B. DeAndrade, "Automated Logic Design Techniques Applicable to Integrated Circuit Technology." *Proc. 1966 FJCC*, Vol. 29, pp. 247–265.

[100] P. Weiner and T. F. Dwyer, "Discussion of Some Flaws in the Classical Theory of Two-Level Minimization of Multiple-Output Switching Networks." *IEEE Trans. on Computers*, Vol. C-17 (February, 1968), pp. 184–186.

[101] H. P. Williams, "The Synthesis of Logical Nets Consisting of NOR Units." *The Computer Journal*, Vol. 11 (August, 1968), pp. 173–176.

[102] S. S. Yau and C. K. Tang, "Universal Logic Circuits and Their Modular Realizations." *Proc. 1968 SJCC*, Vol. 32, pp. 297–305.

APPENDIX A
Modified Sharp-Product Algorithm [25]

Let $C = \{c^1, c^2, \ldots, c^p\}$. Then $U_n \#' C$ is a prime implicant cover for the complex $K^0(U_n) - K^0(C)$. $U_n \#' C$ is computed according to the following procedure.

Step 1: Set $D(0) = xx \ldots x$.

Step 2: Set $D(1) = D(0) \# c^1$.
Step 3: For $r = 2, 3, \ldots, p - 1$, form $D(r + 1)$ from C and $D(r)$ as follows:
Let $D(r) = \{d_1, d_2, \ldots, d_{s(r)}\}$. Set $U(r + 1, d_j) = d_j \# c_{r+1}$ for $j = 1, 2, \ldots, s(r)$. Set $D(r + 1, 0) = D(r)$. For $j = 1, 2, \ldots, s(r)$ set $D(r + 1, j) = \{D(r + 1, j - 1) - d_j\} \cup \{u \mid u \in d_j \# c_{r+1}$ and $u \# \{D(r + 1, j - 1) - d_j\} \neq \varphi\}$. Set $D(r + 1) = D(r + 1, s(r))$.
Step 4: $U_n \#' C = D(p)$.

APPENDIX B
Generalization of #-Product
for Multiple-Output Functions [84]

Let $\hat{c} = ce$ and $\hat{c}' = c'e'$ be two cubes. Then $\hat{c} \# \hat{c}'$ is defined as follows:
1. If $c \cap c' = \varphi$, then $\hat{c} \# \hat{c}' = \hat{c}$.
2. If $c \Rightarrow c'$, then $\hat{c} \# \hat{c}' = ce''$, where $e_i'' = 1$ if $e_i = 1$ and $e_i' = x$; otherwise, $e_i'' = x$ for $i = 1, 2, \ldots, m$.
3. If $c \cap c' \neq \varphi$ and $c \Rightarrow c'$, then
   (a) there is one cube ce'' in $\hat{c} \# \hat{c}'$ where $e_i'' = e_i$ if $e_i' = x$; otherwise $e_i'' = x$ for $i = 1, 2, \ldots, m$;
   (b) there is one cube corresponding to each cube c'' in $c \# c'$ obtained by concatenating on to c'' the output cube e.

Let 1_m consists of the m-tuple consisting of all 1 elements. If C_0 and DC are initial connection covers for the true and don't care complexes of a multiple-output function f, then

$$Z = (U_n 1_m \# (U_n 1_m \# (C_0 \cup DC)))$$

is the set of prime implicants of f.

chapter three

GATE-LEVEL LOGIC SIMULATION

BENSON H. SCHEFF

Assistant Department Manager
Software Department
Digital Systems Laboratory
Raytheon Company
Bedford, Massachusetts

STEPHEN P. YOUNG

Senior Member Technical Staff
Design Automation Department
Computer Systems Division
RCA Corporation
Marlborough, Massachusetts

3.1 INTRODUCTION

Logic simulation consists of exercising a realistic model of a digital system's logic with sets of input stimuli to generate signal-versus-time behavior. A logic simulator operates upon: (a) representations of logic elements; (b) connections of levels of these elements; (c) combinations of Boolean expressions; and (d) statements of timing relationships.

At the logic level, a simulator can assist digital system development by:
1. checking logic before commitment to hardware;
2. evaluating logic design alternatives;
3. verifying fault test procedures;
4. obtaining detailed circuit operational reference data.

The initial impetus for computer-based simulation at the logic level was the need to thoroughly check the set of logic equations characterizing a digital system. Even partial manual evaluations were time-consuming due to the complexity of the logic. Simulation on a digital computer permitted a comprehensive evaluation of logic to be performed automatically.

It became apparent that simulation techniques developed for testing pure logic could be extended to provide a basic modeling tool to help ensure that the fabricated system would operate as the designer intended. Not only could logic equations be checked, but also the designer could be provided with insight into the behavior of his system. The simulation provided access to all logical elements, permitting the intermediate behavior of the logic to be observed and the effects of a single cycling of the logic to be recorded. The simulation model could also be made responsive to probabilistic failures to permit hardware behavior to be analyzed. Logic design conditions such as the failure to invert the output of a gate, races, hazards, spikes, and logic ambiguities, which would have been difficult to recognize either through analysis or by debugging with actual hardware, could be identified. The output, which describes the operation of the network, forms reference documentation for (a) equipment checkout and (b) subsequent validation and acceptance testing of the manufactured product.

The logic simulation can also provide a mechanism for establishing fault-isolation and fault-detection sequences before hardware availability. Effective fault test procedures can be verified by using the logic simulation model, both with and without the insertion of faults. Diagnostic test sequences can be developed from the model by using algorithms which determine specific input stimuli for activating particular paths in the logic network to relate component failures to output responses. Optimal placement of hardware test points is possible by utilizing empirical fault detection procedures in combination with analytic techniques.

In any application, logic simulation complements, rather than replaces, actual equipment debugging. It minimizes the debugging effort required when hardware becomes avilable. Logic checkout is only a part of the hardware debugging process and is not intended to check the hardware for construction errors and strictly circuit problems such as wiring mistakes, bad connections, and faulty components. Furthermore, even though a logic simulation can provide as extensive a check of the logic as is desired by the engineer, a completely thorough checkout results in inordinate simulation computer execution time.

Logic simulation can assist in reducing the time and cost of digital design and checkout in today's technology, although its use is optional. The cost and long lead time of the digital equipment fabrication cycle places a penalty upon logic changes made after hardware commitment. Yet, successful digital systems, using MSI and IC devices, have been implemented with only minimal recourse to simulation. With Large-Scale-Integrated (LSI) circuit technology, however, the ability to test and analyze logic networks by working with the "breadboard" model provided by logic simulation becomes essential. Because an LSI unit is expensive to manufacture and the high density of logic virtually precludes any changes after fabrication, an exceptionally high premium is placed on eliminating errors during design. As a result, extensive digital system development is likely to become increasingly expensive without an extensive logic simulation capability.

3.2 THE LOGIC SIMULATION PROCESS

The simulation process combines the engineer's inputs describing the logic of the digital system with any additional information built into the simulator, converts the commands defining the conditions of the simulation, and dynamically exercises the model by tracing changes in logic values through the interconnected logic. Hard-copy charts are produced showing the logic values of monitored signals at simulated elapsed times.

This process consists conceptually of two phases: machine logic description and simulation processing. In Fig. 3-1 each box represents a separate step within each phase. The input data statements describing the logic of the digital device are converted and combined with previously stored information to form the replica of the logic network. Previously stored catalogs of standard

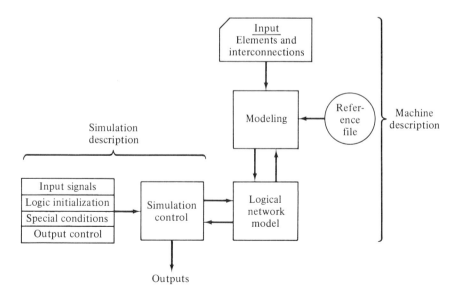

Fig. 3-1. Simulation process.

logic families may be referenced. The simulation commands select the portion of the logic network to be simulated either by direct commands or as a consequence of the particular inputs and outputs specified. Input waveforms can range from explicitly specified time-dependent values to arbitrarily complex signal patterns. The initial logic states and any special conditions (e.g., failures) needed to obtain a realistic environment are also specified as inputs to the simulation process. The simulation control program operates upon the model as directed by the simulation inputs producing the desired output.

The logic network description and the simulation conditions can be changed independently of each other. By changing only the network inputs, alternate logic designs can be evaluated with respect to the same simulation conditions. Similarly, different simulation conditions can be used to test the logic network without restating the machine description. Simulation can be performed as the logic is designed so that the logic specification can keep pace with the design of the hardware.

3.2.1. Logic Simulation

Simulation at the logic level consists of:
1. modeling the network formed by logic elements;
2. exercising the model with a set of discretely changing binary inputs.

The language must permit the designer to describe his logical network completely in a direct manner. Otherwise, regardless of how elegant the linguistics, the simulator will not be used as an engineering tool. For this reason, the input language of most logic-level simulators takes advantage of the convenience of characterizing the elements of digital logic and their interconnections by sets of Boolean equations. With this basic form, digital systems can be appropriately specified. The word appropriate is important; one usually would not describe a backplane (say) entirely in terms of interconnections of flatpacks. Generally, gates in a logic module, cards on a module register, and flip-flop modules would be used. The language may consist of: (a) basic digital types such as gates (AND, OR, NAND, NOR), flip-flops, delay lines, drivers, inverters, either separately or in combination, and more complicated functions expressed as black boxes; (b) the functional behavior of each element, either explicitly or implicitly; and (c) the interconnections between these elements.

The model of the digital network is formed from the source-language input statements either by compiling the input statements into a set of subprograms, each of which performs the functions of a logical element [1], [13], or by translating the input statements into a data structure representing the network [25], [26].

Sec. 3.2   The Logic Simulation Process   105

The simple circuit of Fig. 3-2a, which implements the equations

$$E = (A \cdot B \cdot C)'$$
$$F = (E \cdot D)$$

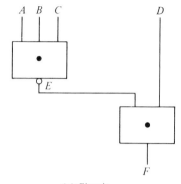

(a) Circuit

1	CAL A	A
	ANA B	$A \cdot B$
	ANA C	$A \cdot B \cdot C$
	COM	$(A \cdot B \cdot C)'$
	SLW E	$(A \cdot B \cdot C)' \to E$
	TRA 2	

(b) NAND gate

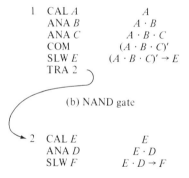

2	CAL E	E
	ANA D	$E \cdot D$
	SLW F	$E \cdot D \to F$

Fig. 3-2. Compiled code model.   (c) AND gate

provides an illustration of these techniques. To depict the operation of the circuit, a compiler would produce computer code similar to that shown in Fig. 3-2b and c. The subroutines representing components are linked to form the logic network expressions.

This logic may also be represented by data tables as in Fig. 3-3. The linkages between logical elements in the table correspond to their structural connections in the network. For a given element, say E, the table defines its input signals (A, B, and C), the type of operation (NAND), and the destination (F).

Exercising the model is analogous to the manual hardware checkout process performed by an engineer. Just as the engineer would set voltages on

106   *Gate-Level Logic Simulation*   *Chap. 3*

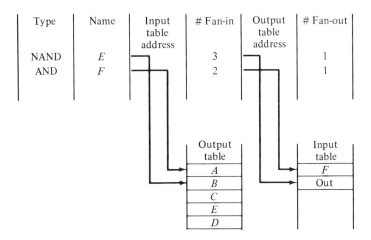

Fig. 3-3. Table-driven model.

various inputs and plot output voltages time on a cathode ray tube (CRT), in a logic simulation the engineer specifies the input signals and can observe the time-varying behavior of the outputs on a computer printout.

In a logical network model formed by compiling input statements into program routines, each collection of routines (or subprogram) represents a Boolean-type expression. The expressions are evaluated by executing the appropriate subprogram. Each component routine in the subprogram performs the indicated operations directly on Boolean variables. The resulting binary output of this evaluation is transferred to the specified signal destinations. For the network shown in Fig. 3-2, the three-legged NAND gate would be evaluated first, using the values of the three system inputs as arguments. The calculated output would be subsequently used, together with the fourth system input, to evaluate the AND gate.

Exercising a logical network model formed by creating a data structure (as in Fig. 3-3) is performed by an interpreter program which follows the linkages from one element to the next. The interpreter utilizes the various entries in the table to perform the required operation. The entry in the type column determines the specific computer code to be referenced, using as arguments the values of the inputs identified by the pointers in the fan-in entry. Unlike the computer model resulting from the compiler-produced code, this program uses only one set of machine codes for each logical operation. This specific subroutine would be referenced each time the interpreter desired to perform that logical operation. The computed output value determines the link to be followed by the interpreter to obtain the next element in the data table. In the network of Fig. 3-2, the value of E is determined by: (a) interpreting the operation; (b) the inputs expressed by the fan-in number; and (c) the input table address which identifies *A*, *B*, and *C* as inputs. After the value of *E* has been computed, its output (*F*) is identified by the fan-out number and the output table address. Its value is computed by interpreting its inputs and its driving operation in a similar manner. Because the interpreter uses the table to drive stored program routines, this modeling approach is denoted as *table-driven interpretive mode simulation.*

The *compiled-code* and the *table-driven* modeling techniques can be combined by replacing the type input address in the table with an address that points to the specific computer code simulating the gate's operation. In this table-driven model, the subroutine would be referenced each time that particular logic element was to be evaluated.

Executing the program code for each transfer function, or moving through the data-base linkage with an interpreter program, in turn takes the simulated digital system from its state at one time to its state for the next time. The term *bit time* is a reference term denoting this time frame.

3.2.2. Comparative Simulation Roles

The role of logic-level simulation in digital system design can be placed in perspective by a brief comparison with circuit analysis and register-level simulation.

At the logic level, the items of interest include gates and flip-flops and their functional relationships. An element in a logic simulation has two discrete states, ONE and ZERO, corresponding to ON and OFF. The next state of the element is determined from its current input state (gates) or from its present and previous input state (flip-flops). This behavior is represented as a transfer function describing the relationship between the outputs and each set of inputs. This is conveniently expressed by the element truth table; the three-legged NAND gate of Fig. 3-4 is typical.

$(A \cdot B \cdot C)' = E$			
A	*B*	*C*	*E*
0	0	0	1
0	0	1	1
0	1	0	1
0	1	1	1
1	0	0	1
1	0	1	1
1	1	0	1
1	1	1	0

Fig. 3-4. Truth table for a three-legged NAND gate.

Comparison of Circuit Analysis and Logic Simulation. At the circuit level, the items of interest include resistors, diodes, capacitors, and transistors.

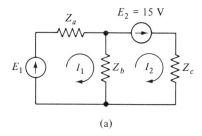

(a)

$$(Z_a + Z_b)I_2 - Z_b I_2 = E_1$$

$$(Z_b + Z_c)I_2 - Z_b I_1 = E_2$$

$$\begin{bmatrix} (Z_a + Z_b) & -Z_b \\ (-Z_b) & (Z_b + Z_c) \end{bmatrix} \cdot \begin{bmatrix} I_1 \\ I_2 \end{bmatrix} = \begin{bmatrix} E_1 \\ E_2 \end{bmatrix}$$

$$\begin{bmatrix} I_1 \\ I_2 \end{bmatrix} = \begin{bmatrix} (Z_a + Z_b) & -Z_b \\ -Z_b & (Z_b + Z_c) \end{bmatrix}^{-1} \cdot \begin{bmatrix} E_1 \\ E_2 \end{bmatrix}$$

(b)

Fig. 3-5. Circuit simulation example.

A circuit program such as CIRCUS [22] or ECAP [27] develops answers in terms of currents and voltages from explicit statements of the values of each element and the connections between them. Based on the electrical properties of each device, the behavior of the system is reduced to a set of loop equations as shown for loops I_1 and I_2 in Fig. 3-5a. Then the simultaneous equations for the unknown currents I_1 and I_2 of Fig. 3-5b are solved.

It is illuminating to compare the processing of a circuit analysis simulator to that of a logic simulator for the same element. Consider, for example, an inverter. A circuit analysis program considers a specific implementation of a general inverter circuit, such as the pnp resistance-coupled inverter shown in Fig. 3-6a. Prior to analysis by ECAP (Electronic Circuit Analysis Program), the transistor must be translated manually into the equivalent circuit contained within the dashed lines (Fig. 3-6b). Points in the equivalent circuit are assigned node numbers, starting with N_0 for the node at ground voltage. The input to ECAP is specified between pairs of nodes as values for DC and AC analysis. Thus, line B5 of Fig. 3-7 indicates that the resistance separating

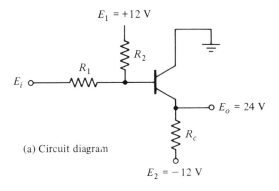

(a) Circuit diagram

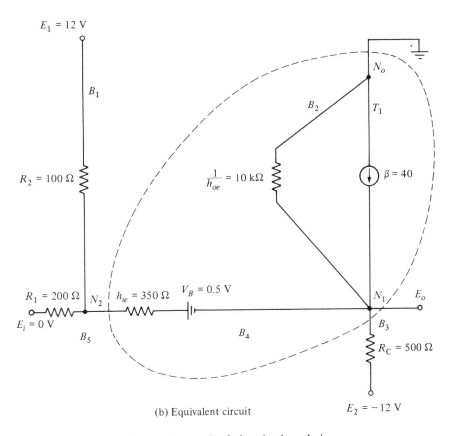

(b) Equivalent circuit

Fig. 3-6. Inverter simulation circuit analysis.

109

110   Gate-Level Logic Simulation   Chap. 3

```
            DC ANALYSIS
B1          N(0,2),R=100,E=12
B2          N(0,1),R=10000
B3          N(0,1),R=500,E=-12
B4          N(2,1),R=350,E=.5
B5          N(0,2),R=200,E=0
T1          B(4,2),BETA=50
            PRINT,VOLTAGES,CURRENTS
            EXECUTE
NODE VOLTAGES
NODES                  VOLTAGES
  1 -   2      .81638552+01      .79462168+01
ELEMENT CURRENTS
BRANCHES               CURRENTS
  1 -   4      .40537831-01      .39520989-01     -.40327710-01     .80674749-
        03
  5 -   5     -.39731084-01
```

Fig. 3-7. DC analysis.

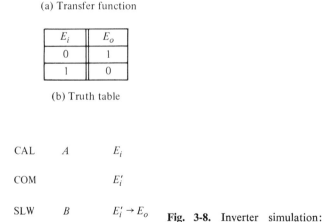

(a) Transfer function

E_i	E_o
0	1
1	0

(b) Truth table

CAL   A E_i

COM E_i'

SLW   B $E_i' \rightarrow E_o$

(c) Computer coding

Fig. 3-8. Inverter simulation: logic simulation.

node N_0 (the ground) from node N_2 (the input) is R_1 or 200 ohms, and the voltage difference is the voltage at E_i or 0 volts. The printout of the voltage at each node and the current between pairs of nodes is shown in Fig. 3-7.

On the other hand, a logic simulation model treats the inverter as a black box (Fig. 3-8a) whose transfer function is completely specified by a truth table (Fig. 3-8b). The computer program (Fig. 3-8c) can be used to simulate a transistor circuit or any other device with the same functional behavior. Thus, a logic simulation program may be considered as describing a more global representation than would the corresponding circuit analysis.

Comparison of Register and Logic Simulation. Although registers are physically composed of groups of logic elements, establishing structural and behavioral characteristics at the register level does not require the same order of specification necessary for defining detailed logic. Registers can be described in terms of their general operation and need not be defined in terms of the operation of their component elements. Because the specifics of logic design are not required, register-level simulations such as LOCS (Logic and Control Sequencing) [33], LOTIS (Logic, Timing, and Sequencing) [24], DDL (Digital Design Language) [6], and APL (A Programming Language) [11], can be used earlier in the design cycle to test design concepts. This may be highly cost-effective; developing a digital design at the logic level might require a significant effort and, especially with LSI modules, may not be required.

The commands for a register-level simulaton specification in APL are illustrated in Fig. 3-9. The process simulated is the decoding of a computer instruction word. The contents of the instruction register and the transfer of appropriate bits to the registers necessary for the actual instruction execution are shown in Fig. 3-9a and b, respectively. In the APL simulation code (Fig. 3-9c), each register is denoted by its mnemonic; the register size (specified elsewhere) is denoted as a binary vector. Actions are specified to the right of an arrow, with the resulting outcome stored in the variable to the left of the arrow. The vector instruction $c \leftarrow a/b$ produces c from b by suppressing each b_i for which $a_i = 0$. For example, the first program statement causes the first three bits of the instruction register to be transferred to the operation-code register.

Conceptually, one simulation program could be used for both register- and logic-level behavior. The source language would be able to express both logic and registers as fundamental descriptors with structural constructions (registers, subsystems, etc.) [18], [19], and control sequences (control-path and data-flow behavior) stated as macro forms based on primitive logic elements and their behavior. However, the practical difficulties of implementing a comprehensive set of register-language definitions in terms of logic-level constructions, coupled with the different objectives of each simulation (e.g., register level for a preliminary design group, logic level for microelectronics facility), have minimized the effort in this direction.

The simulation described in [20] solves a digital system simulation problem using, as components, (a) registers and (b) logic-level simulation (refer to Section 3.6.2). However, essentially two distinct languages are used.

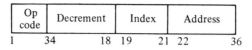

(a) Contents of instruction register

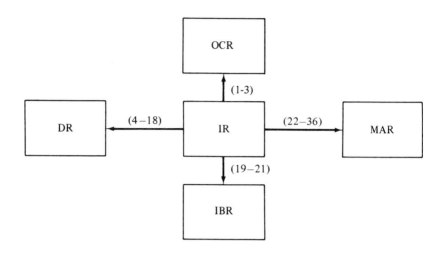

Mnemonic	Register	Register size
IR	Instruction Register	36 bits
DR	Decrement Register	15 bits
IBR	Index Buffer Register	3 bits
MAR	Memory Address Register	15 bits
OCR	Operation Code Register	3 bits

(b) Decoding process

```
OCR ← MASK1/IR      Store bits 1–3 of IR in OCR
DR  ← MASK2/IR      Store bits 4–18 of IR in DR
IBR ← MASK3/IR      Store bits 19–21 of IR in IBR
MAR ← MASK4/IR      Store bits 22–36 of IR in MAR

    MASK1 = 700000000000₈
    MASK2 = 077777000000₈
    MASK3 = 000000700000₈
    MASK4 = 000000077777₈
```

(c) Simulation program

Fig. 3-9. Instruction word decoding: Register level simulation.

3.3 DEVELOPING THE LOGIC MODEL

Modeling the logic network consists of defining the language used to express the design characteristics of this network and building the replica for subsequent simulation processing.
1. *Language:* Grammatical rules establishing the syntactic and semantic description for the logic elements and their basic interconnections.
2. *Model Construction:* Representing the network described by the input-language statements.

3.3.1. Logic Element and Expression Descriptions

Most simulations utilize a language derived from Boolean representations. The basic logical elements are combined in a pseudo-Boolean equation form and either written in a format from which card input, for the simulation, can be produced directly or typed directly via teletype. With the advent of graphic terminals, logic diagrams may become increasingly used as an input form.

Gates. For gates, the AND operation is usually represented by a dot or an asterisk. The OR operation is usually represented by a plus sign, and inversion or negation is indicated by a dash or apostrophe. Alternatively, the function name, followed by its arguments, may be used as a direct subroutine call.

Five different language representations in use for the two-legged NAND gate $C = \overline{(A + B)}$ are

(1) $C = '(A*B)$
(2) $C = *A + *B$
(3) $C = \overline{(A*B)}$
(4) $C = (A.B)*$
(5) $C = NAND (A,B)$

Although the first three representations [25], [18], [17], do not follow directly the exact conventions used by the digital logician, each can be quickly interpreted. The fourth convention [20] is more direct, with operators being written in their normal order. The FORTRAN-like language illustrated in the fifth expression [17] is almost as easy to understand as the pseudo-Boolean equations. This FORTRAN convention names NAND directly and specifies as arguments each leg of the gate, e.g., A and B. If the simulation program were written in FORTRAN, this notation would be translated directly into a FORTRAN statement which calls a subroutine NAND.

A gate output equation defines a signal output as a Boolean combination of input signals, e.g.

CAUT 1 = CY000∗CY060 + CY060∗CAUT11

This logic is shown in Fig. 3-10. Physical placement of this logic to a struc-

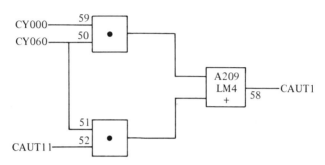

Fig. 3-10. Typical gate circuit.

tural position on a module is needed if the simulation data are to be used as an aid in fabrication. This relationship can be described in a similar manner

LM4-209(58) = 59∗50 + 51∗52

This equation indicates that the circuit (LM4-209) on logic module 4 with output pin 58 has input pins 59, 50, 51, and 52 related as shown [25].

Flip-Flops. The operational characteristics of flip-flops, unlike gates, are not as easily depicted in a notation resembling their logical characteristics. Various conventions are used in different simulations, for example.

(1) TOGGLER/S = /TOGGLER∗T
     TOGGLER/R = TOGGLER∗T
(2) ,AAF = AAF∗DPG∗AAF∗DNG∗AAF∗DNG
(3) JKFLIP (2,A,ABAR,IN1∗IN2,A∗IN4,PULSE,DCSET,DCRSET)
(4) FF1(FT,FF) = XX∗B,AB + Y,CLK,DCR
(5) S1FF1 = XX.B

The first expression denotes a toggle flip-flop which reacts to an input signal of T [18]. The S and R identify the *set* and *reset* functions which are stated in a Boolean form. The second expression for a flip-flop is also in Boolean equation form [28]. In this notation, the set equation is preceded by a comma, and the reset equation is preceded by a period. Because of their similarity, only the set equation is shown. In the third expression, function subroutine notation is used [17]. The subroutine called performs the functions required of a JK flip-flop, using as arguments the number of outputs, the list of outputs, and the inputs for set, reset, pulse DC set, and DC reset.

A somewhat more convenient representation of a flip-flop is shown in expression 4 [25]. This notation, used in the pseudo-Boolean format of expression 4, identifies the true output (FT), the false output (FF), the set signal (XX∗B), the reset signal (AB + Y), the clock signal (CLK), and the DC reset (DCR) of circuit FF1. The simulation in [20] uses a separate equation for each input terminal. A typical input equation for a set input is shown by an $S1$. A reset input would be identified by an R1.

Macro and Black-Box Expressions. To simplify the writing of input expressions for commonly used networks, the language may permit previously defined elements (e.g. a two-stage counter) to be used in macro statements and black-box transfer vectors. As a result, once a circuit is defined explicitly, it can be used by the engineer subsequently in the simulation. Depending upon the hierarchical capabilities of the language, the black box or macro may itself be used in the definition of a larger circuit.

With a macro capability, a network desired by the engineer can be defined in terms of the basic logic-level elements in the laguage. This definition would be made as part of the modeling process. When the macro is used, the subroutine containing the code representing the specified combination of elements is activated. Typically, macros are used to express basic networks used frequently with or without minor variations in the logic being simulated.

A black-box convention is used for logic elements which are too complex to be conveniently expressed in a Boolean equation. If the logic element's functions have been previously programmed as part of this simulation, a black-box expression need only define inputs and outputs. Its use automatically brings in the coded Boolean equations. Flip-flops are commonly represented either as Boolean-type equations or as truth tables. The flip-flop representation described in expression 4, page XX, is a special case of a black-box representation.

A hierarchical application of black-box elements is illustrated in Fig. 3-11. The logic of the NAND flip-flop, denoted RSFF, is defined initially by detailed Boolean equations. This circuit can be used elsewhere by specifying only its inputs and outputs, as in the two-stage shift circuit, SHIFT. The explicit description of the logic flip-flop is not repeated. In turn, SHIFT as well as RSFF can be recursively nested in larger subsystems.

Reference Catalogues. In many logic simulators, libraries defined in terms of any of the building blocks aceptable as input can be stored and referenced directly. These items may range in complexity from a few gates on a module, as shown in Fig. 3-13, to an entire functional unit. The catalog

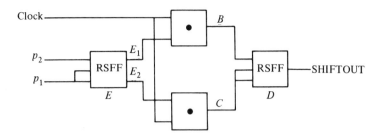

```
*SHIFT:
  OUT/SHIFTOUT;
  INPUT/CLOCK, P1, P2
  LOGIC/
        E (E1, E2) = P2, P1, P1;
        B = CLOCK * E1:
        C = CLOCK * E2:
        D (SHIFTOUT) = B, C, C;
  ALLOCATE/
        E = RSFF:
        D = RSFF:
  END
```

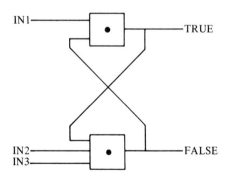

```
*RSFF:
  OUT/TRUE, FALSE
  INPUT/IN1, IN2, IN3;
  LOGIC/TRUE = '(FALSE * IN1);
        FALSE = (TRUE * IN2 * IN3)
  END
```

Fig. 3-11. Macro hierarchical language representation.

facilitates the simulation of a network containing common or repetitive logic blocks such as standard components and manufacturer's catalogs. Once an item is defined and simulated, it can be used in the catalog without further testing.

3.3.2. Time Mechanization

In simulation, time is quantized into discrete time units. Signals may change their values only at these time units. The value assigned to these units is relative, since gate-propagation delays are specified in terms of an integral number of these units.

Delay Considerations. The delay may be specified either as a built-in characteristic of element groupings or as an ad hoc input value. For either definition, an option may permit using a subsequently specified input value to override a standard delay. The delay can be associated with all elements of a given type or, alternatively, set for a single use of an element. To facilitate implementation, a constant value, usually either the nominal delay furnished by the manufacturer or the worst-case delay, is usually used. Although the actual delay varies from this value, it is sufficiently accurate to detect and identify the maximum time for a signal to propagate through the digital system for most simulation purposes.

To fully analyze race conditions, both the largest and shortest delay times are of concern. Using a value which is statistically distributed, but limited to the bounds established by the manufacturer's quality control, gives a more realistic picture of the behavior of the actual hardware. Commonly, the statistical distribution is approximated by a Gaussian curve with a known mean and standard deviation, although any distribution can be assumed [7].

Timing Algorithm: Synchronous or Asynchronous. Timing algorithms are classified as either synchronous or asynchronous, according to whether or not response time is tied to clock time.

In synchronous simulation, each operation takes place under control of a clock, and gated inputs to flip-flops are used only when the clock is ON. This is equivalent to obtaining the element values in a circuit by sampling gate outputs and flip-flop inputs, propagating the current value at the instant when the clock is HIGH. Except at these designated times, the values of the gates are of no interest. In asynchronous simulation, the signal values of each element can be used as soon as state changes occur—not solely at fixed regular clock times.

This does not imply that synchronous simulation should be used for synchronous digital logic while asynchronous simulation should be used for asynchronous logic. The terms synchronous and asynchronous relate to the manner in which timing considerations and responses are modeled. An asynchronous simulator can be used for synchronous circuit logic, e.g., to check if signal pulse changes occurred prior to the onset of the clock signal. The reverse is not true.

In essence, a synchronous simulator models a "zero-delay" network. All state changes are assumed to occur at the time at which a hypothetical clock pulses. The cumulative delay of logic layers may cause circuit behavior to be asynchronous with respect to the clock. Thus, circuit behavior, which is dependent upon the relative delay through different logic paths, may not be realistically depicted. Usually, for example, the clock time is sufficiently slow in relation to gate delays that the output of one gate is always available when needed as input by a succeeding gate. However, a synchronous simulation might fail to detect either the specific instance when too much time was required by an input gate to switch its state, or a race condition.

Small delay variations can cause an action to occur "between clock times" [12], [26]. These conditions, typified by but not limited to races and hazards, cannot generally be detected by synchronous simulation. In a race condition, the output of a gate depends on the precise order of the arrival of its input signals. Consider the simple AND gate in Fig. 3-12a. Normally, if the inputs go from (1, 0) to (0, 1), the output would be expected to be zero. Yet, if IN1 changes to 1 before IN2 changes to 0, for a short space of time both inputs are 1; thus for this period the output goes high, showing a short pulse or "spike." This is shown in the voltage-time chart, Fig. 3-12c. This condition is known as a *static hazard*. If either IN2 changes first or both inputs change simultaneously, the output of the AND gate remains at zero since at no time are both inputs one.

In a dynamic hazard, an output which is supposed to change state only once oscillates before finally settling down. Figure 3-13 exhibits a race condition. The order in which the input values IN1, IN2, and IN3 change from (1, 0, 1) to (0, 1, 0) through the intermediate states of (1, 0, 0) and (1, 1, 0) determines the intermediate values of the output, OUT. The plot of voltage time for signals arriving in the order IN3, IN2, and IN1 shows the oscillation in the output.

It may appear from the foregoing that only asynchronous simulation programs should be designed. Because both synchronous and asynchronous circuits can be analyzed, asynchronous simulation presents a more generally useful test bed environment. Yet, asynchronous simulation may not be required. First, most digital circuits operate synchronously with standard delay gates. Establishing relative delays to elements in the circuit may not yield significantly different results. Second, the ability to detect the presence of a spike analytically would not necessarily indicate a potential error condition. A spike may be inconsequential if the logic it drives depends on voltage

(a) Two-value

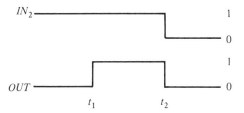

(b) Three-value

(c) Output waveform

Fig. 3-12. Spike condition.

levels. Third, hazards can be detected by simulation of worst-case conditions, although hazards might be indicated which are not actually present in the circuit. Fourth, a synchronous simulator may be simpler and cheaper to implement since the next event can occur only at the next periodic clock time. In an asynchronous simulation, the program must be designed to handle many future event-times. In general, synchronous simulation will consume less computer run time than will asynchronous simulation.

Three-Value Simulation. The behavior of a gate is more accurately described if a value other than zero or one is assigned to represent its transition between states. This third-state value is denoted as X to indicate that it

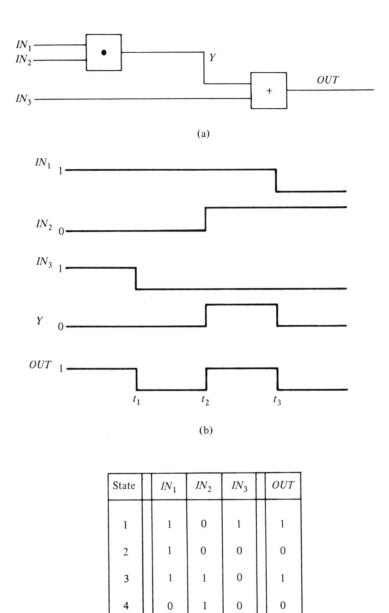

Fig. 3-13. Dynamic hazard example.

represents an unknown value during the transition [12], [25]. The power of this technique with respect to detecting race conditions can be easily shown by considering the simple two-input AND gate shown in Fig. 3-12.

In a two-value simulation (Fig. 3-12a), the output AND gate will remain at zero as the inputs IN1 and IN2 switch from (0, 1) to (1, 0). However, if a third value is used (Fig. 3-12b), both inputs pass through the intermediate X-state as they switch to their opposite values. This is shown simply, by the transition-value sequence in Fig. 3-14. A three-valued truth table of an AND gate (Table 3-1) shows that the gating structure fed by the gate would be affect

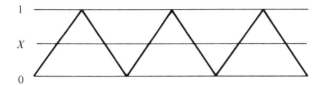

Fig. 3-14. Element transition values.

TABLE 3-1. THREE-VALUE TRUTH TABLE FOR A TWO-LEGGED AND GATE

IN_1	IN_2	OUT
0	0	0
0	X	0
0	1	0
X	1	X
X	X	X
X	0	0
1	0	0
1	X	X
1	1	1

ed by the transient X-state of each input when IN1 has changed to 1 before IN2 has changed to 0. Unlike a two-value simulation, a three-value simulation would detect an output spike for an AND gate and dip for an OR gate under this set of input signal values. The third-state value can also be used to represent any input values unspecified or unknown at the beginning of the simulation. This permits the user to easily observe the extent to which the output of a particular gate is propagated through the logic. Not only does this facilitate developing diagnostic sequences, but, by setting the output of a block of logic to X, it is possible to remove an entire logic block from the simulation.

By avoiding the need for separate specification of the states of 0 and 1 for each element, we greatly reduce the number of different cases which must be simulated. For example, reset logic can be checked in one simulation run by setting the input states to X and noting if the X-value is propagated to all the elements required to be reset. Otherwise, 2^N simulation runs (where N is the number of elements to be reset) would have been required to detect a change. The appearance of the value X reveals that an element was incorrectly eliminated from the simulation. A three-value simulation may be either synchronous or asynchronous.

3.3.3. Model Construction

The actual computer model mechanization is determined by the characteristics particular to the logic to be simulated (number and type of logic elements, logic complexity, timing features), the host computer (computer storage size, special features, processing environment, e.g., time-sharing with engineer interaction or batch processing), and specific applications of the simulation. Thus, no general rule-of-thumb for selecting specific techniques can be made. Rather, this section will examine the basics of some modeling approaches: equation simulation, compilation techniques, table-driven techniques, parallel simulation.

Equation Simulation. An equation simulation checks pure logic. Because it is not concerned with the manner in which the equations are implemented and the related time constants, equation simulation permits rearrangements of the logic equation to simplify the processing.

Two techniques for simulating equations are discussed. The first technique is based on data-dependent subroutine coding; the second technique assumes the ability to examine and reorganize the entire set of logic equations in a large system.

Data-Dependent Code. A given Boolean equation is represented by a set of computer machine code. Generally, the machine-language coding is not unique. The possible routines differ in the memory locations used and execution times. If the sequence of instructions required to evaluate the expression does not depend on the value of the input variables, the compiled routine is data-independent. If the instruction sequence is affected by the input data values, the compiled routine is data-dependent.

In the machine code which simulates a three-legged NAND gate (shown in Fig. 3-2), all five computer instructions are executed regardless of the values of the inputs. Thus, this code is data-independent.

Data-dependent implementation takes advantage of the characteristics of the logic element—in this case, the fact that the output of the NAND gate is 1 if any one of its inputs is 0. Although more host computer instructions might be required, the number of executed instructions and, hence, the execution time, is decreased. This is clearly shown by the data-dependent coding of the three-legged NAND gate:

INSTRUCTIONS	COMMENT
1. SON D	STORE ONE IN D
2. NZT A	A = 0? IF A $\neq$ 0, SKIP NEXT INSTRUCTION
3. TRA OUT	YES
4. NZT B	B = 0?
5. TRA OUT	YES
6. ZET C	C = 0?
7. STZ D	STORE ZERO IN D

For example, if input A is 0, only three instructions (numbers 1, 2, and 3) are executed. As a result of instruction bypassing, the expected number of instructions actually executed is less than four, as is shown by the following computation.

The expected number E is equal to $\sum P_i f_i$ (where P_i is the probability of the ith event, and f_i is the number of instructions executed up to the ith event). Three instructions are executed if A equals 0 ($P_i = \frac{1}{2}$), four instructions are executed if A equals 1 and B equals 0 ($P_i = \frac{1}{4}$) or if A equals 1, B equals 1, and C equals 0 ($P_i = \frac{1}{8}$). If all inputs are 1, five instructions are executed ($P_i = \frac{1}{8}$). Thus the expected number of instructions executed is

$$E = \sum_{i=1}^{4} P_i f_i = \frac{1}{2} \times 3 + \frac{1}{4} \times 4 + \frac{1}{8} \times 4 + \frac{1}{8} \times 5 = 3\frac{5}{8} < 4$$

In this case there is an average processing-time reduction of 27% for the data-dependent coding, assuming all inputs are equally likely.

Although data-dependent techniques are useful in producing compiled computer code, they produce code which is independent of the physical structure of the logic. Thus, they are effective only when the effect of the logic equations, rather than the actual logic, is being simulated.

Data-dependent algorithms can be used to generate reasonably efficient coding. One data-dependent technique [13] produces such code directly from Boolean equations in simplest-sum-of-products form. The algorithm generates a set of logical AND instructions for each term. Thus, if any term has a value of 1, the entire expression is automatically 1. As a result, the number of machine-language instructions (m) needed to simulate a Boolean expression in a computer with a standard instruction complement is

$$m = k + \sum_{i}^{k} n_i$$

where n_i is the number of Boolean variables in the ith term and k is the number of terms in the Boolean expression.

The sequence of instructions which simulates an expression contains k blocks of instructions, one for each term in the expression. In general, block i has $n_i + 1$ working instructions, one instruction to bring the first variable into the accumulator, and $n_i - 1$ logical ANDs. Each block concludes with a conditional transfer instruction which transfers control to the last instruction in the sequence if the accumulator contains 1. The last block, k, does not contain the conditional transfer statement since it must only store the contents of the accumulator in the output variable. The following illustrates the code generated for a sum-of-products expression by this technique.

$$A = BC + CDE$$

BLOCK 1	CAL B		BRING B INTO ACCUMULATOR
	AND C		LOGICAL AND WITH C
	TLQ END		TRANSFER IF TERM = 1
BLOCK 2	CAL C		BRING C INTO ACCUMULATOR
	AND D		LOGICAL AND D
	AND E		LOGICAL AND E
END	SLW A		ACCUMULATOR CONTAINS A, STORE IT

The efficiency of the resulting coding is dependent on the placement of the terms in compilation sequence. Such a placement implies either a manual ordering or a "thoughtful" automatic sorting process. Neither may be feasible with large systems on the order of, say, 10,000 equations.

Generally, the largest terms should be processed first. However, as a result of similarities in some variable expression terms, the minimum code is not always produced.

Simulating an Expression Containing Several Functions of the Same Logic Variable. Each variable is operated upon once in the simulation for each appearance in a Boolean expression. Because a variable may appear several times, the simulation coding may have to be continually reexecuted, unduly slowing the processing. Normally this is not a problem. The Boolean expressions are reduced and converted into several subexpressions for hard-

Sec. 3.3 Developing the Logic Model 125

ware implementation. However, long Boolean equations developed by the computer logician must, at times, be simulated.

Expressions containing several functions of the same input variables are simulated efficiently if the equations can be arranged so that each logic variable is evaluated only once. The equation set

$$F_1 = AB + A\bar{C}$$
$$F_2 = AB + C$$

is organized into a tree structure in Fig. 3-15.

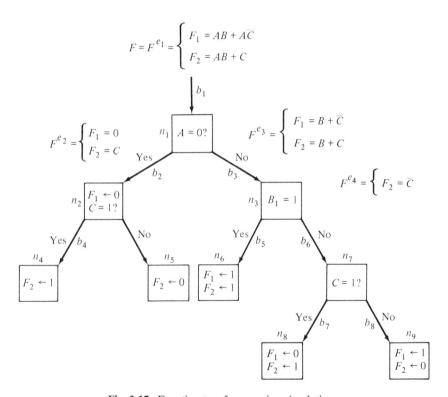

Fig. 3-15. Function tree for equation simulation.

Each node of the tree represents a single variable-state assignment. Thus, each branch from a node corresponds to a different evaluation of the function at that stage. The branches of the tree depict conditional branches in the program that depend on the value of a logic variable. The variable test is performed at the beginning of the program branch, e.g., at the nodes of the tree.

In Fig. 3-15, variable A is tested at n_1. From this node, the left branch is used when $A = 0$, so that $F_1 = 0$ and $F_2 = C$. Variable C is tested at node n_2. If $C = 1$, then F_2 is set to 1; whereas, if $C = 0$, then F_2 is set to 0. The other branches are evaluated in a similar fashion.

Once this tree structure is established, both F_1 and F_2 can be evaluated in three or four steps.

In the generalized program tree created for the function set $F = (F_1 F_2 \cdots F_n)$ of variables $(X_1 X_2 \cdots X_n)$, assume that X_i, a specific variable, is evaluated at node n_j. Node n_j has one input branch, say b_j, and two conditional output branches, b_k (for $X_i = 0$) and b_m (for $X_i = 1$). F_j represents the value of F at node n_j, with the initial variables replaced by the specific values determined with that path through the tree. A set of functions, F^{e_i}, is thereby associated with each node, e.g., if node n has a set of functions F^{e_j} associated with it, then $F^{e_k} = F^{e_j} | X_i = 0$ and $F^{e_m} = F^{e_i} | X_i = 1$. The process ends when each F_i is either 0 or 1.

The tree structure orders the procedures as created during the compilation process. Standard sorting techniques are used coupled with algorithms which attempt to create a structure to minimize the processing time. However, the evaluation process is rather time-consuming and hence may not be feasible either for systems containing many equations, or where a value is repeated in, say, one hundred equations. A simple but effective rule-of-thumb which leads to efficient, if not minimal, processing is to organize the equations to test, first, those variables which appear most frequently in the functions being evaluated. Once the program logic tree has been constructed, the computer code is generated in a straightforward manner.

Compilation Techniques. The machine code executed in the host computer to effect the simulation must correspond with the order in which the logic elements are activated by the signal paths. The order in which the elements in the Boolean expressions are evaluated is determined in the compilation process prior to generation of the machine-language code. After logic levels have been assigned to gates, the gates are sorted by logic levels.

The level assigned to other signals is determined by the number, in general, of cascade levels between an element and the system inputs. With sequential logic [32], the system inputs are assigned level 0, as are the destination sets of feedback lines of cyclic circuits (e.g., flip-flops). All other inputs have a level one greater than the highest level of any input. After the logic elements have been sorted by logic level, sequences of compiled machine code ordered by element level are produced. The compiled code is essentially the computer representation of the input expressions whose elements are the logic elements in the circuit (represented by code similar to that shown for the NAND gate of Fig. 3-2). When the simulation model is executed, the elements are evaluated in order of increasing levels. The elements with the lowest level are processed first. Sequential logic is simulated by performing several simulations of the cyclic logic for each clock time.

As a result of levelizing, updating the logic network usually requires a reassembly of the produced code. In addition, there is no easy provision for selectively evaluating part of the logic. As a result, pure compiled-code modeling is generally limited to small systems or systems in which design changes do not occur frequently.

Table-Driven Techniques. Instead of executable machine-language coding, the input source language can be translated into a data structure representing the network [25], [26]. The circuit network is described by defining the elements' characteristics (primary input, AND gate, etc.), the source of each of its inputs, and the destination of each of its outputs. This data structure is operated upon subsequently by a computer program which interprets the simulation commands to determine the effect on the logic of propagating the specified signals.

Such a data-base representation is shown in Fig. 3-16b, which represents the circuit of Fig. 3-16a. In Fig. 3-16a, the letters a, b, c, d, and e represent the signal name, and the numbers 1, 2, 3, and 4 represent the pins associated with each element.

In the data table structure, three items describe the element. The basic entry states the name given to the element and its circuit type, e.g., system-input code number (-1), flip-flop code number (3), inverter code number (2), and NAND gate code number (1). The pin-list entry distinguishes input pins and output pins, and, for each input, names its source element. The used-on list gives the destination element of each output.

Pointers link the three entries together. In the basic entry, the pin-link entry locations contain the data pertaining to the pins asociated with the element, e.g., the JK flip-flop (element e) has four pins described in pin-list entry positions 8 through 11. In the left-hand column of the pin-list entry, input pins are identified by their source numbers; output pins are represented by a zero.

Pins are denoted by a number with respect to the individual gate. Thus for the JK flip-flop, pins 1 and 2 are output pins, while pins 3 and 4 emanate from element a (item number 1 of the basic entry) and element b (item number 2 of the basic entry), respectively. The output list link points to the destination item. For the JK flip-flop, pin 1's destination is in positions 4 and 5 of the used-on entry; these point to (and hence identify) element c (item number 3 of the basic entry), pin number 3, and element d (item number 4 of the basic entry), pin number 2. Although pin 2 of element e is an output pin, the asterisk indicates that, at the present stage of design, no connection element has yet been assigned. A zero represents a system output to the logic network.

Unlike in the compiled-code model, modifications can be easily effected.

128 Gate-Level Logic Simulation   Chap. 3

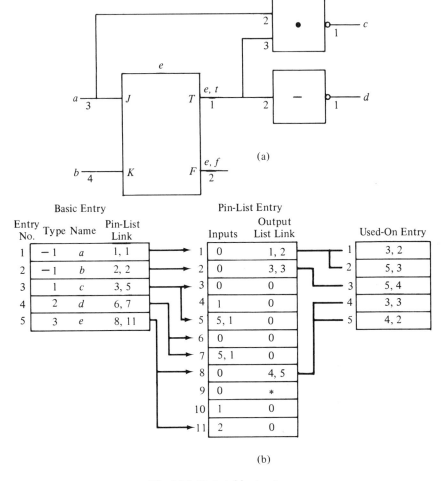

Fig. 3-16. Data table structure.

Only pointers in the data structure may have to be changed to reflect the different propagation routes.

Parallel Simulation. Normally, in the simulation program, an entire computer word is used for a signal value. Only one set of inputs and one machine configuration can be simulated at a time. If each separate bit position in the word is used for a separate input state, as many different conditions as there are bits in the word can be processed simultaneously. This parallel simulation allows a simulated machine to be exercised by several different input patterns simultaneously. Processing multiple conditions simultaneously with the same computer routine is possible because the logical computer in-

Sec. 3.3   Developing the Logic Model   129

structions (AND, OR, etc.) manipulate each bit in a computer word independently [9], [31]. Total computer execution time is reduced because one run testing parallel conditions yields results which normally require many separate runs.

In the program code, the logical operations call for (as operands) a one-bit mask whose active bit is in a different position for each fault.

The input word containing the one-bit representation of the condition is combined with the mask to propagate the desired value as a one-bit digit. An OR instruction, with a one-bit mask containing a 1, causes a 1 to be propagated in that position. For example, suppose that eight test cases were to be processed simultaneously, that the value of the output of gate A was 1 only in the second test case, and that the value of the output of gate B was 1 in the second and eighth test case. The operation $C = A$ OR B would be simulated by ORing 01000000 ... (for A) with 01000001 ... (for B), obtaining $C = 01000001$ .... Similarly, an AND instruction, with one-bit mask contanining a 0, would force a 0 to be propagated. An OR instruction with a 0, or an AND instruction with a 1 causes the desired conditions to be propagated. The bits in the output word for each element are manipulated, as determined by the particular element's logical operation, to determine if an error is to be propagated.

Parallel simulations can be applied to simulating machines with several different input patterns simultaneously. In this case, a system input is represented by a single computer word, with each bit in the word representing a particular input. Thus, bit i in input word INI would represent the condition of system input INI for input pattern i. Because parallel simulation requires data-independent code sequences due to the different data conditions being examined, data-dependent techniques cannot be used.

3.3.4. Output

Useful design and fabrication information describing the logic being modeled can be easily obtained as a by-product of the simulation. This includes up-to-date signal lists, logic fan-out lists, checks for overloaded circuits, checks to ensure that placement of logic to circuit is neither ambiguous nor undefined, and verification that all signals have been used (e.g., a four-input NAND gate used as a three-input NAND). Such data exhibit how the logic is implemented and facilitate subsequent analysis of machine experiences by comparison with simulation results.

Outputs typical of those which could be generated are illustrated by a simple assignment of gates to a module (Fig. 3-17). The function CAUT1 has input functions CY000, CY060, and CAUT11. This gate is on a type LM4 module and is located at A209 on the parent plate. The module contains two separate logic cards. Each input and output pin of every circuit on a

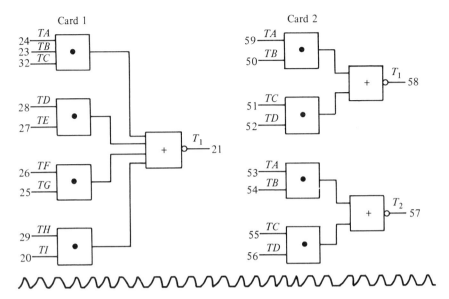

Fig. 3-17. Typical gate module.

flatpack is connected to a connector pin of the module. The first circuit illustrated in card 2 drives connector pin 58.

The circuit resultaing from assigning the gates shown in Fig. 3-10 to this module is shown in Figure 3-18a. This printout shows the assignment of

CIRCUIT   FUNCTION   EQUATION
LM4-58   CAUT1[58]   = '(CY000[59]*CY060[50] + CAUT11[52])

(a) Circuit assignment

GATE TYPE   OUTPUT   INPUTS   DESTINATIONS
NOR   CAUT1   CY000 CY060 CAUT11 D E F

(b) Cross-reference table

Fig. 3-18. Design output.

the logic function CAUT1 to module LM4 at position 58. Cross-reference tables (Fig. 3-18b) show source signals for each output and destination signals for each source signal [10], [25].

A load list for the circuit formed by adding two NAND gates to the circuit of Fig. 3-10 is shown in Fig. 3-19. In this new circuit, one function CAUT1 drives functions AWOB1 and AZR20. Assuming that the functions located at module position A203, pin 24, and position A203, pin 15, each draw 2.0 milliamps, a total of 4.0 milliamps would be drawn by the circuit. This is

Sec. 3.4 *Exercising the Model* 131

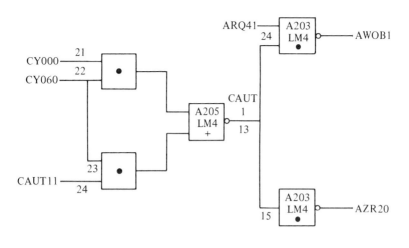

(a)

Signal	Source	Used On	Function	Location	Load	Total Load
CAUT1	CAUT1		AWOB1	A203-24	2	
			AZR20	A203-15	2	
						4

(b)

Fig. 3-19. Load list.

compared to the load driven by CAUT1 to verify that the current drawn does not exceed current capability.

3.4 EXERCISING THE MODEL

Exercising the model requires:
1. setting initial conditions, establishing the initial state values of the network elements, and providing the input waveforms;
2. evaluating the network by performing the programmed operations on the logic sequences;
3. output specification by identifying the outputs to be monitored and types of printout.

3.4.1. Simulation Commands

The input signals used to exercise the model can be of arbitrary complexity, described either implicitly as waveforms at the start of a simulation [25] or explicitly as data values to be inserted as the simulation progresses [18]. Any number of signals may be defined. Consider the specification of the waveforms of the two signals, CLOCK and SIGNAL, shown in Fig. 3-20a. CLOCK is an alternating periodic signal of thirty time units which starts at zero (low); SIGNAL starts at one (high) and is on for ten units, and then alternates with a periodic set of twenty units. The input specification conven-

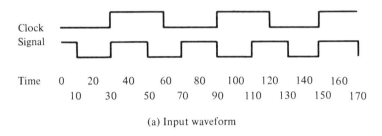

(a) Input waveform

Low/Clock
High/Signal
Schedule/Clock (30), Signal (10, 20, 20, 20, 20, 20, 20, 20, 20)

(b) Implicit specification

TIME	CLOCK	SIGNAL
0	0	1
10	0	0
20	0	0
30	1	1
40	1	1
50	1	0
60	0	0
70	0	1
80	0	1
90	1	0
100	1	0
110	1	1
120	0	1
130	0	0
140	0	0
150	1	1
160	1	1

(c) Explicit specification

Fig. 3-20. Waveform specification

tion shown in Fig. 3-20b describes the entire chart of the waveforms and could be used at the beginning of the simulation. The input specification of Fig. 3-20c describes the same input signals. With this form, however, the specific signals values would be inserted into the simulator at the specific time period specified.

3.4.2. Simulation Algorithms

The algorithm implemented in the computer program describes the simulated behavior of the network. Essentially, algorithms can be classified as: (a) either two or three states depicted per element; (b) either compiled-code or table-driven interpretive mode; and (c) either synchronous or asynchronous. Basic representations of each type are discussed in this section. Many variations of each type have been implemented.

Three-Value. A three-value synchronous simulation has been implemented by alternating two passes over the data at each clock cycle [12]. These passes are denoted as the X pass and the value pass. During the X pass, the changing system inputs are assigned the intermediate value of X. This X is propagated according to the three-value truth table for a two-legged AND gate, as shown in Table 3-1, through the logic. During the value pass, the system inputs are set to their values; these new values are propagated through the logic to complete the simulation of this bit time. An X value which persists from the X pass to the end of the value pass identifies a critical race because the logic has reached a stable value which can be either zero or one.

The problem to be solved in the three-value simulation is shown by the critical static hazard in the circuit of Fig. 3-21a. As the inputs switch from (1, 0, 1) to (0, 1, 1), the output, Z, of the gate which should be 0 could be 1, depending upon which input changed first. In Fig. 3-21b, because X_2 has changed state before X_1, the rising voltage on Y_1 triggers a latch on the output, Z, which does not turn off when the spike is passed, even though Y_1 turns off very quickly. In Fig. 3-21c there is no spike because X_2 has changed state before X_2. Thus, the output remains at its 0 = state.

Figure 3-22 details the process of the three-value simulator for this circuit. The descriptive names *reset*, *switch*, and *hazard* relate simulator actions to circuit actions. In the *reset* cycle, the system is reset, input X_3 is set to zero which makes Y_2 zero which, in turn, makes Z go to zero. During the *switch* cycles, X_1 and X_2 changed state, creating the hazard. In the *hazard* cycle, the hazard is latched onto the system output and held.

In the simulation:

Reset Cycle

134   Gate-Level Logic Simulation   Chap. 3

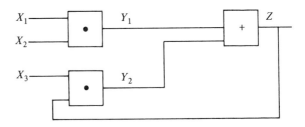

(a) Circuit with a critical race

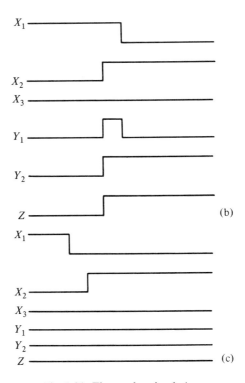

Fig. 3-21. Three value simulation.

X Pass:

All values are set to X. These initial states persist into the X pass of the reset cycle.

Value Pass:

X_1 goes to 1 while X_2 goes to 0, forcing Y_1 to be 0.

X_3 goes to 0 forcing Y_2 to 0. As a result, Z is 0.

Sec. 3.4   Exercising the Model   135

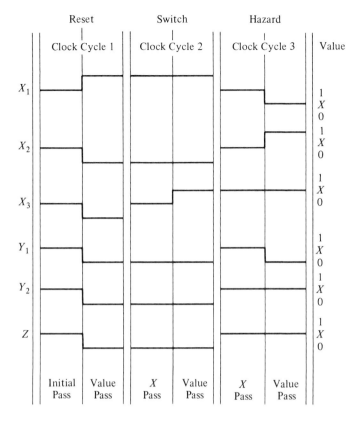

Fig. 3-22. Clock cycle.

The circuit is now in the same condition as shown at the beginning of the graphs of Fig. 3-21.

Switch Cycle

X Pass:

The input X_3 assumes the value of X.

Value Pass:

Input X_3 becomes 1; there is no effect on the other signals.

Hazard Cycle

X Pass:

Input X_1 goes from 1 to X; input X_2 goes from 0 to X.

Since X_1 and X_2 are equal, the output of the AND gate, Y_1, is also X. This forces Z to the value X which, in turn, forces Y_2 to the value X. Notice that Y_1, previously at 0, rises to X in this pass and will subsequently return to

0. This evidences a dynamic hazard in Y_1 which is propagated to Z and represents the possibility that Y_1 might go to 1 temporarily if X_2 goes to the 1 before X_1 goes to 0. There is no evidence as yet as to whether or not this hazard is critical.

Value Pass:

Because Y_2 and Z are still X, the hazard is critical. Thus Z could reach a stable state of either 0 or 1.

Compiled-Code Model. The sorted compiled-code modeling technique results in a comparatively simple and straightforward computer program design. Because the compiled-code model is inherently synchronous, simulating nominal propagation delays and the feedback between logic elements is usually not performed [20].

Such a simulation model utilizes an equation- (or element) ordering algorithm to ensure that output values are computed before they are needed in the subsequent processing. With the nested parentheses used in pseudo-Boolean input expressions, the level of each item is simply determined. The depth of the nesting is determined by scanning the inputs. A level counter is incremented once for each left (open) parenthesis and decremented by one for each right (closed) parenthesis. The engineer-generated input source-language statements can be directly converted as part of the input form to a nested-parenthesis representation. This is shown in Fig. 3-23a which

$$(A + (B \cdot (C + D) \cdot E \cdot (F + G)) + H) = (A' \cdot (B \cdot (C' \cdot D') \cdot E \cdot (F' \cdot G')')' \cdot H')$$

(a) Alternate logic form

Level 1   (A +   + H)
Level 2   (B·   ·E·   )
Level 3   (C + D)   (F + G)

(b) Levels

Fig. 3-23. Levelizing

is a NAND-NAND operation produced by the compiler [20] by removing OR operations.

Typically, the compilation process produces machine-language code routines which perform the desired logic operation, a conditional branch instruction which links the routines together according to their levels, and a test indicator which determines whether a zero or one test will trigger the conditional branch operation. If parentheses are used to establish the level, the conditional branch operation corresponds to the right parenthesis.

After an evaluation is performed, the level identifies the next address (e.g., the next variable) in the processing sequence. The logically equivalent expression of Fig. 3-23a is shown decomposed into levels in Fig. 3-23b. Part of the resulting compiler-produced routine is shown in Table 3-2.

TABLE 3-2.

STEP	CONDITION	PROCESSING	JUMP ADDRESS
1	If A = 0	evaluate B	next element
2	If B = 0	evaluate H	associated with E
	= 1	evaluate C	next element
3	If C = 0	evaluate D	next element
	= 1	evaluate E	associated with D

Next-Event Synchronous Simulation. Only a small percentage of logic in a system is changing at one time in response to a new input. In a conventional, general purpose computer, this percentage is on the order of 2.5% of the total number of logic elements. Obviously, the simulation processing time can be greatly reduced if the inactive logic is not considered in determining the next input state. By considering actions on a *next-event* rather than *next-clock-time* basis, only those elements whose output states are likely to change (because at least one of their inputs has just changed) need to be evaluated.

This process can be implemented by representing current and future events by two pushdown stacks known as the *current-activity stack* and *future-activity stack*. A pushdown stack is a table with the property that the last item entered is the first item removed, as in a pile of cafeteria trays.

All elements whose states can change at a particular time are placed in the current-activity stack (S_0). At the next unit time, each element in the stack is removed and examined. If its state has changed, the elements to which its output signals are propagated and its destination set are placed in the future-activity stack (S_1). As a result, the future-actitity stack contains all the candidate elements whose values might change as a result of new inputs. However, the values of the elements are not updated until the current-activity stack is empty. Consider a two-legged OR gate, j, which is driven by (a) an element, i, whose value changes from 0 to 1 and (b) an element whose value is always 0. If the new value of element i was used, the value of element j would be calculated as 1, an incorrect value. By not evaluating element i until after element j was evaluated, we obtain the correct value regardless of the order in which the gates are evaluated.

When all the elements in the current-activity stack have been examined, the stack functions are reversed. S_1 becomes the current-activity stack to permit evaluation of the candidate elements. The process is repeated, with S_0 being used as the future-activity stack. Each element in S_1 is removed from the stack, its value is evaluated, and its destination set is put into S_0 if its value has changed. Emptying one stack and putting the destination set in the other stack continues until both stacks are simultaneously empty. No more gates remain to be evaluated (a) at the system output or (b) when a stable state has been reached. Unlike the compiled-code model, cyclic logic does not require special-case programming. An oscillation, per se, as might be caused in a circuit in which two gates feed each other, causes no problem in the algorithm. The signal is normally propagated through the network until the simulation ends. It should be noted that an oscillation signal might not be considered by the designer to be an error.

There are two standard computer processes used with pushdown stacks: PUSH, which adds an element to the top of the current stack, and POP, which removes the last element from the current stack. In the next-event simu-

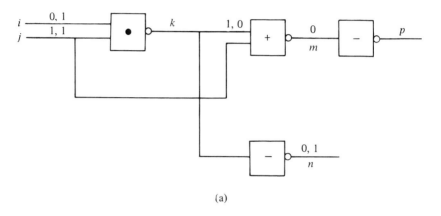

(a)

COMMANDS	ARGUMENTS				COMMENTS		
	Element	Stack	$C(S_0)$	$C(S_1)$	Current-Activity Stack	New Value	Old Value
PUSH	i	S_0	i	0	—	—	—
PUSH	j	S_0	i,j	0	—	—	—
POP	j	S_0	i	0	S_0	1	1
POP	i	S_0	0	0	S_0	1	0
PUSH	k	S_1	0	k	S_0	—	—
POP	k	S_1	0	0	S_1	0	1
PUSH	m	S_0	m	0	S_1	—	—
PUSH	n	S_0	m, n	0	S_1	—	—
POP	n	S_0	m	0	S_0	0	1
POP	m	S_0	0	0	S_0	0	0

(b)

Fig. 3-24. Next-event simulation.

lation, POP also causes the value of the element i, denoted as $v(i)$, to be evaluated. If the new value of the signal differs from the previous value, its destination set is placed (by a PUSH) into the future-activity stack. Both PUSH and POP have two arguments which may be either explicitly or implicitly expressed: the name of the current-activity stack (S_0 or S_1) and the signal name. If not explicitly expressed, the signal name is obtained from the list of signals forming the model of the network.

A next-event simulation is illustrated in Fig. 3-24. The circuit is shown in Fig. 3-24a; corresponding activity stack operation is shown in Fig. 3-24b. For this example, it should be assumed that a list structure describes the linkages between elements so that explicit arguments for PUSH and POP are not required. This list is not exhibited so as not to complicate the discussion.

First, the system inputs are placed into stack S_0. When all of the system inputs (here i and j) have been placed in S_0, S_0 becomes the current-activity stack and S_1 becomes the future-activity stack. Each input, in reverse order, is removed from S_0 and evaluated. Unlike input signal j, the value of input i has changed; so its destination set is placed in S_1. When S_0 is emptied, S_1 becomes the current-activity stack. S_0 contains only signal k. As a result of the change in inputs to the NAND gate, the value of k is now 0; so its destination set (m, n) is placed in S_0. After signal k has been examined and removed from S_1, S_1 is empty, and S_0 becomes the current-activity stack containing only the destination set of k. Although the value of signal n changes, it has no destination set because it is a system output. Signal m does not change so both stacks are exhausted; completing the simulation pass.

Time-Mapping Asynchronous Simulation Model. In the next-event simulation, all future events are modeled as if they were executed and completed at the same time. This assumption is valid in synchronous, but not asynchronous, simulation. Unlike synchronous simulation, asynchronous simulation has many events with different completion schedules in process at the same time. The next-event technique can be easily extended to simulate asynchronous events because each of the time periods at which events are completed is an integral multiple of a time-quantized period. If future activities are represented by utilizing more than one pushdown stack, events can take place in as many time periods as there are future activity stacks [30]. Time is thereby quantized into constant Δt time intervals, e.g., each time period $(t_1, t_2, \ldots, t_i, t_{i+1}, \ldots, t_n)$ at which future activities occur is defined by the rule $t_{i+1} = t_i + \Delta t$.

1. Δt is the greatest common divisor of the individual delays.
2. The time spanned by the Δt intervals is $(d_{\max} + 1) \Delta t$, where $d_{\max}$ is the maximum delay, e.g., there are $n = (d_{\max} + 1)$ intervals.

Any change in the inputs of element x at time t_i is evaluated at time

$t = t + d_x$, where d_x is an integer multiple of Δt. The new time, t, specifies the stack (S_t) into which the destination set of x is placed.

A pointer (P) points at the current-activity stack (S_t), advancing to the next stack to denote the next time period at which an event occurs. The Δt time periods form a circular loop so that future-activity stack S_i is $S_i = (t_i + d_x)$ mod n, and P successively points to $S_1, S_2, \ldots, S_i, S_{i-1}, \ldots, S_n, S_1, S_2, \ldots$. Each item in the new current-activity stack (S_{t+1}) is removed from the stack and evaluated. Each component of the destination set of each signal whose inputs have changed is put into the appropriate stack. The pointer advances

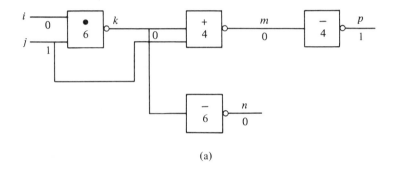

(a)

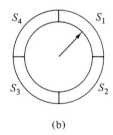

(b)

		S_1	S_2	S_3	S_4	T_1	New Value	Old Value
1	POP (j, S_1)	i				1	$v(j) = 1$	1
2	POP (i, S_1)	o				1	$v(i) = 1$	0
3	PUSH (K, S_4)				k	1		
4	POP (K, S_4)				o	4	$v(k) = 0$	1
5	PUSH (n, S_3)			n				
6	PUSH (n, S_2)		m	n		4		
7	POP (m, S_2)	o	n			2	$v(m) = 0$	0
8	POP (n, S_3)					3	$v(n) = 1$	1

(c)

Fig. 3-25. Time-mapping simulation.

to the next stack in the loop when the current-activity structure is emptied.

The time-mapping technique is described in Fig. 3-25 using the same circuit that was used to illustrate next-event simulation (Fig. 3-24) for comparison.

Each gate's delay is noted in the logic diagram. Since all gates have either a four- or six-unit delay, the value of the greatest common divisor is two units (since $\Delta t = 2$, $n = \frac{6}{2} + 1 = 4$). Thus four pushdown stacks are sufficient to store the actions taking place in the circuit.

Assume that the processing has progressed so that signals i and j are in stack S_1 and that P has just been advanced to point to that stack. Signal j is removed from S_1 and evaluated; because its value is unchanged, its destination set is not of interest. On the other hand, signal i's value has changed from 0 to 1; so its destination set (k) is placed into the S_4 stack ($t + d_k = 1 + 3 = 4$). Since S_1 is empty, the pointer advances to S_2 time period. Because S_2 and S_3 are both empty, the next element evaluation occurs at t_4. No evaluation is required for either t_2 or t_3. Stack S_4 contains signal k, whose value has now changed, causing its destination set (m, n) to be placed into the appropriate stacks, S_2 and S_3, respectively. The process continues until there are no signals left to be evaluated in any of the stacks, completing the simulation.

As an alternative to quantizing time into a specific number of Δt time slots according to the requirements of the specific simulation, events could be linked in a list and ordered by their "time-to-be-executed." This general time list replaces the time slots represented by the pushdown stacks. Each item on the list points to the next item to be evaluated. As a result, time is not incremented by a fixed constant value. Simulation time is advanced according to the time associated with the next element to be evaluated and, because it takes on more of continuous domain, avoids the necessity of specifying a basic time unit. The simulations in [19] and [25] use this approach.

As discussed previously, assigning a standard time unit for simulation poses no difficulty. The choice between the two approaches is determined on the basis of implementation ease and resulting computer processing time for the particular simulation situation.

3.4.3. Output Descriptions

The usefulness of the simulation is determined by its outputs. Thus the designer must be able to specify:

1. a list of signals to be monitored;
2. output formats;
3. the time when output is desired.

The logic behavior is essentially summarized in the changes in internal and output states of selected signals. This output may be of interest at dif-

142   *Gate-Level Logic Simulation*   Chap. 3

ferent time periods, depending upon the purpose of the simulation: (a) for each bit-time; (b) at selected or periodic time intervals [25]; (c) when the selected gate (*s*) changes state; and (d) if the simulation permits the values to be compared automatically with a provided standard whenever the calculated output disagrees with the standard value [12]. Because of its convenience, a chart of signal- and state-values time forms the basic logic simulation output. A typical timing sequence (Fig. 3-26) depicts a waveform in an exceptionally convenient and widely used format. The values of the signals are shown in a table whose columns are the specified signals and the rows are the specified time periods. The logic simulation output displayed in this illustration is based on the circuit configuration described in Fig. 3-10

SIMULATION BEGINS

	C Y 0 0 0	C Y 0 6 0	C A U T 1 1	C A U T 1
1	0	0	0	1
2	0	0	1	1
3	0	1	0	1
4	0	1	1	0
5	1	0	0	1
6	1	0	1	1
7	1	1	0	0
8	1	1	1	0
9	0	0	0	1

Fig. 3-26. State vs. time behavior.

assuming that CY000, SY060, and CAUT11 started out at a low voltage and that an alternating pattern four clock cycles in duration was used on signal CY000.

The term waveform is deliberate. In the printout, the 1's and 0's are staggered so that, when this line printer output is turned on its side, as in Fig. 3-27a, the chart resembles an oscilloscope output. The periods of high state, shown by sequences of 1's, appear at a high level. Spikes and dips are isolated on the page and easily noticed. Other simulators [20] use periods (.) and M symbols as an alternative gimmick to denote low and high signals in a printer format in which rows represent signals. This symbology presents a picture similar to a waveform output and is shown in Fig. 3-27b.

Other types of outputs are useful. Although a waveform diagram clearly presents the signal pattern, the offset limits the number of signals that can be printed on a page. A signal table permits one signal to be assigned to each print position; almost as many signals can be monitored as there are print positions across a page. A special printer chain can be utilized to permit

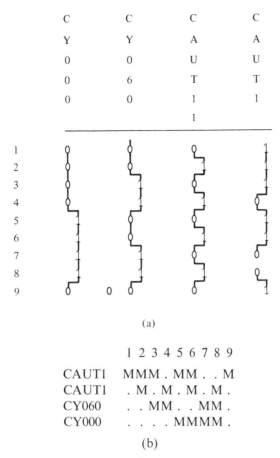

Fig. 3-27. Waveform output example.

characters to be printed sideways in small type so that more signals can be printed out.

3.5 FAULT SIMULATION

The problem of automatically generating test patterns for combinatorial networks has commanded considerable interest. Few solutions (mostly those based on Roth's D-Algorithm [23]) have had any degree of acceptance. Although beyond the scope of this chapter, a brief discussion of the problem is included because existing and proposed solution mechanizations utilize a logic simulation base. There are in effect two problems:

1. to generate an input pattern which tests a portion of a network;
2. to generate the list of single faults (stuck-at-1, stuck-at-0) whose presence is detected by the test pattern.

Gate level simulation is generally used to satisfy the latter requirement. First the fault-free network is simulated. Then, simulation is performed with each of the possible faults inserted. Comparison of the latter simulation with the fault-free case identifies those faults which produce a change in output values. Considerable computer time may be required since the test generation procedure must be repeated for each fault not covered by a previously generated pattern.

However the ability to accept both solid and intermittent fault conditions greatly extends the application of a logic simulation for fault-analysis studies. The output of the failure-free model normally simulated is a useful reference during hardware testing to determine if a malfunction has occurred. But if, in addition, faults can be selectively inserted, the response of the digital system to various types of faults can be determined. The resulting failure-response table associating specific faults with output bit sequences [9], [32] not only reduces the effort for post-fabrication fault-isolation diagnostics, but also permits establishing the sensitivity of the digital system to component failures.

Associating simulated component failures with observable results assumes that:

1. the fault alters the system's behavior;
2. the effects of the fault can be observed in the actual hardware;
3. the logical path between the inputs and outputs is unambiguous;
4. other failures producing the same responses can be identified.

Even with a logic simulator, comprehensive fault-analysis studies are handicapped by the large number of cases which must be evaluated to determine the logic's failure mode behavior characteristics. Each separate combination of element failures requires some degree of modification of the model. Consequently, current simulation approaches attempt to develop relatively efficient, rather than optimum, solutions.

In this section, we examine first the manner in which a component failure may be introduced into a logic simulation. Second, the logic simulator's use in generating detection and diagnostic test sequences is considered. Finally, we discuss the application of the information provided by a logic simulator in a digital test facility.

3.5.1. Fault Insertion

The behavior of the failed element, such as a diode that acts in a particular manner, can be simulated directly by changing the logic model. The normal program code which simulates the correct operation of the element is replaced by a new program code which simulates the element's fault behavior. In an interpretive program system, the links in the data structure are modified to force the desired conditional branching. The frequency and type of failure can be represented statistically to depict either a permanent or transient error.

Changes made to the computer model to depict each failure mode may have to be reset before the next simulation. This modification process is time-consuming. The same effect has been achieved by using fault-simulation parameters, either as constants in the programs or as operands to the logical instructions in the program code. These parameters can be externally set and reset easily by an ancillary program. Rather than being treated as an overlay to the "clean" simulation process, fault and normal conditions are treated as different cases of the same model [9], [31].

Consider the simple two-legged AND gate whose coding is exhibited in Fig. 3-28. The normal coding is shown in Fig. 3-28a. This code has to be revised for each different component-failure simulation. Yet, with the

```
        CAL     INPUT 1
        ANA     INPUT 2     AND BOTH INPUTS
        SLW     OUTPUT
```

(a)

```
        CAL     INPUT 1
        ORA     MASK 1      FAIL/PASS INPUT 1
        SLW     TEMP
        CAL     INPUT 2
        ORA     MASK 2      FAIL/PASS INPUT 2
        ANA     TEMP
        ANA     MASK 2      FAIL/PASS AND GATE
        SLW     OUTPUT
```

(b)

BIT NUMBER	CONDITION	MASK 1	MASK 2	MASK 3
1	FAULT FREE	0	0	1
2	IN1 failed on	1	0	1
3	IN2 failed on	0	1	1
4	OUTPUT failed off	0	0	0

MASK 1   0100
MASK 2   0010
MASK 3   1110

(c)

Fig. 3-28. Parallel fault-insertion coding and gate.

coding shown in Fig. 3-28b, where suitable values are placed in the three masks, several conditions can be simulated simultaneously in parallel. These faults are introduced by ANDing or ORing the state of the applicable input signal with the mask. Each bit in the output word corresponds to a particular failure mode. Forcing an appropriate bit to 1 simulates the related condition of the associated gate being stuck ON. This is accomplished by ORing a mask with a 1 in the proper bit position with the signal word. Similarly, a gate can be forced to 0 by ANDing the corresponding bit position with a mask containing 0. In Fig. 3-28b, each bit position represents a different masking condition as shown in Fig. 3-28c. The masks are arranged so that each bit is unchanged by the masks except for the bit which is failed. Thus, bit 2 represents the case of IN1 being stuck ON. When the state of IN1 is ORed with MASK2, bit 2 is ORed with 1; so it is forced ON. OUTPUT is ANDed with MASK3, which contains a 1 in every bit position except the fourth. As a result, only the fourth bit position is affected, and since it is forced to be 0, it simulates the gate being OFF.

Inserting faults as an ancillary process is more efficient than changing the machine logic description. Conceptually, a separate ancillary program is not necessary. The simulation commands control the values of system input stimuli as a function of time. Extending this mechanism to control internal state values allows the failure to be specified. As a result, the failed condition can be set and reset according to any desired time schedule without directly affecting the logic model.

Different failure conditions, as well as failed element values, can be simulated in parallel. Each mask position set to 1 represents a separate failure, e.g., diode open, diode closed. Both single and multiple faults can be represented. With single faults, one mask represents a failure condition for any one bit. Multiple faults may use several failure masks for one bit.

When the element simulation has been completed, the results with the mask-enabled failure condition can be compared with the mask-disabled normal simulation. If this is done, differences in particular bit positions are used to determine which failures are to be propagated further.

The processing advantages of parallel simulation, when compared to several separate single-failure event simulations, are to a large extent lost in event-driven algorithms. In event-driven simulation, the processing sequence follows only those elements whose states have changed. At any one time, this represents only a small percentage of the total logic. However, because each failure changes the behavior of elements, when several faults are simulated at one time, different flow sequences must be followed. As a result [9], in practice, normal operation is followed until the first path which diverges is encountered. This may result either from differences between inputs or from failure modes. Consequently, after completing each path, the simulator must

Sec. 3.5   Fault Simulation   147

backtrack to the point of divergence, resetting all internal elements to their previous values.

3.5.2. Diagnostic Generation

The high speed processing capability of the computer to generate and execute a large number of trial runs permits the logic simulation model to be used as a basis for developing change in state value under failure conditions. If, for example, the normal output of a gate were 1, a malfunction which fails the gate to 1 would not be detected. With specific faults introduced into the model, input stimuli are generated exhaustively, randomly, or according to some algorithmic procedure. The effect of each run on output and intermediate values is monitored and selectively recorded, usually when a change of state occurs.

After different trial runs have been made, ineffective input sets are weeded out using simple criteria such as (a) maximizing the number of failures de-

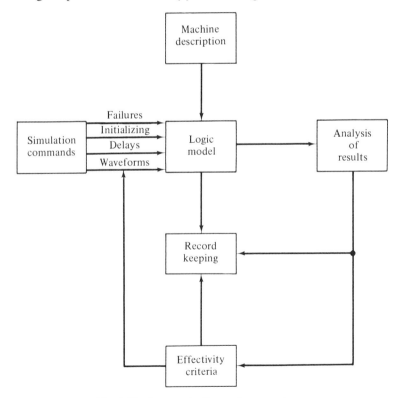

Fig. 3-29. Automatic diagnostic generation.

tected for each input set or (b) minimizing the number of different input sets required to exercise every path in the network. The remaining input sets are used to create the diagnostic sequences. The process is depicted in Fig. 3-29.

Test-generation algorithms [21] attempt this process to avoid the burdens of manual analysis of the large number of computer runs required. For each different input set, the internal state values in a failure-free system are compared with those in a system containing the given faults. The difference between the two systems is used by the algorithm to compute the next input set to be used.

In contrast to the normal simulation in which the inputs are known and the outputs calculated, a backward simulator [23] is given a gate's output and determines the inputs necessary to produce that output. The logic is traced backward from the faulty element to find the system inputs which test this condition.

3.5.3. Logic Testing

Digital modules and digital subsystems are commonly tested in a computer-controlled test facility. The facility is used in design and fabrication phases for troubleshooting detected hardware errors. In the test facility, the digital equipment being tested is connected directly to a digital computer. The software programs resident in the test facility computer control all testing, including error detection, fault detection, and fault isolation. The input stimuli to the digital device under test are transmitted by the computer under program control. Outputs from the device, in response to these stimuli, are returned to the computer. To facilitate subsequent fault-analysis studies, records are automatically kept of the failure occurrences, including transient errors, sequence sensitive errors, and multiple faults.

For logic testing, the logic-level simulation complements the digital test facility (Fig. 3-30). Testing digital logic consists of applying a sequence of inputs and verifying that the correct outputs are produced. Typical benchmark sequences are established during design through simulation by analyzing the effects of time-varying bit input waveforms upon the logic model. The actual hardware outputs from the hardware tests are compared with the simulated outputs. In practice, process testing is used instead of logic testing whenever the logic description becomes too complex for efficient diagnostic testing, as with groups of LSI models, or when functional testing of a subsystem is sufficient to pinpoint erroneous replaceable logic blocks.

Recording all test sequences generated by the test facility computer permits both the stimuli, which triggered a hardware error, and all preceding test sequences to be repeated subsequently in a simulation. As in the failure-free

Sec. 3.5   Fault Simulation   149

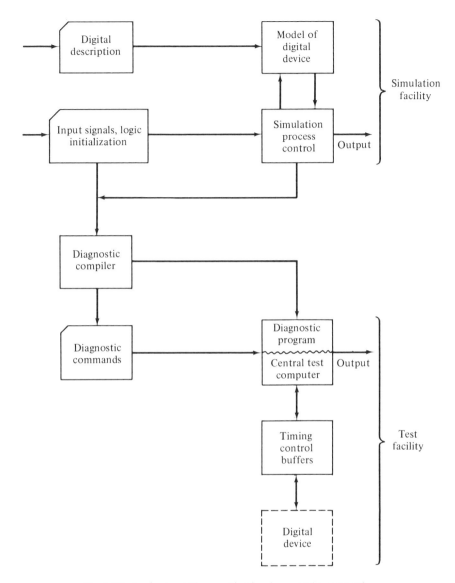

Fig. 3-30. Logic simulation application in central test Facility.

case, the simulation output is compared with the hardware testing results to analyze hardware filures.

The simulation commands, together with the logical and structural description of the digital device in the model, contain sufficient information to define test programs. Consequently, the diagnostic test sequences developed during the simulation studies prior to hardware availability could be automatically converted for the test facility by a diagnostic compiler. The conversion process would utilize the duality between logic simulation and hardware testing. In logic simulation, the commanded input waveforms are translated into a set of computer program instructions which exercises a model of the digital system. In the logic test facility, the input stimuli exercise the digital hardware whose logic was modeled in the simulation.

Implementation of a diagnostic compiler is not simple; but because known test sequences are used, the conversion process is not as complex as the production of a compiler for automatic test program generation. The cost of the compiler could be recovered by reducing equipment checkout expenses in any facility designing and fabricating a large number of different digital modules and subsystems. As a result, several diagnostic compilers are in the planning stage; none, to the authors' knowledge, has yet been implemented.

3.6 SPECIAL SIMULATION TECHNIQUES

Digital logic simulation can be regarded in a more general context than the implementation of a specific program to be executed on a standard host computer. On one hand, advantage can be taken of the already existing general purpose software programs to aid the analysis activities; on the other hand, other computer configurations can be utilized to provide a more efficient simulation vehicle.

3.6.1. Special Software Techniques

Two programming systems, the general purpose system simulators and PERT, have been used to assist the logical design process. For each, a large body of checked-out and well-documented software is commonly available.

General Purpose System Simulators. Logic behavior can be simulated with any general purpose simulator that models a discrete environment. Although usually applied to study system level behavior, they are directly applicable to modeling digital logic. Using the general simulation language, it is possible to express the structure of the network, its behavior, and the simulation conditions. By so doing, the designer can take advantage of the features such as timing controls, output formats, and diagnostics which have been built into the programming system. A thorough comparison of six simulation systems can be found in the [29]. Of the currently available simulators, **GPSS** and **SIMSCRIPT** are the most widely used.

GPSS [7], for example, defines as its basic entities;
1. transactions which move through the system, generating prescribed actions;
2. facilities which are operated upon by transactions;
3. blocks which provide the rules by which transactions flow;
4. statistical queues or tables.

Direct analogies are made between these entities and those in the system to be modeled. For logic simulation, these concepts represent signal pulses, the digital elements (gates, flip-flops), and the signal paths of the network.

Based on these assumptions, the logic simulation rules are defined using the GPSS language structures. From these statements, GPSS generates and monitors the flow of digital pulses through the model of the network. A signal must be acted upon by a gate before it can continue to move through the system. Gates are put into an event queue when their inputs change; they are replaced when their outputs switch state. The GPSS clock is free-running, so that timing of events is asynchronous, and it advances according to the time at which the next event is to occur. Delay values can be expressed as a statistical distribution. The output and the report generation describe both system flow and state changes and provide summary statistics.

There are several drawbacks to creating a logic simulation tool for production application based on a general purpose simulator. None of these limitations necessarily preclude application either as an experimental aid having a limited use or in a situation where a logical simulator must be quickly implemented.

1. General purpose simulators depict a statistical model of a system. The technique is sufficiently detailed to provide a valuable assist in conceptual design activities. Their main value is to complement rather than replace logic-level simulation by facilitating the examination of large blocks of logic. It is, of course, possible to represent digital logic gate-by-gate, with individual GPSS blocks. However, the large number of resulting blocks for a system of reasonable size would not be practical with respect to the time required for either modeling or computer execution.

2. Using a general simulation system requires a reasonable facility with the syntax and semantics of its language and an understanding of the simulation concepts which have been built in by their implementers. For the nonprogramming engineer, the time consumed by learning these rules may out weigh the facility benefits. Whereas GPSS, due to its flow chart orientation, is easy to use in this respect, the logic designer may have difficulty in expressing his model with the sentence-like structure of SIMSCRIPT. In addition, some programming experience (knowledge of FORTRAN) is necessary to use SIMSCRIPT, and only limited source-language diagnostics may be present [29].

3. The fidelity of the general purpose simulation in depicting realistically the specific logic behavior of the model cannot be assumed a priori. A general simulation system models a general process. The processes of specific

application may conflict with the processes built into the simulator. Any inaccuracies or ambiguities between the general purpose simulation's language concepts or the use to which they are put could invalidate the simulation.
4. The general system simulator provides a simulation vehicle and associated program structure for a wide range of applications. In any specific situation, however, these features may not be needed. The implementation of a general approach can easily result in a relatively inefficient processing scheme for a specific application, as logic simulation, when compared to a specially implemented program. A slow simulator, besides the resultant computer execution time costs, could consume so much computation time that, for all practical purpose, it would be relatively useless as a tool in identifying logic errors and race conditions.

PERT. PERT (Project Evaluation and Review Technique) was initially developed to identify critical paths and provide statistical estimates of end dates in large projects. A modification of this technique has been used to obtain worst-case delay information which would not be easily available from a logic simulator [14]. In a logic simulation there is an implicit assumption that fast switching elements will cancel out the effect of slow switching caused by using nominal delay values. As a result, a statistical estimate of the probability of a complex path being a critical path is obtained without having to resort to using the worst-case values. Once identified, this specific path can be studied in detail with a logic simulator.

A PERT network is a directed flow graph between completed events represented by nodes. No events can take place until all the input events connected to it have been completed. The branches between the nodes represent the time required to achieve the next event, given that the events on all connected input branches have occurred. Three time values are assigned: the shortest time (best case), the most likely time (expected time), and the longest time (worst case). An estimate of the most likely time is computed from these three values. In the simple PERT diagram of Fig. 3-31a, events are denoted by letters $A, B, C, \ldots$, and time periods are denoted by numbers. It can be observed that event G may occur at any time from 9 to 28 time periods, and it requires events A, B, C, D, E, and F to have occurred previously. The output is a statistical probability of (a) the expected delay in the system and (b) achieving a given event by a specific time. The critical time path and slack time in noncritical paths are identified.

The validity of the PERT estimates is affected by:
1. the credibility of the three input figures;
2. the correctness of the assumption that the mean-time delay can be computed by an approximation to a normal distribution;
3. an implicit error caused by the statistical bias resulting from using three input values to estimate the distribution.

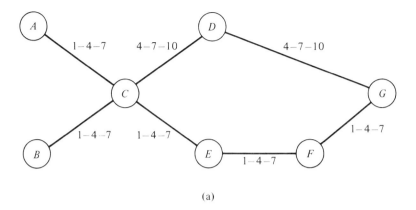

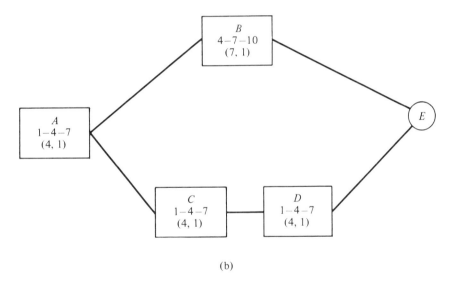

Fig. 3-31. PERT application.

When the PERT is used as an aid to logic design, the nodes are replaced by logic blocks. Time delays are associated with each block rather than with branches. The branches represent signal flow. Two sets of delay figures are normally associated with each block, one identifying the time required for the element to turn ON, the other identifying the time required for the element to turn OFF. A logic AND is represented by using the longest time delay of any input branch (worst case). An OR is represented by using the earliest time of any input branch sequence.

This PERT analysis is illustrated with the PERT logic block diagram of Fig. 3-31. Only one set of delays is shown for simplicity. The mean-time delay (t_e) and the variance (σ^2) are computed as in a project management application.

$$t^e = \frac{s + 4m + w}{6}$$

$$\sigma^2 = \left(\frac{w - s}{6}\right)^2$$

where s is the shortest time, w is the worst-case time, and m is the mean or expected time. These values are shown as ordered triplets in the diagram. The path A, C, D, E has the longest delay, T_e ($T_e = \sum t_e$), in the system. The expected delay of this critical path is 18 time periods. To calculate the probability of achieving this output by 22 time periods, the normal distribution curve is used. For example

$$Z = \frac{T - T_e}{\sum \sigma^2} = \frac{4}{1.732} = 2.31$$

With reference to the standard normal distirubiton function tables, the probability of propagating input signals through the network within 22 time periods is determined to be 98.96%.

PERT was effectively used in designing a ferrite core storage system [14] in this situation; a worst-case analysis indicated that the core output would be too slow. However, the PERT analysis indicated that 98% of the cores could meet the access-time requirement.

3.6.2. Specialized Computer Techniques

While most logic simulations operate on a standard, general purpose computer, other host configurations have been suggested and, at least in several instances, can be implemented to create a dedicated system.

In any except the simplest digital network, logic activity takes place along parallel paths. For instance, many gates may be switching in parallel. A simulator designed to utilize a conventional computer as the host must deal with events sequentially.

Computer systems with a parallel structure have been investigated for use in logic simulation to decrease the execution time. In the parallel computer configuration, there are several elements operating simultaneously. Each processing element acts as a small computer, with a central processor and a bus structure for communication between processing elements. These processing elements operate independently under centralized control.

For logic simulation, each processor contains the routine required to evaluate simple gates and maintains the status of the signals for each of the gates. The controlling processor contains the system clock and maintains the list of future events to communicate with the appropriate cell. The controller activates the proper cells and provides input values for the computations. The processors inform the controlling processor whether or not the elements under their cognizance actually switched. The controlling processor updates the clock and rearranges the events list.

A combined logic and register which is extremely effective in simulating larger systems has been achieved with a host computer configuration consisting of four special purpose parallel programmable hardware elements and a general purpose computer [19]. The special purpose hardware performs logic-element simulation, while the general purpose computer performs register-level simulation as well as all input statement and output format processing and hardware setup and control.

Each special purpose computer permits logic simulation to be effected using both computer code and actual element operation. The parallel operation of these devices results in the logic simulation being performed at high speed. The ability to model registers at the same time greatly extends the scope of the simulation by permitting input/output and interface units to be simulated without defining their logic. Thus the logic of sections of a large system can be simulated, with respect to their operation within a total system framework, without having to define the entire system at the detailed logic level. In addition, the effectiveness of simulation is increased because results from the logic simulation and register simulation of the same device can be compared easily.

3.7 SIMULATOR EXAMPLE

The logic simulation concepts and techniques described in this chapter are illustrated in the following description of a hypothetical simple three-value asynchronous simulator based on [25]. Of significance is the depth of simulation permitted with only a few commands.

3.7.1. Simulator Characteristics

With this simulator, the logician can:
1. initialize any input signal and any logic element state to HIGH or LOW or unspecified;
2. set individual gate delay times, set delay times for groups of gates or set gate times both individually and for groups;
3. schedule pulses, levels, and clock times as periodic functions;

4. select observation intervals for printout of signals when a state change occurs or at predetermined times;
5. control the format and frequency of output;
6. reinitialize the simulation to allow multiple simulation runs with optional respecification of input, delay, and characteristics with the following control commands:

(a) *Simulation Control*

SIMULATE	Initiates the simulation.
EXIT	Signifies the end of the simulation commands.
RESTART	Reinitializes the simulation program and the circuit characteristics.

(b) *Signal Specifications*

HIGH	Sets listed signals to logic-level 1.
LOW	Sets listed signal to logic-level 0.
SCHEDULE	Causes: (1) inversions of listed input signals at successive time increments; (2) printouts at the time periods indicated in parentheses after PRINT statement; (3) conclusion of the simulation at the time given in parentheses after STOP statement.

To illustrate:
1. DATA10 (100, 300, 200) produces the waveform

```
 0      100           400      600
```

assuming DATA10 had been initialized by a HIGH/DATA10 statement.

A *p* symbol causes the sequence to be repeated periodically throughout the simulation from the initial action to the last time listed prior to the *p* symbol, e.g.:

2. PRINT (25,P) causes a printout every 25 time units.
3. CLK (50,P), using a nanosecond (ns) time unit, generates a 10 MHz clock called CLK.

DELAY	Sets the delays of listed gates (ns). If unspecified, gates are set to their nominal values.

(c) *Output*

PRINT	Prints the listed signals. Optional RESET command restarts the printout list.
CHART	Specifies timing-chart format on simulation printout (maximum of fifteen signals across printout page).

LIST   Specifies list format on simulation printout (maximum of one hundred signals across printout page).

CHANGES   Indicates a full-line of simulation printout is to occur every time any signal on the output list changes.

NOCHANGE   Opposite of CHANGES; printout occurs only when explicitly specified.

It should be noted that this simplified simulation has limitations: No language capability exists to allow faults to be inserted with the simulation commands, and no capability exists for performing any type of inverse or backward simulation, e.g., specifying the output(s) of an element(s) and obtaining a set of input stimuli.

3.7.2. Simulation Examples

For a circuit that delays a slowly varying input waveform by two clock cycles and generates a synchronizing pulse on the overlap of the first and second shifted outputs, the simulation commands illustrated in Fig. 3-32, would generate the following input stimuli:

```
HIGH/HIGH1,PULS0,RCLCK$
LOW/GND,PULS$
SCHEDULE/RCLCK(125,P)
PULS(1250,1250,2500,750,2500,1250,500)
PULS0(1250,1250,2500,750,2500,1250,500)
STOP(12000)
PRINT/HIGH1,GND,RCLCK,PULS,PULS0,
F4311,F6730,A1413,F6758,A1420,PLSS$
CHART/
SIMULATE/
EXIT
```

Fig. 3-32. Simulation input commands.

1. HIGH initially sets the listed signals to 1. HIGH1 enables the transfer gates in the flip-flop packages. PULS0 is the inverted value of the slowly varying input pulse, and RCLCK is the clock signal to the JK flip-flops.
2. LOW initially sets the listed signals to ZERO; GND is a permanent ground; and PULS is the true value of the slowly varying input pulse.
3. SCHEDULE schedules timing information. RCLCK(125,P) generates a clock with 125-ns sections (alternately high and low), the resulting period being 250 ns. PULS(...) and PULS0(...) generate waveforms, each of whose sections (alternately high and low), are the lengths in nanoseconds of the numbers in parentheses.

The resulting waveform patterns are:

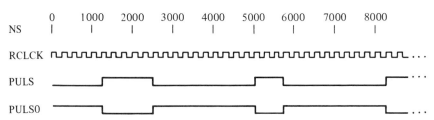

The usefulness of the simulation is shown with the circuit in Fig. 3-33. This logic diagram describes two series transfer flip-flops packages implemented using NOR logic. The input DIN sets the input gate A2. When the clock line CLK falls from high to low, the input is enabled and passed through gates A3 and A4, thus setting the back-to-back gates A5 and A6. The output of A5 takes on the value of DIN two or three gate delays after the fall of the clock. The circuit latches itself, and no further inputs are transferred until the next clock transition from high to low. The second stage sets the output signal DOO on the next falling edge of the clock. The device performs a simple shift function on a one-bit input DIN, which varies slowly with respect to the clock.

The following statements define a simple simulation

    RESTART
    HIGH/DIN,A5,A11
    LOW/CLK
    SCHEDULE/DIN (125,100) CLK(50,P)PRINT (10,P) STOP (400)
    PRINT/A3,A4,A5,A9,A10,A11,DIN,CLK,DOO
    CHART/
    DELAY/ (5)
    SIMULATE/
    EXIT/

The signals DIN and CLK are initialized at high and low levels, respectively. The two gates, A5 and A11, are set at high state to preset the back-to-back NOR configuration. All other unspecified signals are undefined at simulated time 0 and assume high or low level values after appropriate signal combinations have occurred during simulation. The clock line is a 10-MHz clock, that is, with a period of 100 ns. The delays of all gates are set to 5 ns. The varying input waveform on DIN changes to a low level after 125 ns and returns to high level at 225 ns. The input and output pulses, the clock, and

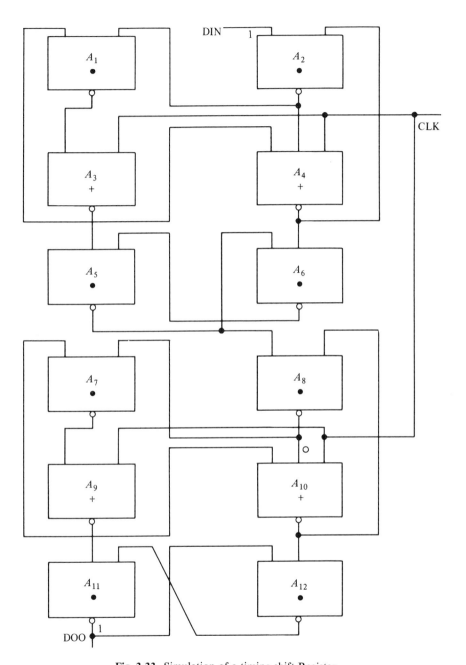

Fig. 3-33. Simulation of a timing shift Register.

159

outputs of several gates are printed out periodically every 10 ns or whenever any signals being monitored change level. The simulation terminates after 400 ns.

In the resulting timing chart, Fig. 3-34, the rows represent simulated time, beginning at simulated time 0. CLK inverts every 50 ns. The input DIN varies comparatively slowly, changing at 125 and 225 ns. The output of gate A5 changes two or three gate delays after the falling clock edge, depending on

	A3	A4	A5	A9	A10	A11	DIN	CLK	DO
0	U	-	-	1	-	-	1	1	0	1
10	U	-	-	1	-	-	1	1	0	1
20	U	-	-	1	-	-	1	1	0	1
30	U	-	-	1	-	-	1	1	0	1
40	U	-	-	1	-	-	1	1	0	1
50	U	-	-	1	-	-	1	1	1	1
55	U	U	0	1	0	0	1	1	1	1
60	U	U	0	1	0	0	1	1	1	1
70	U	U	0	1	0	0	1	1	1	1
80	U	U	0	1	0	0	1	1	1	1
90	U	U	0	1	0	0	1	1	1	1
100	U	U	0	1	0	0	1	1	0	1
105	U	0	1	1	0	1	1	1	0	1
110	0		1	1	0	1	1	1	0	1
120	U		1	1	0	1	1	1	0	1
125	U		1	1	0	1	1	0	0	1
130	U		1	1	0	1	1	0	0	1
140	U		1	1	0	1	1	0	0	1
150	U		1	1	0	1	1	0	1	1
155	U	0	1	0	0	1	0	1	1	
160	U	0	1	0	0	1	0	1	1	
170	U	0	1	0	0	1	0	1	1	
180	U	0	1	0	0	1	0	1	1	
190	0	0	1	0	0	1	0	1	1	
200	U	U	1	0	0	1	0	0	1	
205	1	U	1	0	1	1	0	0	1	
210	1	0	0	0	1	1	0	0	1	
220	1	U	0	0	1	1	0	0	1	
225	1	0	0	0	1	1	1	0	1	
230	1	0	0	0	1	1	1	0	1	
240	1	0	0	0	1	1	1	0	1	
250	1	0	0	0	1	1	1	1	1	
255	0	U	0	0	0	1	1	1	1	
260	0	0	0	0	0	1	1	1	1	
270	U	U	0	0	0	1	1	1	1	
280	0	0	0	0	0	1	1	1	1	
290	0	0	0	0	0	1	1	1	1	
300	U	0	0	0	0	1	1	0	1	
305	U	1	0	1	0	1	1	0	1	
310	U	1	0	1	0	0	1	0	0	
315	U	1	1	1	0	0	1	0	0	
320	0	1	1	1	0	0	1	0	0	
330	0	1	1	1	0	0	1	0	0	
340	0	1	1	1	0	0	1	0	0	
350	0	1	1	1	0	0	1	1	0	
355	0	U	1	0	0	0	1	1	0	
360	U	0	1	0	0	0	1	1	0	
370	U	0	1	0	0	0	1	1	0	
380	0	0	1	0	0	0	1	1	0	
390	0	0	1	0	0	0	1	1	0	

Fig. 3-34. Simulation output.

the direction of change. The waveform of DIN appears at A5, shifted by one clock interval. The output of gate A11, DOO, shows the same signal shifted by the second clock period.

The dashes for gates A4, A6, A10, A12 indicate that their states are unknown.

3.7.3. Relationship Between the Data Base and the Simulation Algorithm

In the simulation program, the model of the logical network is represented as a structured data base. Each equation is linked by pointers to its predecessors (input signals) and successors (fan-out). Equations may be grouped together to form ICP or plate groupings, and any such block may be used as a logical unit by higher level structures. In Fig. 3-11 RSFF was embedded in a larger circuit denoted as SHIFT by the allocation statements of A and D. The definition of RSFF as an entity could have been stored, initially, in this system in a separate reference file.

The simulation mechanization is shown in Fig. 3-35. The pointer structure permits elements to be linked dynamically as their definitions are called for during simulation. The list linkage permits a complete logical path from input to output to be quickly tracked. A change in any input signals causes the equations on which the signal is used to be reevaluated in order to establish if new outputs were generated. Thus, propagating a signal consists of following changes through a path of successor pointers until the change has no additional effects on the system.

Timing of pulses and wave propagation through the interconnected logic is handled by a version of the time-mapping simulation algorithm. When recomputed, any change in output state is noted, and the output change is scheduled to occur at a time determined by the gate delay. Through a chaining mechanism, the time schedule controls the relative times at which future changes in signal states will occur. A threaded list links the logical equation of the events occurring at the same time.

Processing at a given time consists of looking up all scheduled signal changes for the current simulated time and stacking references to the input combinations they affect in the appropriate time list. Changes in any input state causes the destination elements to be put in the time schedule according to their propagation delay. When the event list for the current time has been completely processed, the timing schedule is recorded. The clock is advanced to the time associated with the next earliest event chain on the time schedule, and processing commences with that event.

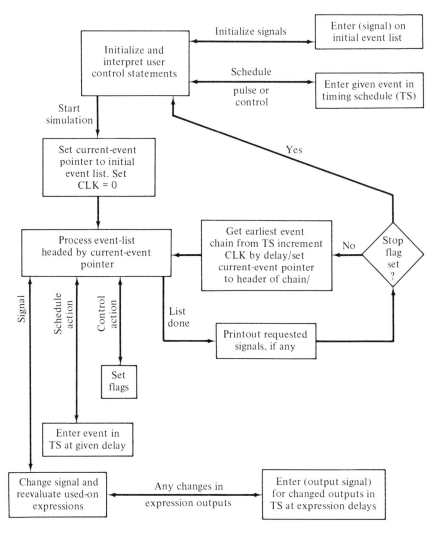

Fig. 3-35. Simulation algorithm.

The tables of Fig. 3-36a would be produced by simulation commands

HIGH/A2,A3

SCHEDULE/A3(25)STOP(100)

At clock time 0, the current time pointer identifies A2 and A3 on the threaded list whose successors identify A1 for processing. The new state of A1 is computed and put in the timing schedule at the appropriate 5-ns delay time (delay time of A1). The clock is updated to the next scheduled time (5 ns).

Sec. 3.7 Simulator Example 163

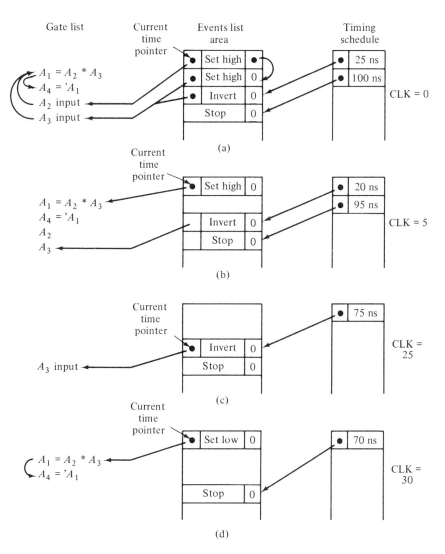

Fig. 3-36. Intermediate states of data tables.

At this time (refer to Fig. 3-36b), the change to A1 causes a change to A4, scheduled for 5 ns (delay time of A4); hence, at CLOCK = 10, the tables have a similar appearance. The current clock time pointer heads a list which consists only of setting A4 low. At CLOCK = 25 (Fig. 3-36c), it points to A1, which is reevaluated to set an event pointing to A4, as shown in Fig. 3-36. At CLOCK = 30, the tables are as shown in Fig. 3-36d. A1 is set low, A4 is reevaluated, and an event to change it to high 5 ns later is generated. At CLOCK = 35, the change to A4 is performed. No new events are generated, and the CLOCK increments to 100 ns where the stop event causes the simulation to terminate.

3.8 SIMULATOR SELECTION CRITERIA

Selecting simulation techniques involves trade-offs between various objectives to determine the most useful and least expensive simulator which satisfies the needs of the designer. Cost-effectiveness ratios for basic algorithms would provide a base-line reference for associating expenses with differing capabilities. The relative trade-offs with respect to each simulation algorithm would permit a potential user, after evaluating his application, to establish his own figures of merit. However, there are many factors which make an accurate evaluation of the implementation costs and expenses of computer execution time of implemented algorithms difficult. First, as implemented in each simulation, an algorithm may be associated with at least one of several methods for reducing computation time:

1. evaluating only those gates which are likely to change at a given time;
2. simulating many different inputs by packing many states into one computer word;
3. organizing the simulation program to take advantage of input characteristics;
4. utilizing an undetermined state of an initial value to minimize the number of input combinations.

Second, any evaluation must make some general simplifying assumptions to account for the variation in the execution capabilities of various host computers. Third, further complications are introduced by the wide variation of output produced by different simulations for the same simulation conditions. These factors could lead to obvious anomalies. For example, a simulator in which all output was suppressed would rank very high in any comparative processing-time evaluation. Fourth, the statistics recorded for each program are based on different numbers of gates being simulated. Obviously, the more gates that are simulated, the longer will be the running time. However, this relationship is not linear and does, in fact, differ from one simulator to another.

The only statistics available are those in the articles describing the particular simulator. Such statistics generally apply to only a specific set of cases, and are highly dependent upon the characteristics of the particular configuration of the host computer (e.g., type and quantity of peripheral units), the specific logic of the network, the particular conditions being simulated, and the amount of output produced. In any event, these figures must be carefully interpreted to be meaningful. Thus, the reader is cautioned to use any printed figures, including those in this text, as a guide rather than as absolute invariant facts.

Comparative figures obtained from the literature for eight implemented simulators are given in Fig. 3-37. Those characteristics of the simulator which affect execution time are listed in columns 1–7. Column 8 is the number of bit time cycles that are quoted as being simulated in one second of computation time. This statistic is analogous to the slow-down ratio (SDR) which is occasionally used to state the ratio of simulation time to device time. One should note that the SDR is measured against the speed of the target machine; hence, a fast target machine can make a simulator appear to be inordinately slow.

The simulation execution figures have been converted to an estimate of running time on a standardized computer for gross comparison (column 9). An explanation of the columns follows:

1. Synchronous vs asynchronous; whether the output of a gate is sampled only at regular clock pulses, or whether the output is available as soon as the gate becomes stable.
2. Compiler vs interpreter; whether the simulator produces a model in the form of executable code, or whether the model is in the form of tables which are processed by a prewritten program.
3. Whether the simulator evaluates all gates at each step or only the gates that can change their value because at least one input changed.
4. Whether several faulty jachines can be simulated simultaneously.
5. Whether the simulator allows signed values of only 0 and 1, or whether there is an additional value of X which can represent an initial unknown state, a signal which is of no interest, or a possible hazard condition.
6. The maximum number of gates that can be simulated at one time.
7. The host computer, the machine on which the simulator runs.
8. The number of machine cycles that can be simulated in one second on the host computer.
9. The number of cycles per second that can be simulated on an IBM 7094, which has enough core storage to contain the other simulators. This does not mean that all the simulators could run on an IBM 7094; some require more core storage than is available in the 7094. The higher the figure, the faster the simulator.

The fastest simulator is McKay's special purpose computer which uses dedicated hardware [19]. Hardie and Suhocki's method derives its speed from its simultaneous simulation of thirty-three cases and by evaluating only those gates that are likely to change [9]. A range of figures is listed for the Shafer and Scheff simulation because of the unusually large amount of input and output processing required in their application [25]. The higher figure should be used for comparison with the other simulators in the table.

	Simulator Characteristics							Time Characteristics	
	1	2	3	4	5	6	7	8	9
Author	Asynchronous Synchronous	Compiler Interpreter	Evaluates Only Gates That Change	Parallel Fault Simulation	Three Value	Maximum No. of Gates	Host Computer	Cycles/Sec on Object Computer	Cycles/Sec on Standard Computer
Ulrich	A	I	YES	NO	NO	10,000	IBM 7094	0.33	0.33
McClure	S	C	NO	NO	NO	1,200	CDC 1604	100	100
McKay	A	I	YES	NO	NO	36,000	SPECIAL PURPOSE	2500	NA
Hays	S	I	NO	NO	NO	4,000	UNIVAC 1108	0.8	0.3
Hardie and Suhocki	A	C	YES	YES	NO	—	IBM 7090	1100	1100
Shafer and Scheff	A	I	YES	NO	YES	2,000	UNIVAC 1108	1–50	0.3–15
Case et al.	A	I	YES	NO	NO	3,000–4,000	IBM 7090	.02	.02
McKinney	S	C	NO	NO	NO	6,000	UNIVAC 1108	2.5	0.8

Fig. 3-37.

Typically, many practical simulation efforts are unreported. The authors are aware of other working logic simulators which are in the "company proprietary" classification. Yet, although these simulations may be a part of a more comprehensive design automation package, the experinces described in the literature are indicative of both the techniques and the relative processing time obtained in practice.

PROBLEMS

1. Discuss the implications contained in the statement: "Logic simulation permits the exhaustive evaluation of a digital design in hours; the same task on hardware might take months."
2. Describe applications for simulation at:
   (a) circuit level;
   (b) logic level;
   (c) register transfer level.
3. Describe applications in which the use of (a) PERT and (b) GPSS for logic simulation would be advantageous.
4. What are the advantages and disadvantages of allowing logic simulation to be interactive with the engineer in, say, a typical time-sharing environment?
5. Develop a logic-level language capable of describing a simple shift register.
6. What are the basic language elements in a register of subsystem-level language and how would they be represented using only logic level constructions?
7. (a) Devise an input expression which permits concise semantic representations for coding of a network involving a JK flip-flop.
   (b) Devise a data table structure for this circuit.

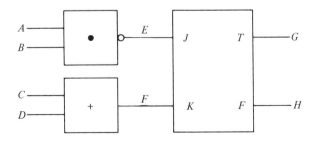

8. Design a simple translator or compiler which produces (a) data-independent code and (b) data-dependent code, for

$(A \cdot B \cdot C)'$

$(A + B)$

$(A + B + C + \cdots)$

$(A' \cdot (B \cdot (C' \cdot D') \cdot E \cdot (F' \cdot G')')' \cdot H')$

using, where appropriate, the following instructions

CAL	Z	Bring contents of Z into accumulator.
ORA	Z	Logical OR contents of Z with accumulator and put results in accumulator.
ANA	Z	Logical AND contents of Z with accumulator and put results in accumulator.
STN	Z	Store logical word in Z.
COM		Complement accumulator.
STZ		Store zero in Z.
STO	Z	Store accumulator in Z.

9. For a five-legged AND gate, compare the number of instructions required for data-dependent and data-independent coding. Compute the expected value of the number of instructions executed by a data-dependent subroutine.

10. Draw a detailed flow chart of the logic of the time-mapping simulation algorithm in which a self-contained subset describes the next-event algorithm. Identify this subset.

11. Would it be advantageous for a logic designer to be able to simulate the delay of a gate switching from 0 to 1 (rise time) separately from its delay in switching from 1 to 0 (fall time)? How might this simulation be implemented effectively?

12. Under what conditions would an asynchronous simulation consume less computer running time than for a synchronous simulation?

13. Give an example of a circuit exhibiting a static hazard which latches the output to remain at the spurious level.

14. Give a three-value truth table for a three-input NAND gate. Compare it with Fig. 3-4.

15. Describe an efficient algorithm for automatically identifying race and hazard situations without the use of a three-value simulator.

16. Describe an algorithm which provides automatic detection of races and hazards to be incorporated into an asynchronous logic simulator. Assume that the purpose of the algorithm is to relieve the logic designer from having to check pages of listings and to protect against designers who might ignore timing and race outputs.

17. Compare the program design considerations in developing a parallel fault-logic simulator with a fault insertion capability as opposed to a simulator in which only one case can be simulated.

18. Write the coding for simulating a five-input NAND gate under the failure mode conditions in which each input fails OFF and where each input fails ON.
19. List techniques for minimizing total computer execution time by either speeding up simulation runs or decreasing the total number of runs required.

REFERENCES

[1] M. A. Breuer, "Techniques for the Simulation of Computer Logic." *Comm. ACM*, Vol. 7, No. 7 (July, 1964), pp. 443–446. Discusses the possibility of increasing a simulator's efficiency by a method analogous to the relay switching tree, i.e., simultaneously simulating many functions of the same inputs as well as many inputs with the same set of functions.

[2] M. A. Breuer, "General Survey of Design Automation of Digital Computers." *Proc. IEEE*, Vol. 54, No. 12 (December, 1966). Gives an overview of the entire field of design automation with a section on each aspect. Contains a very extensive bibliography.

[3] P. W. Case, H. H. Graff, L. E. Griffith, A. R. Leclerg, W. B. Murley, and T. M. Spencer, "Solid Logic Design Automation." *IBM Journal of Research and Development*, Vol. 8 (April, 1964), pp. 127–140. Good example of a simulation program that fits into a generalized design automation system.

[4] T. A. Connolly, "Automatic System and Logic Design Techniques for the RW-33 Computer System." *1960 IRE Conv. Rec.*, Part 2, pp. 124–132. Describes a gate-level simulation program used to simulated entire programs for a small airborne computer.

[5] R. L.. Davis, "The ILLIAC IV Processing Element." *IEEE Trans. on Computers*, Vol. C-18, No. 9 (September, 1969), pp. 800–815. Describes the processing element of the ILLIAC IV parallel computer. The organization can be used to aid in simulating simultaneous results in a logic network.

[6] J. R. Duley and D. L. Dietmeyer, "A Digital Design System Language (DDL)." *IEEE Trans. on Computers*, Vol. C-17, No. 9 (September, 1968), pp. 850–860. Example of a register transfer level language which can be used for logic synthesis.

[7] R. Efron and G. Gordon, "General Purpose Digital Simulation and Examples of Its Applications." *IBM Systems Journal*, Vol. 3 (1964), pp. 22–34. Describes the GPSS (General Purpose System Simulator) language, including the block diagram notation, and gives a detailed example of the simulation of a computer with three disc drivers.

[8] D. F. Gorman and J. P. Anderson, "A Logic Design Translator." *Proc. Fall Joint Computer Conference* (1962), pp. 251–261. Describes a register transfer level language which can be used for design evaluations, simulation, and implementation.

[9] F. H. Hardie and R. J. Suhocki, "Design and Use of Fault Simulation for Saturn Computer Design." *IEEE Trans. on Electronic Computers*, Vol. EC-16 (August, 1967), pp. 412–429. Presents a method of fault insertion and parallel error simulation as well as some very interesting statistical results about diagnostics which were deduced by running the simulator on the Saturn computer.

[10] G. G. Hays, "Computer-Aided Design: Simulation of Digital Design Logic." *IEEE Trans. on Computers*, Vol. C-18, No. 1 (January, 1969), pp. 1–10. Describes a simulation program which mentions, but does not describe, several interesting simulation techniques.

[11] K. Iverson, *A Programming Language*. New York, John Wiley & Sons, Inc., January, 1962. The original description of APL, a language used for representing a computer at the register transfer level.

[12] J. S. Jephson, R. P. McQuarrie, and R. E. Vogelsberg, "A Three-Value Computer Design Verification System." *IBM System Journal*, No. 3 (1969), pp. 178–188. Describes and discusses a three-value synchronous simulator.

[13] J. H. Katz, "Optimizing Bit-Time Computer Simulation." *Comm. ACM*, Vol. 6, No. 11 (November, 1963), pp. 679–685. Describes several techniques for optimizing the simulation time required for a long Boolean Expression.

[14] T. I. Kirkpatrick and N. R. Clark, "PERT and an Aid to Logic Design." *IBM Journal of Research and Development* (March, 1966), pp. 135–141. Discusses the use of PERT in calculating delay time through a circuit. Results include the probability of a given circuit having a delay falling within the delay specifications.

[15] L. J. Koczela and G. Y. Wang, "The Design of a Highly Parallel Computer Organization." *IEE Trans. on Computers*, Vol. C-18, No. 6 (June, 1969), pp. 520–529. Describes the concept of a parallel computer which is divided into relatively independent groups of relatively independent cells. Facilitates simulation of parallel events in a computer.

[16] R. P. Larsen, "Logic Simulator Program." 1962 AIEE Design Automation Workshop. Describes one of the earliest logic simulation programs. Each Boolean expression is converted into Polish notation.

[17] R. Marsh, "A General Logic Simulation Program." *Software Age* (November, 1969) pp. 28–35. Describes the input and output of a logic simulator. Program purchasable directly from author.

[18] R. M. McClure, "A Programming Language for Simulating Digital Systems." *J. ACM*, Vol. 12 (Janvary, 1965), pp. 14–22. Describes a logic simulator which contains declarations for registers as well as for flip-flops and gates.

[19] A. R. McKay, "Comment on 'Computer-Aided Design: Simulation for Digital Design Logic'." *IEEE Trans. on Computers*, Vol. C-18 (September, 1969), p. 862. Describes a combination hardware–software system which can simulate a computer at the register level or gate level and can intermix the descriptions.

[20] J. G. McKinney, "Synchronous Logic Simulation." *Proceedings of the SHARE-ACM Design Automation Workshop* (June, 1967), pp. 19–22. Describes a simple two-value synchronous simulator based on sorted compiler code, primarily from the viewpoint of a programmer.

[21] Y. Metze and S. Seshu, "Proposal for a Computer Compiler." *Proc. Spring Joint Computer Conference* (1966), pp. 254–263. Describes a program for converting a description into detailed logic. The input language is similar to FORTRAN and there are subcontrols which correspond to FORTRAN subroutines.

[22] L. G. Millman, "*CIRCUS, a Digital Computer Program for Transient Analysis of Electronic Circuits.* Report 346-2, Harry Diamond Laboratries, Washington, D.C. (January, 1967). Describes a circuit analysis program which emphasizes the transient response of a circuit.

[23] J. P. Roth, W. C. Bouricius, and P. R. Schneider, "Programmed Algorithms to Compute Tests to Detect and Distinguish Between Failures in Logic Circuits." *IEEE Trans. on Electronic Computers*, Vol. EC-16, No. 5 (October, 1967), pp. 567–580. Describes D-algorithm which is the basis for backward simulation.

[24] H. P. Schlaeppi, "A Formal Language for Describing Machine Logic, Timing, and Sequencing (LOTIS)." *IEEE Trans. on Electronic Computers*, Vol. EC-13, No. 8 (August, 1964), pp. 439–448. Describes a register transfer level simulator which can handle several levels of description.

[25] L. Shafer and B. H. Scheff, "Efficient Simulation Within a Comprehensive Design Automation System." *Proc. Joint Conference on Mathematical and Computer Aids to Design*, October 30, 1969. Describes a three-value asynchronous table-driven simulator. Hierarchical input structures are permitted so that a backboard or parent plate simulation can call in a description of a module which, in turn, calls in a description of an integrated circuit chip.

[26] L. Shalla, "Automatic Analysis of Electronic Digital Circuits Using List Processing." *Comm. ACM*, Vol. 9, No. 5 (May, 1966), pp. 372–380. Describes a simulator which translates computer logic into a list structure rather than into a machine-language program. This approach permits a very ingenious approach to the analysis of race conditions and instabilities.

[27] *Sixteen-Twenty Electronic Circuit Analysis Program (ECAP) (1620-EE-02X) Application Description.* IBM Technical Publications Department, 112 East Post Road, White Plains N.Y., 10601, 188 pp. Manual describing the use of the ECAP circuit analysis program.

[28] G. N. Stockwell, "Computer Logic Testing by Simulation." *IRE Trans. on Military Electronics*, Vol. MIL-6 (July, 1962), pp. 275–282. Describes a logic simulation giving considerable information about the simulation of a magnetic drum attached to the simulated computer.

[29] D. Teicherow and J. F. Lublin, "Computer Simulation-Discussion of the Technique and Comparison of Language." *Comm. ACM*, Vol. 9, No. 10 (October, 1966), pp. 723–741. Discusses simulation as a technique and

gives a detailed comparison of six of the most widely used general purpose discrete-flow simulators.

[30] E. G. Ulrich, "Time-Sequenced Logical Simulation Based on Circuit Delay of Active Network Paths." *Proc. ACM 20th Nat. Conf.* (1965), pp. 437–448. Introduces and discusses in a simplified form the concepts of a time-mapping simulation and the Δt-loop.

[31] R. E. Wolfe, "Logical Simulation Techniques Using the IBM-709." *AIEE Conference Paper CP-603*. Gives a good explanation of the operation of an early synchronous compiler.

[32] I. H. Yetter, "High Speed Fault Simulation for UNIVAC 1107 Computer Systems." *Proc. 1968 ACM Nat. Conf.*, pp. 265–277. Valuable discussion of (a) fault insertion, (b) parallel simulation of different faults, and (c) levelizing, assigning the order in which variables are to be evaluated.

[33] M. S. Zucker, "LOCS: An ECP Machine Logic and Control Simulation." *1965 IEEE Conv. Rec.*, Part 3, pp. 28–51. Describes a register-level simulator.

chapter four

PARTITIONING AND CARD SELECTION

UNO R. KODRES

*Naval Postgraduate School
Monterey, California*

4.1 INTRODUCTION

In this chapter we shall assume that the logic design has been completed and that the description of the design is specified in some unambiguous form. The documentation may be in algebraic-equation form, functional notation (such as the so-called Polish notation), or in the form of drawings which are either prepared by hand or computer-generated.

The circuit element assignment problem is viewed in this chapter as the decision process which assigns circuits to replaceable modules and the modules to a hierarchy of subassemblies, so that the total cost of the resulting assembly is minimized.

Section 4.2 gives a historical sketch of how changes in technology have affected the assignment problem. Section 4.3 describes how the logic design is represented by a directed bipartite graph. Section 4.4 gives a decision procedure which is used to determine which modules should be put into the library of standard replaceable modules.

Section 4.5 defines a partitioning problem in a mathematical form. How an assembly is constructed from subassemblies is viewed as the partitioning problem at a given level. Various solution approaches which have appeared in the literature are reviewed and compared qualitatively. Section 4.6 describes the selection problem, i.e., the problem of choosing the minimal number of standard modules in order to build the logic circuits that are found in one subassembly. This problem is formulated as an integer linear programming (ILP) problem as well as a restricted network flow problem. Unsolved problems and concluding remarks are given in Section 4.7.

4.2 THE INFLUENCE OF BUILDING MATERIALS AND TECHNIQUES

The logic design problem of digital circuits is relatively independent from the layout design problem. The logic design problem depends on the electronic properties of the circuits rather than on the physical properties and manufacturing techniques.

The layout design is greatly influenced by some seemingly unimportant design details:

1. the size of the transistors, resistors, capacitors, etc.;
2. the size of the connections;
3. the size of the replaceable units;
4. the number of input/output connections to a replaceable unit;
5. the type of printed wiring surfaces used to achieve the interconnections.

The broad aim of a layout procedure is to be able to build, with a given set of components and with the given interconnection techniques, a system specified by the logic design. The system, when built, must usually satisfy additional constraints:

1. It must operate reliably in the prescribed physical environment.
2. It must be serviceable in case of failures of the components.
3. It must satisfy some upper bound criteria on the cost of manufacturing and servicing.

We are not interested here in tracing the influence of each manufacturing technique, but rather we would like the reader to get the impression that the layout problem does radically change with changes in manufacturing techniques. Therefore, layout techniques which were used to construct the first generation computers are not generally applicable to designing fourth generation computers.

The components of the first generation computers can be grouped into two categories:
1. the vacuum tubes;
2. the remaining circuit elements: resistors, capacitors, and inductors.

The vacuum tube is by far the most expensive, the most space-consuming, and the most heat-producing. It is also the most common cause of failure of a circuit. Because of these reasons, we minimize the number of tubes in the circuits, and we make the tube pluggable for easy removal.

The introduction of a transistor to replace the tube changes the space requirement and the heat production of the second generation computer. The transistor is now the most expensive element; thus we are interested in minimizing the number of transistors in the circuits. Increases in the density of the circuits introduce interconnection problems which are solved by the printed circuit techniques. The fundamental replaceable unit is a printed circuit, but now the service engineer needs a more extensive inventory of replaceable units because the circuit card usually contains more than one basic circuit. Such an inventory of replaceable units becomes a costly item which cannot be ignored.

On some third generation computers, we replace the circuit card by a wafer which contains several basic logic circuits. Because transistors, resistors, diodes, etc., are manufactured simultaneously, their individual costs are not identified separately. The logic design problem changes from a transistor minimization problem to a basic circuit minimization problem. The replaceable unit consists of a printed circuit card with many wafers soldered onto the card. The space required by the pluggable connector is a significant percentage of the total space required by the circuits and their interconnections.

The foregoing technique of building computers increases the inventory of replaceable units substantially. Two conflicting design aims become apparent. If many circuits are placed on one replaceable unit (to increase density and thus reduce total cost), then replaceable units tend to become unique. Design and inventory costs are consequently increased.

Large scale integration (LSI) accentuates this problem. Because a few pluggable connections require as much space as a large number of interconnected circuits, it is highly desirable to minimize pluggable connections. In the extreme case, each replaceable unit may become unique; hence the design cost and the inventory costs become prohibitive.

The architecture, logic design, and manufacturing techniques all have an important bearing on the problem of repetitive replaceable units. In a subsequent section, we shall deal with this problem in a more quantitative fashion.

From this brief historical sketch, the reader can see the increasing importance of two problems:
1. How should the replaceable units be designed so that the number of distinct units remains within reasonable bounds?

176 Partitioning and Card Selection Chap. 4

2. How should the replaceable units be selected so that a particular design can be implemented most economically?

These two problems are viewed as two aspects of the so-called assignment problem, i.e., how should logic circuits be assigned to modules so that the total cost of the resulting assembly is minimized?

4.3 GRAPH THEORETIC REPRESENTATION OF THE LOGIC DESIGN

After the detailed logic design is completed, the digital system exists in some documented form. The design is frequently documented in Boolean equation form, in functional notation form, or in the form of a block drawing as illustrated in Fig. 4-1a, b, and c. The plus and product symbols refer to Boolean operations rather than the usual arithmetic operations. The notation

$$z = (x_1 * y_1 + x_2 * y_2) + (x_1 * y_2 + x_2 * y_1)$$

(a)

$$z = x_1 y_1 * x_2 y_2 * + x_1 y_2 * x_2 y_1 * + +$$

(b)

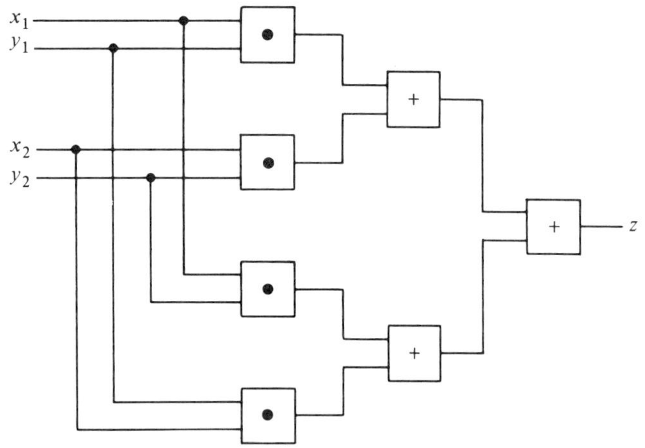

(c)

Fig. 4-1. Documentation of logic.

in Fig. 4-1b is parenthesis-free functional or reverse Polish notation and is also frequently used during the translating process of formula-oriented computer languages (FORTRAN, ALGOL, etc.) to the machine language. The symbols * and + refer to AND and OR logic circuits as well as to Boolean operations.

Because the layout problem concerns itself with the physical properties of the circuit elements and their interconnections, representation in Fig. 4-1c is most natural for our purposes.

The AND, OR, and INVERT circuits, which are sufficient to describe any logic design, are somewhat more complicated in reality than described

$$z = (x_1 * y_1 + x_2 * y_2) + (x_1 * y_2 + x_2 * y_1)$$
(a)

$$z = x_1 y_1 * x_2 y_2 + x_1 y_2 * x_2 y_1 * + +$$
(b)

in the Fig. 4-1c. Depending on the circuit family used, an AND circuit may consist of various configurations of resistors, capacitors, transistors, and diodes. In most designs there are at least three connections—a ground and two voltage levels—connected to each of the fundamental circuits. A typical example is given in Fig. 4-2, where the circuit functions either as an OR gate or as an inverted AND gate, depending on which output is used. In the logic drawing, the voltage and ground connections are understood and are therefore omitted from the drawings.

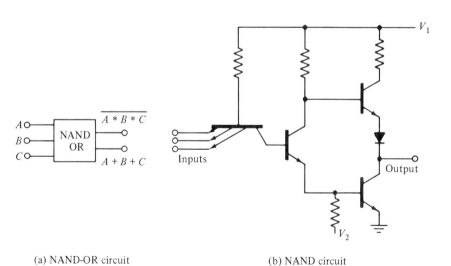

(a) NAND-OR circuit   (b) NAND circuit

Fig. 4-2. Typical gate circuits.

178 *Partitioning and Card Selection* *Chap. 4*

For someone who is familiar with the basic definitions in graph theory, it is quite natural to think of a digital system as a graph. One must be a little careful, however, about associating logic elements with nodes and interconnections with edges. Because interconnections or signals generally connect more than two distinct logic elements together, it is preferable to think of both the logic elements and the interconnections as nodes. However, because logic elements are interconnected by means of signals, and signals in turn must have interposing logic elements, the nodes corresponding to logic elements and the nodes corresponding to signals form distinct classes. Such graphs occur frequently and they are known as bipartite graphs. For basic definitions the reader may consult texts such as Berge [1], Harary [9], and Ore [21].

In some problems in layout, it is also important to distinguish between inputs and outputs of a logic element. This can be very conveniently accomplished by assigning directions to the edges of a graph. An output signal of a logic element would be directed away from the corresponding node, whereas an input signal would be directed into the node.

Figure 4-3a and b illustrates a small part of a digital system with the

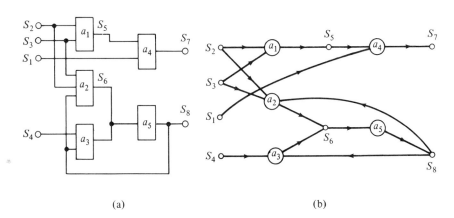

Fig. 4-3. A circuit and its bi-digraph.

corresponding bipartite directed graph. Harary [9] calls directed graphs *digraphs*; thus we shall call the bipartite directed graph a *bi-digraph*.

Associated with bi-digraphs is a matrix, B, which describes the incidence relation of the logic elements to the signals. If with each logic element a_i we associate a row, and with each signal s_j, a column, then the element

$$b_{ij} = \begin{cases} 1 & \text{if } s_j \text{ is an output signal of } a_i \\ 0 & \text{if } s_j \text{ does not touch } a_i \\ -1 & \text{if } s_j \text{ is an input signal of } a_i \end{cases}$$

Figure 4-4 gives the matrix B for the bi-digraph in Fig. 4-3b.

	s_1	s_2	s_3	s_4	s_5	s_6	s_7	s_8
a_1	0	1	1	0	-1	0	0	0
a_2	0	1	1	0	0	-1	0	1
a_3	0	0	0	1	0	-1	0	1
a_4	1	0	0	0	1	0	-1	0
a_5	0	0	0	0	0	1	0	-1

Fig. 4-4. Incidence matrix for bi-digraph in Fig. 4-3.

Because there seems to be a considerable amount of confusion about the representation of digital circuits as graphs, it may be worthwhile to elaborate a little more on the foregoing definition.

The source of confusion appears to be related to the branching of logic signals. Because an edge or an arc of a graph is a binary relation among the nodes, i.e., either an unordered or an ordered pair of nodes, a graph cannot be used to represent a relation which involves a subset of more than a pair of nodes. Because many theoretical results are already known about bipartite as well as directed graphs, the bi-digraph representation seems to be not only adequate but appropriate as well.

One must be careful not to misinterpret the meaning of the two sets of vertices. In printed circuit embedding problems, the vertices corresponding to the signals, together with the edges connected to these vertices, constitute the interconnection, i.e., the electrically common path.

Although the bi-digraph representation of a digital system appears to be new, Mamelak [17] has used an incidence matrix of the type in Fig. 4-4 to represent logic circuits. Ramamoorthy [22] has used a directed graph to represent discrete systems and to study some of their properties.

Although certain characteristics of bi-digraphs which correspond to digital systems can be readily identified, much more information must be gathered from existing systems to aid designers in the future.

The characteristics about which we do know something are:

1. The degree of the signal node, i.e., the number of edges which come together at a signal node, is bounded (usually by fifteen or less). The bound is sometimes specified by the sum

$$d^+ + d^- \leq 15$$

where d^+ and d^- refer to the number of arcs leaving and entering a node respectively.

2. The degree of a basic circuit element, such as an AND, OR, or NOR, is bounded (usually by six or seven), i.e.,

$$d^+ \leq 2, \quad d^- \leq 5$$

3. The distribution of the degrees of both signal nodes and basic logic-element nodes is known: the averages over sizeable collections of circuits range 2.0–4.0 for signal nodes and 3.0–5.0 for logic elements.

The characteristics we would like to know more about would be the cyclic nature of the bi-digraph. A simple *cycle* is an alternating sequence of nodes and directed edges incident on the nodes, with no repetitions of nodes except the first and last nodes, which are the same. We would like to know the distribution of cycles with respect to length.

We would like to be able to characterize arithmetic, decode-encode, and control circuits. Are there simply computed characteristics which distinguish between these three major subcategories of digital computer circuits?

Before we leave this section, we would like to suggest a method of recording a bi-digraph in a computer. Because the matrix B is sparse, that is, relatively few elements are nonzero, it is inefficient to keep the information in matrix form. A form which consists of two lists has been found effective in many applications where large sparse matrices must be manipulated. Only the nonzero elements of the matrix are recorded. For purposes of matrix operations, it is convenient to represent the nonzero elements in either row-by-row or column-by-column order. For a row-by-row representation, the column indices of all nonzero elements in the first row of the matrix are recorded in consecutive locations on a list, K. For example, the incidence matrix in Fig. 4-4 has three nonzero elements in columns 2, 3, and 5 in the first row. We record these column indices on the list K as follows

$$K \quad \begin{array}{cccccccc} 1 & 2 & 3 & 4 & 5 & 6 & 7 & 8 \\ 2 & 3 & 5 & & & & & \end{array}$$

An auxiliary list, L, is used to indicate the position of the first nonzero element in each row. Thus

$$L \quad \begin{array}{ccc} 1 & 2 & 3 \\ 1 & 4 & \end{array}$$

indicates that the column indices of the nonzero elements in the first row are found in locations starting with 1 on list K. The column indices of the nonzero elements in the second row start at location 4 on list K. If the elements in the matrix have only three possible values 1, -1, or 0, then we may use the positive and negative sign to record this on list K. Thus $K(3)$ becomes -5. The compact representation of the incidence matrix in Fig. 4-4 appears in Fig. 4-5.

	1	2	3	4	5	6									
L	1	4	8	11	14	16									
	1	2	3	4	5	6	7	8	9	10	11	12	13	14	15
K	2	3	−5	2	3	−6	8	4	−6	8	1	5	−7	6	−8

Fig. 4-5. Compact representation of the incidence matrix.

For a matrix that contains nonzero elements which may assume an arbitrary value, a third list similar to K may be used to store the values of the elements.

Note that if the degree of the basic circuit is bounded by 7, we need at most $n + 7n = 8n$ memory locations to store what normally would be an n by m matrix, where m is of the same order as n.

4.4 THE STANDARD LIBRARY OF REPLACEABLE UNITS

The logic design of digital circuits is usually expressed in terms of a fundamental set of logic functions. One such basic set consists of the AND, OR, and INVERT blocks, where the number of inputs to the AND and OR blocks generally ranges between two and five.

Historically, the first electronic computers were built with replaceable units which corresponded to the basic circuits. In addition to the logic connections, there were other connections including those for powering the tube, etc. The number of distinct units (differing in the number of inputs) was small, namely, four AND gates, four OR gates, and one INVERT circuit. Thus nine distinct standard replaceable units were sufficient to implement all possible logic functions.

Theoretically, one distinct replaceable unit containing one five-input AND and OR unit and an INVERT block would be sufficient to construct all configurations. This is because a two-input OR or AND circuit may be constructed by using any of the corresponding circuits with the number of inputs greater than or equal to two. This relationship, in mathematical terminology, is known as *partial order*. A relation $\geq$ in a set X is said to be a partial-order relation if:

1. $x \geq x$, $x \in X$;
2. $x \geq y$, $y \geq x$ implies $x = y$;
3. $x \geq y$, $y \geq z$ implies $x \geq z$.

A convenient way to display a partial-order relation is by using a corresponding directed graph. For the above AND, OR, and INVERT circuits, this graph is given in Fig. 4-6. Some useful properties of graphs corresponding to partial orders is given in Chapter ten of Ore [21].

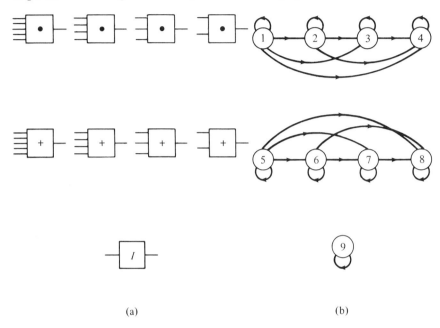

(a) (b)

Fig. 4-6. Circuits and the corresponding partial-order graphs.

In a graph, G, which corresponds to partial order, there exist certain nodes, called *basis nodes*, from which directed edges emanate. In Fig. 4-6b, nodes 1, 5, and 9 constitute the basis elements. More precisely, a set of nodes B in a partial-order graph G constitutes a basis if:

1. $b \in B$ is not the terminal node of any directed edge from some other node;
2. Every "source" node (node with the above property) in G is in the set B.

In terms of this concept, we can now define the concept of a *complete library* of replaceable units.

In logic design certain combinations of circuits occur frequently. Such circuits we shall call *recognized circuits*. Corresponding to such recognized circuits, we may construct a partial-order graph similar to the one in Fig. 4-6b. For a standard library of replaceable units to allow the construction of any recognized circuit, the basis elements of the recognized-circuits graph must have corresponding circuits in the standard library. For example, if circuit 1 of Fig. 4-6b were not in the library, and if such a circuit occurred among the circuits to be implemented, it would be impossible to build the given combination.

Even though a complete library allows the construction of any collection of circuits, a library that contains more elements may allow a more efficient way of doing it. For example, if the library consisted of only one replaceable unit containing circuits 1, 5, and 9 in Fig. 4-6b, and if we wished to implement a grouping which contained only circuits of type 4, we would waste all the INVERT and OR circuits as well as three input connections on each AND circuit.

In the days when the vacuum tube was the most costly circuit element, such a waste of OR and INVERT circuits would have been intolerable. The waste of input connections on each AND circuit would not have been as serious.

With today's integrated circuits, a waste of transistors would not be as serious because the cost of making one on a substrate is not much different from the cost of making several hundred. The same manufacturing steps must be followed in both cases. However, it would be intolerable to waste the input connections to each of these circuits because we could conceivably have constructed a large part of the entire computer on that substrate.

In order to decide what should be included in the standard library of circuits, we propose the use of the procedure described next.

A formula given by Kodres [12] provides a new way of deciding whether or not a new circuit card should be added to a given standard circuit-card library.

Notz et al. [20] propose the following equation to account for the cost of a system

$$CS = \sum_{i=1}^{XN} \left[CM(i) + CT(i) + \frac{CG(i)}{V(i)} \right] + B \cdot CMF + CPK \qquad (4\text{-}1)$$

where

$CM(i)$ = manufacturing cost for the ith part
$CT(i)$ = test cost for the ith part
$CG(i)$ = generating cost for the ith part
$V(i)$ = the number (volume) of parts of the ith part number
B = parameter which depends on field maintenance strategy
CMF = field maintenance cost
CPK = cost of higher level packaging including cooling and power supply
N = number of distinct parts needed for the system

X = increase in systems parts to allow for duplications in systems part numbers

The Equation (4-1) can be simplified considerably if further assumptions are made.

Assumption 1: The manufacturing and testing costs are relatively independent of the particular circuit card. By this we mean that the average value

$$\overline{CM} = \frac{1}{N} \sum_{i=1}^{N} CM(i)$$

gives a good approximation to the cost of each manufactured circuit. That is

$$\max_i \frac{\overline{CM} - CM(i)}{\overline{CM}} \leq 0.1, \quad i = 1, \ldots, N$$

or the largest departure of the actual cost from the average cost is at most 10%.

Assumption 2: The design cost of a card is relatively independent of a particular card. One would expect more variance in this case because particularly dense cards are difficult to design and may have to undergo several design cycles before a satisfactory design is produced. An average value or minimum cost value may be associated with a particular card.

Assumption 3: To analyze similar alternatives, we can assume that the field maintenance costs and strategies remain the same and that the higher level packaging cost is not changed significantly by small changes in the standard library.

Equation (4-1) presents the generating costs of a new part number somewhat awkwardly. An equivalent way would be to let

$$\sum_{i=1}^{XN} \frac{CG(i)}{V(i)} = \sum_{i=1}^{N} \sum_{1}^{V(i)} \frac{CG(i)}{V(i)} = \sum_{i=1}^{N} CG(i)$$

The entire equation could be expressed more clearly by

$$CS = \sum_{i=1}^{N} \{[CM(i) + CT(i)] \cdot V(i) + CG(i)\} + B \cdot CMF + CPK$$

where $V(i)$ is the number of parts of the ith part number used in the system. Let us compare two alternatives to produce a system.

Alternative 1: Produce the system with the currently defined standard library.

Alternative 2: Introduce a new card with an estimated generating cost, CG, which replaces C previously defined library cards. The amount C may be fractional in that k new cards may replace p previously designed cards. In this case

$$C = \frac{p}{k}$$

If we compare the old system's cost, CS, to the new system's cost, CS', under Assumptions 1, 2, and 3, we get

$$CS = \sum_{i=1}^{N} \{[CM(i) + CT(i)] \cdot V(i) + CG(i)\} + CF$$

where $CF = B \cdot CMF + CPK$. Then

$$CS = \sum_{i=1}^{N} \{(\overline{CM} + \overline{CT}) V(i) + \overline{CG}\} + CF$$

$$= (\overline{CM} + \overline{CT}) \sum_{i=1}^{N} V(i) + N \cdot \overline{CG} + CF \quad (4\text{-}2)$$

$$= (\overline{CM} + \overline{CT})V + N \cdot \overline{CG} + CF$$

where

$$V = \sum_{i=1}^{N} V(i)$$

If we estimate that r new cards can be used in the system and each such card replaces p previously defined cards, then by introducing the new card we get

$$CS' = (\overline{CM} + \overline{CT}) \cdot [(V - rC) + r] + (N + 1)\overline{CG} + CF$$

Now

$$CS - CS' = (\overline{CM} + \overline{CT}) \cdot (rC - r) - \overline{CG}$$

In order to realize a saving

$$(\overline{CM} + \overline{CT})(rC - r) - \overline{CG} > 0$$

or

$$r(C - 1) > \frac{\overline{CG}}{\overline{CM} + \overline{CT}}$$

According to Thompson [24], the ratio of design cost to manufacturing-testing cost, d, for circuits employing LSI technology is approximately

$$d = \frac{\overline{CG}}{\overline{CM} + \overline{CT}} = \begin{cases} 3000 & \text{for custom designed LSI chips} \\ 1000 & \text{for general purpose chips} \\ 300 & \text{for standard arrays} \end{cases}$$

186 *Partitioning and Card Selection* Chap. 4

The critical values of C, d, and r, which define the point at which a saving is realized, are thus given by

$$C = 1 + \frac{d}{r} \qquad (4\text{-}3)$$

Table 4-1 gives the C values in Equation (4-3), where d and r assume a sequence of likely values.

TABLE 4-1. VALUES OF C AS FUNCTIONS OF d AND r IN EQUATION (4-3)

r	\multicolumn{9}{c}{d}
	10	50	100	200	300	500	1000	2000	3000
1	11	51	101	201	301	501	1001	2001	3001
2	6	26	51	101	151	251	501	1001	1501
3	4.3	18	34	67	101	167	334	667	1001
5	3	11	21	41	61	101	201	401	601
10	2	6	11	21	31	51	101	201	301
20	1.5	3.5	6	11	16	26	51	101	151
30	1.3	2.7	4.3	7.7	11	17	34	67	101
50	1.2	2.0	3.0	5.0	7	11	21	41	61
100	1.1	1.5	2.0	3.0	4	6	11	21	31
200	1.05	1.25	1.5	2.0	2.5	3.5	6	11	16
300	1.03	1.17	1.3	1.67	2.0	2.7	4.3	7.7	11
500	1.02	1.10	1.20	1.40	1.6	2.0	3.0	5.0	7
1000	1.01	1.05	1.10	1.20	1.3	1.5	2.0	3.0	4.0
2000	1.005	1.025	1.05	1.10	1.15	1.25	1.5	2.0	2.5
3000	1.003	1.017	1.03	1.07	1.10	1.16	1.3	1.5	2.0
10000	1.001	1.005	1.01	1.02	1.03	1.05	1.1	1.2	1.3

To be clear about what the numbers in the table mean, let us take a particular example. If the ratio of design cost to manufacturing testing cost is 3000, and if the number of repetitions of the proposed circuit throughout one computer is 10, we look at the entry in row 10, column 3000, which is 301. This means that, to save anything at all, at least 301 standard library circuit cards must be replaced by a single card of the new design. Quite clearly, it is next to impossible to achieve this kind of increase in density. Therefore, for one computer we could not afford to design such a new card. If we produced 100 such cpmputers, then the repetition number would be 1000, and the corresponding 3000 column entry would be 4.0. If the proposed new card increases the density at least fourfold, we will realize a saving.

One interpretation of interest in the LSI technology is this:

1. The standard library will consist of "independent" chips that contain logically unconnected basic circuits.

2. Any other chips, which satisfy the criteria in Table 4-1, must be exceedingly dense, highly functional chips.
3. To increase the repetition number r, the architectural design should perhaps consist of a collection of identical small computers under the direction and command of a master computer. Such a design would make the servicing problem easier also.
4. The design cost unfortunately increases as the density we wish to achieve on a given card increases. With successful automated techniques, the design cost would be reduced.
5. The ratio of the costs is a deciding factor in the cost effectiveness of a standard library. If the manufacturing costs increase but the design costs remain the same, it may become advantageous to introduce new library cards.

4.5 THE PARTITIONING PROBLEM

A partitioning problem in the most general form may be stated as follows.

Given a detailed description of the logic design of the entire system, subdivide the system into a hierarchy of subsystems so that at:

Level 1. The subsystem could be built in a cabinet of convenient size.

Level 2. The intermediate packages which constitute the contents of the cabinet can accommodate the assigned subsystem.

Level 3. The fundamental, replaceable, standard units can be used to build the circuits assigned to that subsystem.

Associated with any partitioning problem are parameters which allow us to measure the solutions of a partitioning problem. Two parameters which must always be considered as part of any partitioning problem are space and external connection requirements of circuits. Neglecting for the moment any other parameters, we may define a partitioning problem as follows.

Given is a bi-digraph of a digital system. With each circuit node, a, of the graph, we associate two numbers, $S(a)$ and $E(a)$, corresponding to the amount of space and the number of external signals, respectively, required by a.

With each subset A_i of circuit nodes we also associate two numbers

$$S(A_i) = \sum_{a \in A_i} S(a)$$

and

$$E(A_i) = \sum_{a \in A_i} E(a) - \sum_{s \in I(A_i)} \bar{d}(s) - \sum_{s \in B(A_i)} (\bar{d}(s) - 1)$$

The set of internal signal nodes of A_i is denoted by $I(A_i)$ and is defined to consist of all these signal nodes in the graph which are adjacent only to circuit nodes belonging to A_i. The set of boundary signal nodes of A_i is denoted by

$B(A_i)$ and consists of these signal nodes which are adjacent to at least one node of A_i and are either input or output nodes or adjacent to a node not in A_i. The notation $\bar{d}(s)$ refers to the number of nodes in A_i adjacent to s. If $s \in I(A_i)$, then $\bar{d}(s) = d(s)$, where d is the degree of the signal node.

We define a partition of the set A of all circuit element nodes to be a family of subsets of A

$$\{A_j\}, \quad j = 1, 2, \ldots, m$$

such that $A_i \cap A_j = \Phi, i \neq j$

$$\bigcup_{j=1}^{m} A_j = A$$

A feasible partition is one which satisfies the two criteria

$$S(A_j) \leq \bar{S} \qquad (4\text{-}4)$$
$$E(A_j) \leq \bar{E} \quad \text{for all } j \qquad (4\text{-}5)$$

$\bar{S}$ and $\bar{E}$ are the corresponding space and external connection limits for a particular electronic package.

A partitioning problem is to find a *feasible partition* with the minimum number of partition elements.

The preceding formulation is somewhat different from the one proposed by Lawler [14] and Habayeb [8]. The objective in their proposals is to minimize the total number of electrical connections between subassemblies, rather than the number of subassemblies. Either of these formulations gives rise to a difficult combinatorial design problem.

Before we discuss the attempted solutions of some partitioning problems in detail, some additional parameters must be introduced. Certainly, the manufacturing and design costs of the partition elements cannot be ignored. In most applications the cost parameters are of the greatest importance.

It is intuitively clear that the optimal solution of the partitioning problem, which results in the least number of partition elements, leads to partition elements which are unique. The likelihood of identical logic elements is remote in problems of any realistic size.

It is also clear from the economic point of view that any system which consists of distinct replaceable units must be completely reliable, so that replacement is unnecessary, or the number of such systems at a location is large, so that one complete system of spare parts remains economically feasible.

For the immediate future it seems likely that systems must be built with replaceable units which have a high degree of repetition throughout the system. Our analysis in Section 4.4 indicates that new designs for replaceable parts become feasible when they increase density or have a high degree of repetition or both.

We are led to the conclusion that replaceable units must consist of logically independent, multi-use units, or highly dense, functional multi-use units. Thus any practical partitioning scheme must be able to cope with this added restriction.

The partitioning problem we formulated remains unchanged if we assume that the bi-digraph which represents the system consists of circuit nodes which correspond to the so-called recognized combinations of basic circuits. For example, a highly complex functional combination, such as a section of an adder, might be identified by a single node. Because of the assumed dichotomy of the replaceable units into either highly complex functional units or very simple logic circuits, the identification of groups of nodes of the original bi-digraph with nodes which represent recognized combinations does not pose a serious problem. A design engineer can do this even in the initial design phase. Incidentally, a design engineer is presumed to at least influence, if not define, the recognized circuits.

The space and external connection limits are usually associated with all variants of partitioning problems. There may be added requirements of a more complicated nature.

Lawler et al. [15] consider a variant of the problem which minimizes the maximum delay through the circuit. They assume that each external connection contributes a unit of delay, whereas internal connections cause negligible delay.

Because the performance characteristics may outweigh economic considerations for many systems, some added side conditions might be:

1. Feedback loops must be in one partition.
2. Certain signals must be externally available for monitoring purposes.

A concept which changes the partitioning problem into a covering problem is also brought out by Lawler et al. [15]. Recreation or replication of certain signals may be useful in avoiding delay caused by external connections. Replication changes logic design which, in some instances, might be so critical that the entire logic design must be reviewed. In other instances replication causes only localized change which is routinely accomplished.

Next we present some of the algorithms which have been proposed to solve partitioning problems. For the purpose of discussion, we classify these algorithms into three groups. In the first group are algorithms which generate partition elements one element at a time. Algorithm 1 incorporates the typical features of such algorithms. The second group, typified by algorithm 2, contains the procedures which generate the partition elements simultaneously. In the third group are the procedures which take a more global view of the problem but, at the same time, oversimplify it to an unrealistic extent.

The reader who hopes to find an algorithm which solves his particular partitioning problems among the algorithms to be described will probably be disappointed. All of the algorithms to be described have serious shortcomings. We are presenting them to provide a deeper understanding of the problems and to review the published work to date.

We present each algorithm in a semiformal way first. Then we use the example problem in Fig. 4-7 to illustrate the operation of the algorithm. After the algorithms are presented and illustrated, we compare and analyze the strengths and weaknesses of the procedures.

Some general concepts must be defined before we can proceed. We shall assume that the digital system is represented by a bi-digraph such as the one shown in Fig. 4-7. The circuit nodes represent the recognized circuits which, in this example, are the AND, OR, and INVERT circuits. Each recognized circuit a has associated with it a space parameter $S(a)$ and an external connection parameter $E(a)$. The parameter values are so chosen that they accurately reflect the construction requirements.

In our example we take the space requirement and the external connection requirement to be

$S(a) = E(a) =$ number of external connections of circuit a

The reason for choosing $S(a)$ to be the same as $E(a)$ is that our circuit module library consists of modules which contain independent circuits, in which case space limitation is dictated by external connections. For logically complex recognized circuits, $S(a)$ may be considerably larger than $E(a)$.

We must also point out that the parameter values $\bar{E}$ and $\bar{S}$, which describe the external connection and space bounds, must be so chosen that the resulting feasible partitioning element may be actually constructed. In terms of our terminology, a feasible solution of the partitioning problem should give rise to an optimal solution of the selection problem which satisfies construction requirements. We assume that the partitioning problem is solved first and then appropriate modules are selected from the library of modules to implement the design.

The term *conjunction* of two circuit nodes, i and j, is defined as the number of signal nodes which are simultaneously adjacent to i and j. More generally, the *conjunction* of two disjoint *subsets of nodes* V_i and V_j is defined as

$$\text{Con}(V_i, V_j) = |B(V_i) \cap B(V_j)|$$

i.e., the number of signal nodes which are simultaneously connected to a node in V_i and V_j. In Fig. 4-7, for example

$$\text{Con}(\{1\}, \{2\}) = 2$$
$$\text{Con}(\{1, 2\}, \{3\}) = 1$$
$$\text{Con}(\{1, 3\}, \{2, 9\}) = 2$$

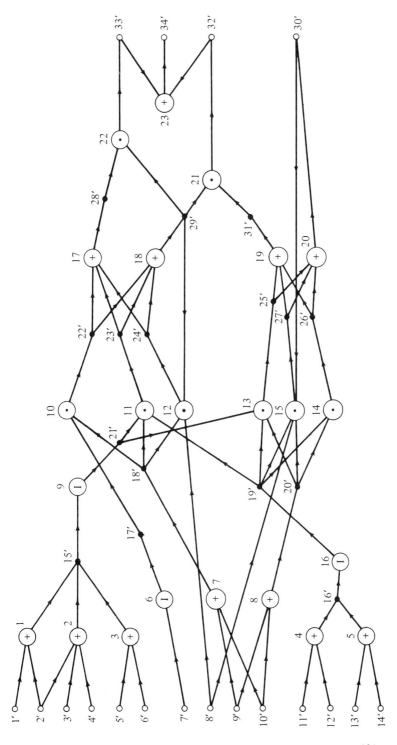

Fig. 4-7. A bi-digraph of a digital system.

The term *disjunction* is also defined for two disjoint subsets as

$$\text{Dis}(V_i, V_j) = |B(V_i) \cup B(V_j)| - \text{Con}(V_i, V_j)$$

i.e., the total number of external signals adjacent to V_i or V_j, minus the conjunction of V_i and V_j.

For example

$$\text{Dis}(\{1\}, \{2\}) = 5 - 2 = 3$$
$$\text{Dis}(\{1, 2\}, \{3\}) = 7 - 1 = 6$$
$$\text{Dis}(\{1, 3\}, \{2, 9\}) = 8 - 2 = 6$$

With these definitions we now state algorithm 1, which typifies the algorithms of Gamblin et al. [6], Haspel [10], and Nidecker and Simon [19].

Algorithm 1: A_0 is the set of circuit nodes to be partitioned. Let $A = A_0$.

Step 1: Let $i = 1$.

Step 2: Consider a given subset A of the totality of circuit nodes V. Let the complement of A be denoted by $-A = V - A$, and let $\bar{a} \in A$ be such that $\text{Con}(\bar{a}, -A)$ is maximum, i.e., $\bar{a}$ is most connected externally. If more than one such $\bar{a}$ exists, select from among these nodes the ones for which $\text{Con}(\bar{a}, \{A - \bar{a}\})$ is minimum. An arbitrary selection of the node with the lowest index is made if more than one candidate node remains.

Step 3: Assign selected node $\bar{a}$ to partition element P_i. If $A - P_i = \varnothing$, we are finished. If not, then consider $\bar{a} \in A - P_i$ such that $\text{Con}(P_i, \bar{a})$ is maximum. Among such nodes, select those for which $\text{Dis}(P_i, \bar{a})$ is minimum. Again an arbitrary selection is made if more than one candidate remains.

Step 4: Check if $(P_i \cup \{\bar{a}\})$ is a feasible partition element by checking if

$$S(P_i \cup \{\bar{a}\}) \leq \bar{S} \quad \text{and} \quad E(P_i \cup \{\bar{a}\}) \leq \bar{E}$$

If so, return to step 3. If either limit is exceeded, then we check all other nodes for feasibility. If a feasible candidate is found, return to step 3. If none is found, then we let

$$A \longleftarrow A - P_i \quad \text{and} \quad i \longleftarrow i + 1$$

and return to step 2.

Example 1: As an example, we use algorithm 1 to partition the nodes of the graph in Fig. 4-7, subject to the bounds $\bar{S} = 20$ and

$\bar{E} = 10$. At the conclusion of step 2, $i = 1$, $\bar{a} = 2$. Step 3 results in

$$P_1 = \{2\} \qquad \bar{a} = 1$$

In step 4 we check

$$S(\{1, 2\}) = 7 \leq 20$$
$$E(\{1, 2\}) = 5 \leq 10$$

We return to step 3 and obtain two candidates $\{3, 9\}$. We select 3, check space and external connection limits, and return to step 3.

The next candidate $\bar{a} = 9$.

$$S(\{1, 2, 3, 9\}) = 12 \leq 20$$
$$E(\{1, 2, 3, 9\}) = 7 \leq 10$$

We obtain $P_1 = \{1, 2, 3, 9\}$.

At the next step 3

$$\text{Con}(P_i, \{11\}) = \text{Con}(P_i, \{13\}) = 1$$
$$\text{Dis}(P_i, \{11\}) = 10 - 1 = 9$$
$$\text{Dis}(P_i, \{13\}) = 10 - 1 = 9$$

Choosing $\bar{a} = 11$, we next check feasibility in step 4.

$$S(\{1, 2, 3, 9, 11\}) = 16 \leq 20$$
$$E(\{1, 2, 3, 9, 11\}) = 10 \leq 10$$

We return to step 3 and obtain

$$P_1 = \{1, 2, 3, 9, 11\}$$
$$\bar{a} = 13$$

Step 4 gives

$$S(P_1 + \{13\}) = 19 \leq 20$$
$$E(P_1 + \{13\}) = 12 > 10$$

Checking all other nodes for feasibility indicates that adding any node to P_1 will violate condition (4-5). Therefore

$$P_1 = \{1, 2, 3, 9, 11\}$$
$$i \longleftarrow 2$$
$$A \longleftarrow A - P_1$$

and we return to step 2.

Continuing this process with a selection rule which selects the node with the lowest index (in case several candidate nodes appear), we get the following set of partition elements

$$P_2 = \{23, 22, 21, 17\}, \quad S(P_2) = 14, \quad E(P_2) = 10$$
$$P_3 = \{4, 16, 5, 14\}, \quad S(P_3) = 11, \quad E(P_3) = 9$$
$$P_4 = \{7, 8, 10, 6, 12, 18\}, \quad S(P_4) = 19, \quad E(P_4) = 10$$
$$P_5 = \{20, 15, 13\}, \quad S(P_5) = 13, \quad E(P_5) = 8$$

Algorithm 1 has thus produced a solution to the partitioning problem with five partition elements.

Among the shortcomings of algorithm 1, the most obvious is the strictly sequential process which determines the partition elements. If the partition elements were allowed to "grow" simultaneously, a more satisfactory result might be expected.

Algorithm 2 determines the partition elements simultaneously, thus removing a serious shortcoming of algorithm 1. A decision as to how many partition elements there should be is made prior to a "seeding" operation. This operation selects from among the largest components those which are far removed from each other, and it chooses one such component as a "seed" component for each partition element. A measure which determines the size of components may be chosen as $S(c)$, the space parameter associated with component c. The process we now formalize resembles one used by Weindling [25] to solve the component placement problem.

Algorithm 2: A_0 is the set of circuit nodes to be partitioned. The assumed number of partition elements is m. Let $i = 1$ and $A = A_0$.

Step 1: Calculate $S(a)$ for all circuits a in the subset A of circuit nodes V.

Step 2: Let $\bar{a}$ be a circuit node which requires the most space, i.e.,

$$S(\bar{a}) \geq S(a), \quad a \in A$$

If more than one such circuit node $\bar{a}$ exists, select the one with the lowest index. Let $P_i = \{\bar{a}\}$, and $A \leftarrow A - \{\bar{a}\}$. If $i = m$, we go to step 5; otherwise continue with the next step.

Step 3: $j = 1$, $N_o(\bar{a}) = \{\bar{a}\}$. Construct the j-neighborhood $N_j(\bar{a})$

$$N_j(\bar{a}) = \{b \mid \text{Con}(N_{j-1}(\bar{a}), b) > 0, b \in A\} \cup N_{j-1}(\bar{a})$$

If $A - N_j(\bar{a}) = \phi$, we let $j \leftarrow j - 1$ and go to step 4. If

$$S(N_j(\bar{a})) < \bar{S} \quad \text{and} \quad N_j(\bar{a}) \neq N_{j-1}(\bar{a})$$

we let $j \leftarrow j + 1$ and return to compute the j-neighborhood of $\bar{a}$. If the space restriction is exceeded or two successive neighborhoods coincide, we go to the next step.

Step 4: Remove from the remaining candidate nodes A the j-neighborhood, i.e.,

$$A \leftarrow A - N_j(\bar{a})$$
$$i \leftarrow i + 1$$

Return to step 2.

Step 5: Having chosen the seed components for each of the partition elements P_i, $i = 1, 2, \ldots, m$, we now proceed to assign the remaining elements. We let

$$A \leftarrow A_o - \bigcup_{i=1}^{m} P_i$$

We formulate a sequence of linear assignment problems which result in a complete assignment of circuit nodes to partition elements, subject to the space and external connection restrictions. The solution techniques for the assignment problem are described elsewhere; Munkres [18], Rutman [23], and Kurtzberg [13] explain the theoretical and practical aspects of solving the assignment problem. The Appendix of Chapter five gives solution algorithms.

Step 6: We form a matrix consisting of the same number of rows as we have partition elements and the same number of columns as we have candidate nodes in A. The entries d_{ij} in this matrix are calculated as follows

$$d_{ij} = \text{Con}\,(P_i, a_j), \quad a_j \in A$$

If either

$$S(P_i \cup \{a_j\}) > \bar{S} \quad \text{or} \quad E(P_i \cup \{a_j\}) > \bar{E}$$

then

$$d_{ij} = -\infty$$

Step 7: We solve the assignment problem where each partition element P_i will be assigned a candidate circuit a_j so that the sum

$$\sum_{ij} d_{ij} X_{ij}$$

is maximized. The assignment variable

$$X_{ij} = \begin{cases} 1, & \text{if } a_j \text{ is assigned to } P_i \\ 0, & \text{otherwise} \end{cases}$$

We require that

$$\sum_{j=1}^{|A|} X_{ij} = 1, \quad \text{for each } i$$

which means that each partition element will receive exactly one circuit node.

We update the contents of each partition element and remove the corresponding circuit nodes from the group of candidates, i.e., $P_i \leftarrow P_i \cup \{a_j\}$, where $X_{ij} = 1$ and $d_{ij} \neq -\infty$

$$A \leftarrow A - \bigcup_{i=1}^{m} P_i$$

If $A = \Phi$ we are finished. If $A \neq \Phi$, we return to step 6, unless $d_{ij} = -\infty$ for each entry. In the latter case, no solution can be found unless the number of partition elements is increased. We would thus return to step 1 with a larger m.

Example 2: We apply algorithm 2 to the circuitry in Fig. 4-7. Let $\bar{S} = 20$, $\bar{E} = 10$, and $m = 4$. In step 2 the circuit node 15 will be chosen and placed into partition element P_1. In step 3 we calculate the 1-neighborhood

$$N_1(15) = \{8, 11, 12, 13, 14, 15, 16, 19, 20\}$$
$$S(N_1(15)) = 3 + 4 + 4 + 4 + 3 + 5 + 2 + 4 + 4 = 33 > 20$$

After step 4

$$A = \{1, 2, 3, 4, 5, 6, 7, 9, 10, 17, 18, 21, 22, 23\}$$

We return to step 2, which concludes with the selection of node 2. Thus

$$P_2 = \{2\}$$

Step 3 yields a 1-neighborhood

$$N_1(2) = \{1, 2, 3, 9\}$$

and a 2-neighborhood

$$N_2(2) = \{1, 2, 3, 9\} = N_1(2)$$

Step 4 updates A, and we return to step 2 with

$$A = \{4, 5, 6, 7, 10, 17, 18, 21, 22, 23\}$$

In step 2

$$P_3 = \{17\}$$

and

$$N_1(17) = \{10, 17, 18, 22\}$$
$$N_2(17) = \{6, 7, 10, 17, 18, 21, 22, 23\}$$
$$N_3(17) = N_2(17)$$

We return to step 2 with

$$A = \{4, 5\}$$

and hence

$$P_4 = \{4\}$$

We continue with step 5.

Table 4-2 illustrates the assignment process. The columns correspond to the circuit nodes. The last two columns tabulate the space and external connection parameters. The consecutive four rows, 1–4, indicate the stage of the partitioning process. Other tabular entries correspond to

$$d_{ij} = \text{Con}(P_i, a_j)$$

which are checked for feasibility. If the feasibility criterion is not satisfied, an entry $-\infty$ is recorded.

The last four rows contain three columns with enties $-\infty$. This indicates that the algorithm was unable to find a four partition element solution. We can either construct a five partition element solution by simply placing the remaining circuit nodes into a new partition element, or we can start all over again by assuming that $m = 5$ and returning to step 1.

Complex combinatorial problems frequently defy analysis. A much used tool is to oversimplify the problem so that some theoretical results may give useful heuristic ideas which can be exploited to solve the original problem. One such approach by Luccio and Sami [16] defines a minimal collection of circuits to be such that any non-null subcollection requires more external connections. That is, $\bar{A}$ is minimal if and only if, for any proper subset

TABLE 4-2. THE ASSIGNMENT TABLEAU

	1	2	3	4	5	6	7	8	9	10	11	12	13	14	15	16	17	18	19	20	21	22	23	S	E
1	0	X	0	0	0	0	0	1	0	0	1	1	2	2	X	1	—	0	1	2	1	0	0	5	5
2	2		1	X	0	0	0	0	1	0	0	0	0	0		0	—	0	0	0	0	0	0	4	4
3	0		0		0	0	0	0	0	1	1	1	0	0		0	X	3	0	0	0	1	0	4	4
4	0		0		1	0	0	0	0	0	0	0	0	0		1	—	0	0	0	0	0	0	3	3
1						0	0	1	1	0	0	1	0	2					2	3	1	0	0	9	7
2						0	0	0	1	0	2	0	0	0					0	0	0	0	0	7	5
3						0	0	0	0	1	0	2	0	0					0	0	1	2	0	7	5
4						0	0	1	1	0	1	0	0	0					0	0	0	0	0	5	5
1						0	0	0	1	0	2	1	3	3					3	0	1	0	1	13	8
2						0	0	0	0	1	0	0	0	0					0	0	0	0	0	10	7
3						0	0	1	0	0	1	2	0	1					0	0	1	0	0	11	6
4						0	0	0	0	0	0	0	0	0					0	0	0	0	0	8	5
1						0	0	0	0	0	1	2	0	3							0	0	0	17	6
2						0	0	1	0	1	0	0	0	0							2	0	0	12	7
3						0	0	0	0	1	1	0	0	0							1	0	1	15	6
4						0	1	0	0	0	0	0	0	1										12	8
1						−∞	1	−∞													−∞	−∞	−∞	20	6
2						−∞		−∞													1		−∞	15	10
3						−∞		−∞													−∞		−∞	19	8
4						−∞		−∞													−∞		−∞	15	10

$A \subset \bar{A}, A \neq \phi$

$$E(A) > E(\bar{A})$$

The main theorem of this paper establishes an interesting property, namely, that minimal sets cannot partially overlap, i.e., either

$$A \supset B \quad \text{or} \quad B \subset A \quad \text{or} \quad A \cap B = \phi$$

A procedure based on this property is then used to partition a given collection of circuits into minimal subcollections.

Unfortunately this technique enables one to identify only very strongly interconnected subcollections. In the problem we used as an example, the original circuits (defined to be minimal collections) are the only minimal collections. The procedure thus fails to identify any of the clusters which are of importance in partitioning.

Another approach by Lawler et al. [15] makes the following oversimplification: The graph describing the system is assumed to be a directed tree. A labeling algorithm is then used to decompose the graph into clusters which satisfy the space and external connection constraints. The clusters are so constructed that the longest delay in the circuit is minimized. (A delay of one unit is encountered whenever a signal leaves a cluster.) The procedure thus produces feasible partition elements which minimize the maximum delay through the circuit in case the original circuit is a tree. In applying the procedure to acyclic graphs, which appear more frequently in practice, the acyclic graphs are transformed into directed trees. The transformation process implies that circuits must be replicated whenever necessary. The logic design must therefore be reconsidered and many circuits duplicated in order to use the clustering technique. In many applications replication is not possible; hence this procedure would not be useful. Replication is particularly difficult to achieve with high speed circuits, where delays due to wire lengths start to play an important role.

To conclude the description of various approaches used to solve partitioning problems, we present the procedures used by Kernighan and Lin [11]. Their objective is to minimize the weighted sum of connections between partition elements that are subject to size constraints.

First a simpler problem of partitioning a graph of $2n$ nodes into two subsets of n nodes is considered. Two initially constructed sets A and B, $A \cup B = V$, are used to start the process. If A^* and B^* are a pair of sets constituting the optimal solution, then

$$A^* = A \cup (B \cap A^*) - (A \cap B^*)$$
$$B^* = B \cup (A \cap B^*) - (B \cap A^*)$$

If $X = A \cap B^*$ and $Y = B \cap A^*$ can be determined without the knowledge of B^* or A^*, then A^* may be constructed simply by exchanging the set X with Y in A, and B^* may be constructed by exchanging Y with X in B.

The following procedure is devised to approximately determine X and Y. If the cost of an external connection is assumed to be unity, then we define an external cost

$$E_a = \text{Con}(\{a\}, B)$$

and an internal cost

$$I_a = \text{Con}(\{a\}, A - \{a\})$$

A difference of external cost and internal cost is defined as

$$D_a = E_a - I_a$$

A completely analogous definition is made for D_b.

If vertex $a \in A$ is exchanged with vertex $b \in B$, then

$$g = D_a + D_a - 2\,\text{Con}(a, b)$$

measures the number of external connections gained by the interchange of a and b. If $g > 0$, then the interchange is beneficial, that is, the total number of connections between A and B is reduced.

Let a_1 and b_1 be the nodes which lead to the largest gain, g_1, when exchanged. A and B will then be replaced by $A - \{a_1\}$ and $B - \{b_1\}$, and the process repeated until A and B are reduced to the null set. This given sequences $a_1, a_2, \ldots, a_n$; $b_1, b_2, \ldots, b_n$; and $g_1, g_2, \ldots, g_n$. We choose a value, k, for which the partial sum $\sum_{i=1}^{k} g_i$ is maximum.

By letting

$$X = \{a_1 a_2 \ldots a_k\}$$
$$Y = \{b_1 b_2 \ldots b_k\}$$

we have constructed two sets whose interchange will yield a smaller number of interconnections between the resulting sets

$$A_1 = A \cup Y - X$$
$$B_1 = B \cup X - Y$$

The same process is now repeated, starting with A_1 and B_1 as the initial sets, until no further positive gains can be made. The resulting sets are considered to be approximations to A^* and B^*.

This process can be generalized to sets of unequal size simply by starting with unequal sets. The process can also be generalized to k-way partitions by treating pairs of partition elements sequentially or by treating the k

partition elements as pairs of unequal size, p and $k\text{-}p$ partition elements, respectively. The reader is referred to Kernighan and Lin [11] for the results of experimentation with the various alternatives.

We cautioned the reader earlier not to expect too much. It should be quite clear that the techniques we have discussed have many shortcomings. No one has experimentally compared the various techniques; consequently little can be said about which procedure gives the most satisfactory result.

It is easy to find fault with all of the procedures. One might therefore consider that procedure the best which appears to have the smallest number of serious objections.

Algorithm 1 lacks foresight in the partitioning process. Its tendency is to discover tightly connected clusters at first. The last partition elements, however, are composed of relatively disconnected circuits. The overall solution is therefore far from optimal. The main advantages of algorithm 1 are its speed and simplicity.

Algorithm 2 was designed to overcome the main objection to algorithm 1. The seeding process still ignores the overall structure of the graph. However, the seed elements are well distributed throughout the graph, and experimentation with the parameter $\bar{S}$, which determines the maximum size of a j-neighborhood, may further improve their dispersion. The solution of the sequence of assignment problems, which may be solved exactly or approximately, does give rise to an overall partitioning with nearly all partition elements containing connected circuit clusters. Although algorithm 2 is not quite as easy to program as algorithm 1, the process still requires a relatively small number of basic operations and therefore is applicable to problems of practical size. The most serious shortcoming of algorithm 2 is that it adds circuits to partition elements one circuit at a time. As a consequence, the external connection bound $\bar{E}$ may be exceeded if only one circuit were added to the existing partition element; however, if a group of several circuits were added, the partition element would remain feasible. Fortunately this does not occur frequently in practice.

The Kernighan–Lin [11] procedure is designed to solve a different problem, namely, to minimize the total number of external connections of a given number of partition elements. The procedure may be adapted to solve the partitioning problem we stated by iteratively reducing the number of external connections below the bound $\bar{E}$ for each partition element. If all partition elements have become feasible, then the procedure may be continued to determine the near-minimal feasible solution.

The most serious shortcoming of this procedure is also its inability to exchange more than a pair of circuits at a time. Another fault is the procedure's inability to escape a local optimal solution. One might think that algorithm 2 followed by the Kernighan–Lin procedure should give the best solution. Unfortunately, algorithm 2 usually finds a solution which is locally optimal with respect to pairwise interchange. One further disadvantage of this procedure is its length in comparison to algorithms 1 and 2.

It should be clear that the partitioning problem is a difficult combinatorial problem which has not been studied extensively. Without experimental comparisons, it is folly to try to say anything conclusive. Because all procedures outlined require the same set of basic subroutines to calculate connectivity, our suggestion is to develop all three procedures and to compare the results.

4.6 THE SELECTION PROBLEM

The solution of the partitioning problem consists of subsets

$$A_i = \{a_{i_1}, a_{i_2}, \ldots, a_{i_k}\}, \quad i = 1, 2, \ldots r$$

of the set of all nodes of the bi-digraph, G, which represents the digital system. Each node in the subset A_i represents a recognized circuit $c_k, k = 1, 2, \ldots, n$, in the catalog of recognized circuits. Let r_k denote the number of circuits c_k required by the subset A_i.

The decision procedure in Section 4.4 allows us to construct a standard library of replaceable modules m_j, each of which contains a_{ij} circuits of type $c_i, i = 1, 2, \ldots, n$. A directed graph Γ may be used to display the relationship which allows us to implement certain circuits in terms of others. Let this graph Γ be represented by the matrix (g_{ij})

$$g_{ij} = \begin{cases} 1, & \text{if } c_i \text{ may be used to implement } c_j \\ 0, & \text{otherwise} \end{cases}$$

We are interested in selecting a set of standard replaceable units with minimal total cost, so that all circuits in the partitioned set A_i can be implemented.

Let the cost of each standard module m_j be h_j, and let x_j be the number of modules of type m_j. Let y_{ki} represent the number of circuits of type c_k used to implement circuits of type c_i.

The cost of x_j modules of type m_j is $h_j x_j$. The total cost is

$$z = \sum_{j=1}^{m} h_j x_j$$

The preceding collection of modules supplies

$$s_i = \sum_{j=1}^{m} a_{ij} x_j$$

circuits of type c_i.

$$\sum_{k=1}^{n} y_{ik} g_{ik}$$

circuits of type c_i are used to implement others, whereas

$$\sum_{k=1}^{n} y_{ki} g_{ki}$$

circuits of other types are used to implement circuits of type c_i.
Since the requirement for the circuits of type c_i is r_i, we can state

$$\sum_{j=1}^{m} a_{ij} x_j - \sum_{k=1}^{n} y_{ik} g_{ik} + \sum_{k=1}^{n} y_{ki} g_{ki} \geq r_i, \quad i = 1, \ldots, n \quad (4\text{-}6)$$

The selection problem can thus be formulated as an integer linear programming problem (ILP): minimize

$$z = \sum_{j=1}^{m} h_j x_j$$

subject to inequalities (4-6), where the variables

$$x_j \geq 0 \quad \text{and} \quad y_{ik} \geq 0$$

are integers.

Breuer [2] and Ching and Keenan [4] have formulated the selection problem as an integer linear programming problem for the special case where the graph Γ consists of chains. The ILP problem involves fewer variables for that special case.

Another perhaps more intuitive interpretation of this problem is in terms of a restricted minimal-cost flow problem. We use the digital system in Fig. 4-7 as an example to demonstrate the formulation and solution of the selection problem. A reader unfamiliar with network flow problems may wish to read the appropriate sections in Ford and Fulkerson [5], Busacker and Saaty [3], or Berge [1].

Consider a graph Γ whose nodes represent the recognized circuits c_i. A directed edge or arc (i, j) corresponds to the statement that recognized circuit c_i can be used to implement recognized circuit c_j. Figure 4-6 is a simple example of a graph Γ.

To simplify the problem, we consider a subgraph $\bar{\Gamma}$ of Γ from which we remove all these directed edges (i, j) for which a path connecting (i, j) remains in $\bar{\Gamma}$. Loops are removed also. Figure 4-8 displays the nodes and edges of $\bar{\Gamma}$ with heavy lines for the graph Γ in Fig. 4-6b.

We augment the graph $\bar{\Gamma}$ by introducing a source node s, a terminal node t, and an excess-flow node e. If the requirement for a circuit c_i is $r_i > 0$, we introduce a directed edge from node i to the terminal node t. We associate with this edge the minimal capacity constraint r_i. In this formulation, the maximal capacity constraint will be r_i also. The flow cost associated with this edge is zero.

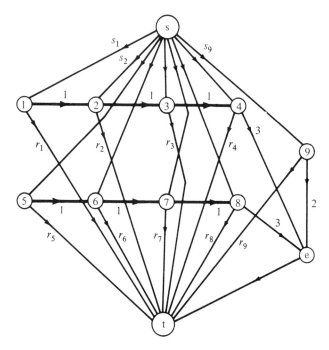

Fig. 4-8. Network flow graph.

The nodes in $\bar{\Gamma}$ from which no directed edges leave are connected to the excess-flow node e. The minimal capacity of this edge is zero, the maximal capacity is infinity, and the cost is nonzero.

The availability of circuits c_i, which are found on some standard circuit modules m_j in the library of circuits, is indicated by arcs leaving s and terminating at i. We associate with each of these arcs the minimum capacity zero, the maximum capacity infinity, and flow cost zero. The flow in these arcs, however, is constrained by the following equalities

$$s_i = \sum_{j=1}^{m} a_{ij} x_j, \qquad i = 1, 2, \ldots, n$$

where x_j represents the number of circuit modules, m_j, and a_{ij} represents the number of circuits, c_i, available on module m_j.

The arcs in the subgraph $\bar{\Gamma}$ will all have a minimum capacity of zero and a maximum capacity of infinity. The flow cost, however, will be nonzero.

If the cost h_i of each circuit module is distinct, then the flow cost in each arc cannot be a constant, and the problem therefore is not a restricted minimal-cost flow problem. If, however, the costs h_i do not differ substantially, then one may consider

$$h_i \approx h, \quad i = 1, 2, \ldots, m$$

In that case, h may be taken to be the number of external connections on a module. The total cost C becomes

$$C = \sum_{i=1}^{m} h_i x_i = h \sum_{i=1}^{m} x_i$$

which may be separated into the cost of connections used and cost of connections wasted. For any feasible solution, the cost of used connections is a constant C_u, where

$$C_u = \sum_{i=1}^{n} r_i d_i$$

and where d_i is number of connections of circuit type c_i. The cost C_w of wasted connections varies, and to minimize total cost C, C_w must be minimized.

A unit flow in arc (i, j) in Γ represents the use of circuit c_i to implement circuit c_j which has one less input/output. Each circuit so used would give rise to one wasted connection. The cost is thus conveniently represented by associating a unit cost with each arc in Γ. A completely wasted circuit is represented by a unit flow in the arc (i, e), which terminates at the excess-flow node e. If the cost per unit flow for each such arc is defined to be the number of connections of circuit c_i, then the total cost C_w of wasted connections is given by

$$C_w = \sum_{(i,j) \in \Gamma} f(i, j) + \sum_{(i,e)} f(i, e) d_i,$$

where $f(i, j)$ represents the total flow in arc (i, j).

Figure 4-8 displays a flow network for the selection problem associated with the digital network in Fig. 4-7. The standard library is given in Table 4-3. The AND, OR, and INVERT circuits are labeled by $\cdot$, $+$, and I, respectively, in Fig. 4-7.

If the flow from the source s were not restricted by the equations

$$s_i = \sum_{j=1}^{5} a_{ij} x_j, \quad i = 1, 2, \ldots, 9$$

the selection problem represented in Fig. 4-8 would be a standard minimal-cost flow problem discussed, for example, by Ford and Fulkerson [5]. The foregoing restrictions change the problem from a minimal-cost flow problem, which has several effective algorithms, to a special minimal-cost flow problem, for which the solution techniques constitute a topic of research.

If the circuit card library has a particularly simple structure so that the graph $\bar{\Gamma}$ consists of isolated chains (shown in Fig. 4-8 by the heavy lines), then a new linear integer programming problem may be formulated. The new problem will have fewer variables; consequently, enumeration techniques to solve the integer linear programming problem become applicable.

As an example, we formulate the selection problem in Fig. 4-8 in this new form. We first note that the necessary conditions for a feasible flow are:

TABLE 4-3. STANDARD MODULE LIBRARY AND CIRCUIT REQUIREMENTS FOR THE NETWORK IN FIG. 4-7

Circuit Type	Module Type					r_i	d_i
	1	2	3	4	5		
1	2					0	0
2		2				1	5
3	1		2		1	3	12
4		2				4	12
5						0	0
6			2			0	0
7		2			1	5	20
8			2			7	21
9				4	3	6	

$$s_1 + s_2 \geq r_2 \quad (4\text{-}7)$$

$$s_1 + s_2 + s_3 \geq r_2 + r_3 \quad (4\text{-}8)$$

$$s_1 + s_2 + s_3 + s_4 \geq r_2 + r_3 + r_4 \quad (4\text{-}9)$$

$$s_5 + s_6 + s_7 \geq r_7 \quad (4\text{-}10)$$

$$s_5 + s_6 + s_7 + s_8 \geq r_7 + r_8 \quad (4\text{-}11)$$

$$s_9 \geq r_9 \quad (4\text{-}12)$$

These conditions are also sufficient to provide a feasible flow. This can be shown by constructing a feasible flow, f_{ij}, in each arc joining i to j.

Let $f_{s1} = s_1$, $f_{12} = s_1$, and $f_{s2} = s_2$. Because of inequality (4-7), the flow into node 2 is not less than r_2. Hence make $f_{2t} = r_2$, $f_{23} = s_1 + s_2 - r_2$, and $f_{s3} = s_3$. Inequality (4-8) ensures that the flow into node 3 is

$$f_{s3} + f_{23} = s_1 + s_2 - r_2 + s_3 \geq r_3$$

Again choose

$$f_{3t} = r_3, \quad f_{34} = s_1 + s_2 + s_3 - (r_2 + r_3), \quad f_{s4} = s_4$$

Inequality (4-9) assures that the flow into node 4 is not less than the requirement there. Hence we choose

$$f_{4t} = r_4 \quad \text{and} \quad f_{4e} = s_1 + s_2 + s_3 + s_4 - (r_2 + r_3 + r_4)$$

Similarly, we can construct flows which satisfy the capacity requirements in the remaining arcs in the network.

The optimal solution to the selection problem can be found by minimizing the cost function

$$z = \sum_{j=1}^{5} h_j x_j = h \sum_{j=1}^{5} x_j$$

subject to the Inequalities (4-7) through (4-12). Comparison of this formulation of the problem to the general formulation (4-6) shows that all the intermediate variables y_{ik} have been eliminated. This decreases the number of variables sufficiently to make the use of enumeration techniques for solving integer linear programming problems feasible. The recent work of Greenberg [7] makes this approach look quite attractive.

The graph $\bar{\Gamma}$, which displays how circuits may be used to implement each other, is particularly simple in the foregoing case. As Γ becomes more complex, obtaining the inequalities which constitute the necessary and sufficient conditions for a feasible flow becomes more complicated also. General methods to eliminate the intermediate variables y_{ik} in formulation (4-6) are still unknown. Good near-optimal solutions can be obtained, however, by relaxing the requirement that the inequalities describe necessary and sufficient conditions for a feasible flow. For example, we may replace a highly complex graph $\bar{\Gamma}$ by its subgraph which consists of disjoint chains. This reduces any problem to the form described above. The inequalities then describe only sufficient conditions for a feasible flow. Considerable insight must be used to obtain near-optimal solutions in this case.

We continue with the solution of the integer linear programming problem which originates from Fig. 4-7.

Minimize $16 \sum_{i=1}^{5} x_i$, subject to

$$
\begin{aligned}
2x_1 + 2x_2 &\geq 1 \\
3x_1 + 2x_2 + 2x_3 \phantom{{}+{}} + x_5 &\geq 4 \\
3x_1 + 4x_2 + 2x_3 \phantom{{}+{}} + x_5 &\geq 8 \\
2x_3 + 2x_4 + x_5 &\geq 5 \\
2x_3 + 4x_4 + x_5 &\geq 12 \\
4x_5 &\geq 3
\end{aligned}
$$

Note first that the cost of used connections

$$C_u = 76$$

Because total cost cannot be less than C_u, we know immediately that

$$\sum_{i=1}^{5} x_i \geq 5$$

Exhaustive enumeration of all possible combinations of x_i

$$\sum_{i=1}^{5} x_i = 5$$

shows that no feasible solutions exist with cost 80. Therefore any minimal solution would have cost 96. Among such solutions is

$$x_1 = 1, \quad x_2 = 1, \quad x_3 = 0, \quad x_4 = 3, \quad x_5 = 1$$

The procedure outlined in the example above can be used quite effectively even when the graph Γ is complicated. The simplest way of obtaining sufficient conditions for feasible flow is to decompose the graph into chains. However, the resulting inequalities are unnecessarily restrictive and the minimal solution is frequently missed.

Other methods of decomposing the graph into smaller graphs may be used. The main objective of the decomposition is to make the resulting ILP problem sufficiently small for exhaustive enumeration. We thus reduce the selection problem to a sequence of small ILP problems, each of which is solved optimally. With experience in decomposing the graph, a minimal solution may be obtained in a large majority of cases.

A further benefit of being able to solve the selection problem is that alternative standard module libraries may be tested before final decisions are made about their contents.

The solution of the selection problem consists of a set of standard modules sufficient to implement the circuits in a given partition element. The final assignment of a logic circuit to a particular standard module may be done at this point; however, there are good reasons for delaying this step until the placement procedure.

If we think of the different logic circuits as differently shaped pegs connected appropriately with rubber bands, and if we think of the standard modules as particular arrangements of appropriately shaped holes, then assigning a peg to a hole that can accomodate the peg is the final assignment process. If in the placement problem only entire modules are allowed to move, with everything within the module remaining fixed, then much flexibility is lost. Far better solutions to the placement problem can be found if every peg is allowed to move within the group of appropriately shaped holes no matter to which module the hole belonged. This is especially true when the modules are large, are elongated in one direction, and contain many similarly shaped holes.

Although the placement problem defined in the next chapter assumes that the modules are uniformly shaped, the placement methods can be adjusted to tolerate differently shaped modules. We may therefore think of circuits as modules, and the differently shaped slots as the available positions for the circuits. It is clear that nothing is lost if we view the final assignment process as a placement problem.

4.7 UNSOLVED PROBLEMS AND CONCLUSIONS

The assignment of circuits to replaceable modules was accomplished quite naturally for the first two generations of computers. Changes in technology have provided an opportunity to construct a large number of circuits in the same amount of space required by just a few pluggable connections. To exploit the economic benefits, it is desirable to create highly functional modules, with a high degree of repetition throughout the digital system.

In Section 4.4 we discuss the economics of a standard module library in quantitative terms. The results indicate that the logic designer of the digital system must be cognizant of the standard module library and must influence the make-up of this library. We give a formula which expresses the economic advantage of a new standard module in terms of the degree of repetition and the design-cost to manufacturing-cost ratio.

It is quite clear that the economic benefits brought about by the new technology can be exploited only if the entire architecture of the digital system is designed with the standard module library in mind. How to accomplish this is the major unsolved problem in our new technology.

In Section 4.5 we formulate our version of the partitioning problem and review some of the approaches which have appeared in the literature. Although we cannot claim that our proposed procedures solve the problem from a mathematical point of view, the approximate solutions give sufficiently good results to satisfy most design requirements.

The selection problem, formulated in Section 4.6 as an integer linear programming (ILP) problem and as a restricted network flow problem, presents serious difficulties. For practical applications, the ILP problems are, in general, too large to be solved by existing algorithms. Simplifying assumptions allow the problems to be solved, and the resulting near-optimal solutions are sufficiently good for design purposes.

The circuit assignment problems are treated in this chapter from a mathematical point of view. The proposed solution procedures do not produce optimal solutions to the mathematical problems. Although this is disappointing to the mathematician, it should be remembered that the design problems we are trying to solve can be solved frequently by very poor approximations to the optimal solution. The mathematical procedure often competes with a crude and not very carefully thought-out procedure used by a relatively unsophisticated layout designer. With this in mind, we can say with confidence that the circuit assignment problems can be, and in many cases have been, formalized into successfully operational computer programs.

PROBLEMS

1. The incidence matrix which describes the subgraph of the graph in Fig. 4-7 defined by the circuit nodes {10, 11, 12, 17, 18} consists of five rows and ten columns. Determine this matrix.
2. The compact representation of the matrix in problem 1 consists of a list L of pointer indices and a list K of column indices with positive and negative signs. Determine these lists for the matrix in problem 1.
3. Consider the five circuits:

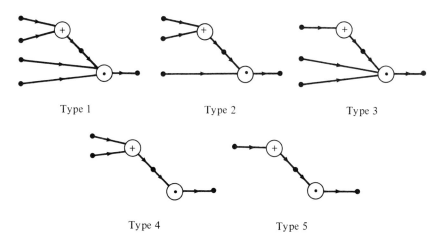

      Type 1   Type 2   Type 3

      Type 4   Type 5

   (a) Determine the graph Γ which describes the partial order in which the circuits may be used to implement each other.
   (b) Identify the set of basis nodes B in Γ.
   (c) If the circuit type 1 were not in Γ, what would constitute the set of basis nodes B.

(d) Which circuits must be in the standard module library for the library to be called complete in parts (b) and (c).

4. Suppose that a standard circuit module can have at most fifteen external connections. Suppose also that the standard module library contains two modules, one consisting of five independent two-input OR circuits, and the other consisting of three independent three-input AND circuits and one two-input AND circuit.

    A new module consisting of three circuits of type 1 is proposed. Using Table 4-1, indicate the values of r and d for which such a new module would be economical.

5. Compute the space parameter $S(A)$ and the external connection parameter $E(A)$ for the subset of circuit nodes in problem 1. Define $S(a) = E(a) = d(a)$, where $d(a)$ is the degree of the circuit node a.

6. Use algorithm 1 to partition the nodes of the graph in Fig. 4-7, subject to the bounds $\bar{S} = 20$, $\bar{E} = 14$.

7. Use algorithm 2 to partition the nodes of the graph in Fig. 4-7, subject to the bounds $\bar{S} = 20$, $\bar{E} = 14$.

8. Formulate the integer linear programming (ILP) problem given by Equations (4-7) to (4-12) with a slightly modified module library in Table 4-3. Let the module of type 1 consist of one circuit of type 3 and three circuits of type 7. The circuit requirements and the remaining modules are given in Table 4-3.

9. Determine a module library which will solve the selection problem in Fig. 4-7 with five modules, assuming that the external connection limit remains sixteen per module and that independent circuits make up each module.

REFERENCES

[1] C. Berge, *The Theory of Graphs and its Application*. New York, John Wiley & Sons, Inc., 1962.

[2] M. A. Breuer, "The Application of Integer Programming in Design Automation." *Proc. SHARE Design Automation Workshop*, 1966.

[3] R. G. Busacker and T. L. Saaty, *Finite Graphs and Networks*. New York, McGraw-Hill Book Company, 1965.

[4] S. W. Ching and W. F. Keenan, "A Note on the Card Assignment Problem." Burroughs Corp., June 6, 1964, unpublished.

[5] L. Ford and D. Fulkerson, *Flows in Networks*. Princeton, N.J., Princeton University Press, 1962.

[6] R. L. Gamblin, M. Q. Jacobs, and C. J. Tunis, "Automatic Packaging of Miniaturized Circuits," in G. A. Walker, ed., *Advances in Electronic Circuit Packaging*, Vol. 2. New York, Plenum Press, 1962, pp. 219–232.

[7] H. Greenberg and R. L. Hegerich, "A Branch Search Algorithm for the

Knapsack Problem." *Management Science*, Vol. 16, No. 5, January 1970.

[8] A. R. Habayeb," System Decomposition, Partitioning, and Integration for Microelectronics." IEEE Trans. on Systems Science and Cybernetics, Vol. SSC-4, No. 2 (July 1968), pp. 164–172.

[9] F. Harary, *Graph Theory*. Reading, Mass., Addison-Wesley Publishing Company, Inc., 1969.

[10] C. H. Haspel, "The Automatic Packaging of Computer Circuitry." *1965 IEE International Conv. Rec.*, Vol. 13, Part 3, pp. 4–20.

[11] B. W. Kernighan and S. Lin, "An Efficient Heuristic Procedure for Partitioning Graphs." *Bell Sys. Tech. J.*, Vol. 49 (1970), pp. 291–307.

[12] U. R. Kodres, "Logic Circuit Layout." *The Digest Record of the 1969 Joint Conference on Mathematical and Computer Aids to Design*, October, 1969.

[13] J. M. Kurtzberg, "On Approximation Methods for the Assignment Problem." *J. ACM*, Vol. 9, No. 4 (October, 1962), pp. 419–439.

[14] E. L. Lawler, "Electrical Assemblies With a Minimum Number of Interconnections." *IEEE Trans. on Electronic Computers* (*Correspondence*), Vol. EC-11 (February, 1962), pp. 86–88.

[15] E. L. Lawler et al., "Module Clustering to Minimize Delay in Digital Networks." *IEEE Trans. on Computers*, Vol. C-18 (January, 1969), pp. 47–57.

[16] F. Luccio and M. Sami, "On the Decomposition of Networks in Minimally Interconnected Subnetworks." *IEEE Trans. on Circuit Theory*, Vol. CT-16, No. 2, May, 1969.

[17] J. S. Mamelak, "The Placement of Computer Logic Modules." *J. ACM*, Vol. 13 (October, 1966), pp. 615–629.

[18] J. Munkres, "Algorithms for the Assignment and Transportation Problems," *J. SIAM*, Vol. 5 (1957), pp. 32–38.

[19] H. A. Nidecker and W. F. Simon, "Logic Partitioning—Component Assignment." *Proc. 1968 ACM Nat. Conf.*, pp. 211–221.

[20] W. A. Notz et al., "Large Scale Integration; Benefitting the Systems Designer." *Electronics* (February 20, 1967), pp. 130–141.

[21] O. Ore, *Theory of Graphs*. Providence, R.I., American Mathematical Society, 1962.

[22] C. V. Ramamoorthy, "A Structural Theory of Machine Diagnosis." *Proc. AFIPS* (Spring, 1967), pp. 743–756.

[23] R. A. Rutman, "An Algorithm For Placement of Interconnected Elements Based on Minimum Wire Length." *Proc. 1964 SJCC*, pp. 477–491.

[24] S. A. Thompson, "CAD Graphics: Circuits Made to Order." *The Electronic Engineer*, August, 1969.

[25] M. N. Weindling, "A Method For Best Placement of Units on a Plane." *Proc. 1964 SHARE Design Automation Workshop*, and Douglas Paper 3108, Douglas Aircraft Co., Santa Monica, California.

chapter five

PLACEMENT TECHNIQUES

MAURICE HANAN
JEROME M. KURTZBERG

*IBM Thomas J. Watson Research Center
Yorktown Heights, New York*

5.1 INTRODUCTION

The physical design of a computer backplane requires the partitioning and assignment of logic to hardware packages, placement of these packages onto larger functional units (which in turn might be part of other packages), and establishment of wiring patterns to interconnect the circuitry in conformance with the logic design. These tasks actually denote a class of problems whose precise definition strongly depends upon the particular technology being employed. The electrical constraints of the circuitry, the physical properties of the package, and even the tools available determine the specific problems. Although the problems of partitioning and assignment, placement, interconnection and routing are intimately related, historically they have been treated separately because of the inherent computational complexity of the total problem.

Other chapters of this text examine the partitioning and wiring problems. In this chapter we explore the placement problem and present algorithms for it and the associated quadratic assignment problem.

Because of the complexity and magnitude of most practical problems, no methods exist which guarantee an optimum solution. Hence methods based upon heuristic rationales are employed. All of these algorithms are either *iterative* or *constructive*. The iterative algorithms seek to improve a placement by repeated modification of it. At every stage there is a complete placement available. Thus an iterative algorithm may be interrupted at the end of any cycle. The constructive algorithms, on the other hand, produce a placement configuration only upon termination.

In our presentation, we have grouped the placement techniques for the purpose of discussion into three classes:

1. *constructive initial-placement;*
2. *iterative placement-improvement;*
3. *branch-and-bound.*

In the main, the algorithms in the first and third categories are constructive; in the second, they are iterative.

In the next section, the placement problem is defined in full and some useful graph theoretic notions applicable to the problem are developed. In Section 5.3, various interconnection rules are discussed, and in Section 5.4 important associated mathematical problems are given. The constructive initial-placement methods are defined in Section 5.5 and the iterative placement-improvement techniques in Section 5.6. In Section 5.7, the branch-and-bound techniques are presented. Known experimental results, conclusions, areas for future research and general remarks are given in the last section. The chapter concludes with problems, references and an appendix which gives two algorithms for the assignment problem for use in one of the iterative placement-improvement methods.

5.2 PLACEMENT PROBLEM DEFINITION AND GRAPH MODELS

Prior to defining the placement problem, we develop appropriate terminology by considering the typical structure of a computer backplane. It consists of an n-level hierarchy of intrawired and interwired packages of circuitry where the specified number of levels is determined by the technology.

The current integrated circuitry approach usually results in three-level backplanes. On the lowest level, sets of *logic elements*[†] form *modules*. Inter-

---

[†] A *logic element* or *circuit element*, or simply an *element*, is a set of one or more preconnected *logic gates*. A gate is the most primitive circuit unit which implements logic functions on the AND, OR, NAND, etc., level. If a logic element is constructed as an independent physical entity, it is usually referred to as a *component*.

connected sets of modules define *boards;* and interconnected sets of boards define the *backplane*, sometimes termed a *plane*. A backplane may be implemented with a multiplane construction. Historically, with pre-integrated circuitry technology, two-level backplanes are the rule. The modules then are termed *cards*, and the "board level" becomes the "backplane level."

For the sake of being concrete, we speak in this chapter of the placement of modules to form a board. It must be recognized that the other levels of backplane formation can be similarly treated, and in general, the algorithms given in this chapter are applicable to virtually any package level.

5.2.1. Definition of Problem

Physically, forming a board involves positioning a set of computer modules into various *slots*† in the board. The various circuit elements on the modules lead to *connection pins* located on the sides of the modules. On the basis of the given logic design, various subsets of these pins, termed *signal sets*, must be interconnected to form *nets*. We speak of these pins as being *electrically common*. Since each module contains a number of connection pins, a module generally belongs to more than one distinct net. Furthermore, because of engineering considerations, some of the nets may be more critical than others, and thus the corresponding signal sets have greater "weight."

Formally, the *module placement problem* consists of finding the optimal placement of modules in the board with respect to some norm defined on the interconnections, such as minimal weighted wire length.

In the real world, there are actually a number of (possibly conflicting) goals which must be satisfied in a placement configuration: wire buildup in the routing channels on the board must be kept within tolerable bounds; signal cross-talk must be eliminated without undue use of expensive shielding techniques; signal echoes must be eliminated; heat dissipation levels must be preserved, i.e., no severe heat source concentrations can be permitted; and so on. Most important of all, the placement configuration must result in ease and neatness of wirability to reduce the expense of construction and maintenance in the field.

For practical purposes, it is impossible to directly incorporate all of the circuit goals. However, minimizing total weighted wire length is an expedient used to satisfy them en masse. In effect, total weighted wire length is an approximation to the numerous restrictions of the circuit designers and desires of the maintenance men and is the usual measure adopted for the placement algorithms. It best reflects all the goals.

†The terms *location* and *position* are used as synonyms for slot. Also, *placement, placement configuration, configuration*, and *layout* are used interchangeably in the text following traditional use of the words. For expositional purposes we assume that the slots in the board are completely filled with modules. If as in realistic problems such is not the case, then dummy modules with no connections are considered to occupy the empty locations.

At times, however, some goal is particularly critical and must be emphasized. In these cases, the measure of total weighted wire length is not used; instead some other norm on the interconnections is adopted. For example, in a technology which permits wires in only horizontal and vertical directions (*rectilinear* or *Manhattan distance*), the total number of noncollinear connections (corner-turns) is frequently of primary importance. Pomentale [47] employs this norm. Also, in certain integrated circuitry, no wire crossovers are permitted. Kodres [29] deals with this. Wire routing congestion is frequently of concern and Clarke [6] attacks placement with the goal of minimizing expected congestion. We stress, however, that all placement goals relate to wirability and that the primary aim of placement is to aid the wire routing.

Finally, we note that the distance function used in the measure of the layout is most usually taken as the rectilinear distance. That is, if (x_1, y_1) and (x_2, y_2) are two points in the plane, then the rectilinear distance, d, is defined to be

$$d = |x_1 - x_2| + |y_1 - y_2|$$

The other distance function that is adopted is the Euclidean,

$$d = [(x_1 - x_2)^2 + (y_1 - y_2)^2]^{1/2}$$

Frequently, the distance function is adjusted by incorporating scale factors for the x and y components.

5.2.2. Graph Models

We conclude this section with a discussion of some graph theoretic representations of modules and signal sets to supply a framework for the placement algorithms. This is done with a small example. For a more thorough discussion of graph theory see Harary [24], Ore [46], Busacker and Saaty [5], and Berge [2].

Consider the following four signal sets† and their corresponding weights p_i, defined on the seven modules $A, B, \ldots, G$

$$S_1 = \{A, B, E\} \qquad p_1 = 4$$
$$S_2 = \{A, B, C, F\} \qquad p_2 = 1$$
$$S_3 = \{C, D, F\} \qquad p_3 = 3$$
$$S_4 = \{B, C, G\} \qquad p_4 = 5$$

†In realistic problems the pin numbers, e.g., A_1, A_2 and B_7, are shown in the signal sets. For clarity these are omitted in the present example, and we take the signal sets to be defined on the modules rather than on the pins.

There are several ways to graphically represent the modules and the signal sets. For the purpose of illustration, assume that for each signal set the first module listed contains the source pin (driver). Then with this convention, the modules and the signal sets defined on them can be represented in the form of a module diagram as shown in Fig. 5-1. This is not a graph in the graph theoretic sense since the joining of lines in this representation does not imply a vertex.

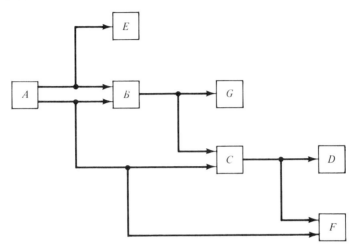

Fig. 5-1. Module diagram.

A useful graph model is the *bigraph* (or *bipartite*† *graph*) shown in Fig. 5-2. In this graph there are two sets of vertices, namely, the module vertices and the signal set vertices.‡ An edge joins a module vertex X to a signal set vertex S_k, or k for short, if and only if X is an element of k.

A third representation is illustrated by the graph in Fig. 5-3, where the modules are the vertices of the graph and an edge (or line) joins two vertices if and only if the modules they represent have a signal set in common. One way of generating this graph is to construct the *complete graph*§ on every signal set and then form the union of these complete graphs. The *adjacency matrix* (or *connection matrix*) for this graph is given in Fig. 5-4. By definition, the diagonal elements are taken to be zero.

Two distinct modules are said to be *adjacent*, if they have at least one signal set in common. A set of modules is said to be *independent* if no two modules

†A graph is *bipartite* if it can be partitioned into two sets of vertices V_1 and V_2 such that there are no edges joining vertices in V_i ($i = 1, 2$).

‡For the purposes of illustration we have shown the module vertices as squares and the signal set vertices as circles. From a graph theoretic view these vertices are mathematical points.

§A graph with n vertices is *complete* if there is an edge joining every pair of vertices.

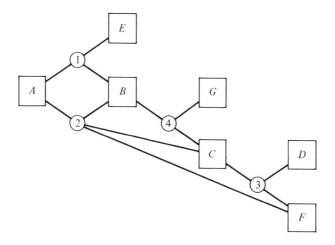

Fig. 5-2. Signal-set bigraph.

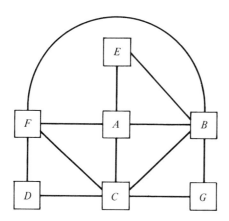

Fig. 5-3. Module-adjacency graph.

	A	B	C	D	E	F	G
A	0	1	1	0	1	1	0
B	1	0	1	0	1	1	1
C	1	1	0	1	0	1	1
D	0	0	1	0	0	1	0
E	1	1	0	0	0	0	0
F	1	1	1	1	0	0	0
G	0	1	1	0	0	0	0

Fig. 5-4. Module-adjacency matrix.

in the set are adjacent. This terminology is motivated by the graph of Fig. 5-3 and the graph theoretic use of these terms. That is, in graph theory two vertices are said to be adjacent if there is an edge joining them, and a set of vertices is independent if there are no adjacent pairs. A module is not considered to be adjacent to itself.

It is easy to see that the two definitions are equivalent upon referring to the module adjacency graph of Fig. 5-3. Note, however, that this is not the case in the signal set graph of Fig. 5-2. In fact, in the signal set graph, the entire set of vertices representing the modules is, by definition, independent.

5.3 INTERCONNECTION RULES

After the module placement is complete, it is necessary to define the interconnection of the pins to form the nets. *Wire routing*, which consists of specifying the geometrical path (i.e., the precise wire channels to follow) for the interconnection of the pins, can then be performed. The interconnection task can be incorporated either into the wire routing program or into the placement program; but, in either case, the norm of the placement algorithm depends upon the type of net used for the interconnections.

The net used to interconnect the points in a signal set depends strictly upon the technology under consideration. For example, for some computers the number of wires on a connection pin is limited to two or three;† for other computers the interconnections are wires from the source pin (driver) to virtually all other pins in the signal set; and for still others, auxiliary junction points (e.g., dummy pins) can be added to the net to aid in reducing the wiring costs.

If we take the J_k pins of the kth signal set to be the vertices of an undirected graph, then the interconnection net is a *tree*‡ defined on these J_k vertices. There are basically three types of trees used:

1. trees of minimum length§ whose vertices are the J_k points of the signal set (a *minimum spanning tree*);
2. trees of minimum length whose vertices include the J_k points of the signal set (a *Steiner tree*);
3. special trees (generally of minimum length) which satisfy certain *degree*‖ requirements on their vertices, or which satisfy other geometrical and topological constraints.

†To illustrate, with the Gardner–Denver wirewrap machines only three wires can be connected to a pin. Usually the third wire is reserved for corrections and modifications to the backplane.
‡A *tree* is a connected graph with no cycles or, equivalently, a tree with J_k vertices is a connected graph with $J_k - 1$ edges.
§The *length* of a tree is the sum of the lengths of all its edges.
‖The *degree* of a vertex is the number of edges incident to it.

In Fig. 5-5 we show a signal set of four points and four different trees connecting them. The encircled vertex represents the source pin and the others represent the sink pins. We use the rectilinear distance defined in Section 5.2 as the metric. Figure 5-5a illustrates a minimum spanning tree. The length, l, is 20 units. Figure 5-5b shows a Steiner tree whose length is 15 units. Note that junction points, i.e., additional vertices, have been introduced. (In general, it is considerably more difficult to find Steiner trees than minimum

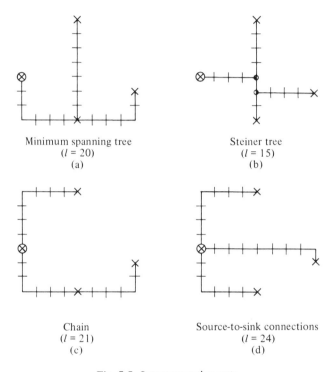

Fig. 5-5. Interconnection nets.

spanning trees.) In Fig. 5-5c and d we illustrate two special trees. In Fig. 5-5c we have found the minimum length tree which has at most two edges incident to any vertex; such a tree is called a *chain*. In Fig. 5-5d we establish a tree which satisfies the restriction that there is a connection from the source to each sink.

For a more thorough discussion of minimum spanning trees see Kruskal [30], Loberman and Weinberger [42], and Prim [49]. For a discussion of Steiner trees see Hanan [20], [21]. Some special trees have been treated by Kurtzberg and Seward [37].

5.4 ASSOCIATED MATHEMATICAL PROBLEMS

We now introduce three other problems, the *assignment*, the *traveling salesman* and the *quadratic assignment* problems, which are related to the placement problem. The assignment problem is a special case of the quadratic assignment problem which, in turn, is a special case of the placement problem. The traveling salesman problem is a constrained assignment problem.

5.4.1. The Assignment Problem

Although this problem appears in many applications, it is usually cast in terms of assigning men to jobs in an optimal fashion, namely, given n men and n jobs and the cost a_{ij} of assigning man i to job j, find an assignment of one and only one man to each job such that the total cost is minimum.

If we let p represent a permutation of the first n integers, then the problem becomes: Find the minimum of

$$F = \sum_i a_{i p(i)}$$

over all permutations p, where $j = p(i)$ is the job assignment of man i.

There are $n!$ permutations on n integers, that is, there are $n!$ distinct ways to assign n men to n jobs. Therefore, any method of solution which depends on an enumeration of all solutions is clearly not feasible. There are, however, well-known methods of solution to this problem, e.g., Munkres [44], which can treat reasonably large problems on the order of 200 men and jobs within five minutes on the IBM 360/65. In addition, approximation methods for the assignment problem have been investigated by Kurtzberg [33].

5.4.2. The Traveling Salesman Problem

In the traveling salesman problem we are given n cities and the distance between every pair. A salesman wishes to start at city c, visit every city once and only once, and return to c, traveling the minimum total distance.

The problem can be regarded as a constrained (subcycle-free) assignment problem, but one whose method of solution is vastly more difficult than that of the unconstrained problem. In fact, no known computationally feasible exact method exists for general problems with more than 20 cities. We now cast the problem more formally to exhibit the similarity between it and the other two problems in this section.

We form a matrix A in which the entry a_{ij} denotes the cost of going from city i to city j, $1 \leq i, j \leq n$. A tour t is a set of n ordered city pairs

$$t = \{(i_1, i_2), (i_2, i_3), \ldots, (i_{n-1}, i_n), (i_n, i_1)\}$$

which forms a cycle of size n. Each (i, j) represents a branch of the cycle, or leg of the trip, in which each city is visited exactly once.

The cost of the tour t under the matrix A is the sum of the elements in A denoted by the couplets in t. We wish to select a tour that minimizes cost; thus the problem becomes: minimize

$$\sum_{(i,j) \in t} a_{ij}$$

Clearly, a brute-force enumerative technique requires examining on the order of $n!$ tours. Dynamic programming approaches [25] reduce the time to an exponential function of n. Branch-and-bound methods [41] also reduce the time for solution but still require an exponential order of steps. Linear programming has been employed [9], but for a general class of problems it is necessary to go to approximation methods. Hence heuristic algorithms are used to find suboptimum solutions, and on occasion even yield the optimum solution. See Bellmore and Nemhauser [1] for an excellent review of the attacks upon the traveling salesman problem and a list of references.

5.4.3. The Quadratic Assignment Problem

This problem is a generalization of the assignment problem in the following sense. In addition to a cost matrix $C = [c_{ij}]$, there is a distance matrix $D = [d_{kl}]$ and the problem is to find the minimum of

$$G = \sum_{i,j} c_{ij} d_{p(i)p(j)}$$

over all permutations p.[†] One possible interpretation of this formulation is as follows. We are given n people (modules) and a matrix $[c_{ij}]$, where c_{ij} is a measure of the affinity (wires) between the two people i and j. Also given are n possible offices (slots) for these people and a matrix $[d_{kl}]$, where d_{kl} is the distance between office k and l. If i is assigned to office $p(i)$ and j is assigned to office $p(j)$, then the cost of this assignment is taken to be $c_{ij} d_{p(i)p(j)}$. We now take as our measure of total cost the sum of the $c_{ij} d_{p(i)p(j)}$ costs over all i and j, and seek that assignment which yields the smallest total cost. If the affinity represents the amount of "face-to-face communication,"

[†]If C and D are symmetric then the quadratic assignment problem is frequently formulated as: Find a permutation p which minimizes $\sum_{i<j} c_{ij} d_{p(i)p(j)}$.

then the assignement we desire is the one which minimizes the total amount of walking (wire distance) for all the people.

As in the assignment problem there are $n!$ ways to assign the n people to n offices. Unlike the assignment problem, however, there are no known methods of solutions that are computationally feasible for a reasonably large problem (n larger than 10–15). The semienumerative methods (branch-and-bound) of Gilmore [18] and Lawler [38] do yield optimum solutions for small n; however, the computation time increases exponentially.

It is considerably more difficult to find a solution to the quadratic assignment problem than to find a solution to the assignment problem. An intuitive reason for this is that in the assignment problem, the cost of assigning man i to job k is independent of the assignment of the other men; whereas in the quadratic assignment problem, the cost of assigning man i to office k depends on the assignment of all men who have a nonzero affinity for i.

We now generalize the quadratic assignment problem in a way to make the distinction between it and the assignment problem even more obvious. In addition to the cost already defined, assume there is a cost a_{ij} of assigning man i to office j so that the function to be minimized is

$$H = F + G = \sum_j a_{ip(i)} + \sum_{i,j} c_{ij} d_{p(i)p(j)}$$

over all permutations p.

We can interpret $[c_{ij}]$ as a connection matrix of a graph with weighted edges, where the men are the vertices of the graph and an edge with weight c_{ij} joins vertices i and j for all nonzero c_{ij}. Hence the quadratic assignment problem is to assign the vertices of an edge-weighted graph to locations such that the function H is minimized. If $c_{ij} = 0$ for all i and j, that is, if the graph is a set of isolated (or independent) vertices, then the second summation in H drops out and the problem reduces to the assignment problem.

This observation that the quadratic assignment problem reduces to the assignment problem whenever the vertices are independent (i.e., no edge between any pair) is exploited by Steinberg [53] for an iterative technique used to find an approximate solution to the quadratic assignment problem. See Section 5.6.2.

Finally we note that the quadratic assignment problem is a special case of the placement problem. In the general placement problem we are concerned with sets of points, whereas in the quadratic assignment problem we deal only with the simpler problem of pairs of points. Therefore, if all signal sets of a placement problem contain exactly two modules, the placement problem reduces to the quadratic assignment problem. In this case the interconnection rule reduces to the trivial case of joining a pair of points with a single line.

Even if the signal sets are not uniformly of cardinality two, an associated quadratic assignment problem can be derived from the placement problem.

This is accomplished by taking as elements of the cost matrix, $C = [c_{ij}]$, the sum of the weights of the signal sets common to modules i and j, $i \neq j$. c_{ii} is defined to be zero. The distance matrix of this quadratic assignment problem is the same distance matrix as in the original placement problem.

In Fig. 5-6 we show the matrix C for the example of Section 5.2.2 (compare with Fig. 5-4), and in Fig. 5-7 we show the corresponding graph. Note that this is the graph of Fig. 5-3 with weights assigned to the edges.

	A	B	C	D	E	F	G
A	0	5	1	0	4	1	0
B	5	0	6	0	4	1	5
C	1	6	0	3	0	4	5
D	0	0	3	0	0	3	0
E	4	4	0	0	0	0	0
F	1	1	4	3	0	0	0
G	0	5	5	0	0	0	0

Fig. 5-6. Connection matrix for associated quadratic assignment problem.

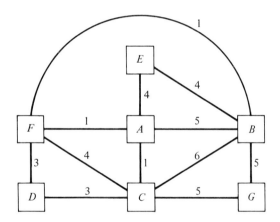

Fig. 5-7. Module-connection graph for associated quadratic assignment problem.

We observe that C transforms any given placement problem into an associated quadratic assignment problem. As an important consequence, any method of solution for the quadratic assignment problem can be used for the placement problem. We note, however, that an optimum solution to this associated quadratic assignment problem is not necessarily an optimum solution to the original placement problem. An illustration of this fact is given by the six-module example defined by the following signal sets

Sec. 5.4   Associated Mathematical Problems   225

$$S_1 = \{A, B\} \qquad p_1 = 1$$
$$S_2 = \{A, C\} \qquad p_2 = 1$$
$$S_3 = \{A, D\} \qquad p_3 = 1$$
$$S_4 = \{A, B, C, E\} \qquad p_4 = 1$$
$$S_5 = \{A, B, E, F\} \qquad p_5 = 1$$

The derived connection matrix for the associated quadratic assignment problem is given in Fig. 5-8a. The optimum configuration for this quadratic assignment problem is shown in Fig. 5-8b, and the otptimum configuration for the placement problem is shown in Fig. 5-8c. A chain interconnection rule with a rectilinear metric is assumed. We leave it as an exercise for the reader to verify that the respective configurations are optimum.

	A	B	C	D	E	F
A	0	3	2	1	2	1
B	3	0	1	0	2	1
C	2	1	0	0	1	0
D	1	0	0	0	0	0
E	2	2	1	0	0	1
F	1	1	0	0	1	0

Fig. 5-8a. Connection matrix.

	E	F
C	A	B
	D	

Fig. 5-8b. Optimum configuration for associated quadratic assignment problem.

E	F	
C	A	B
	D	

Fig. 5-8c. Optimum configuration for placement problem.

5.5 CONSTRUCTIVE INITIAL-PLACEMENT METHODS

In this section we present a general discussion of the class of initial-placement techniques and then illustrate this class with two algorithms due to Kurtzberg [32], [35], the pair-linking method and the cluster-development method.

5.5.1. General Discussion

In constructive initial-placement methods the placement configuration is formed by adjoining modules to a subset (*nucleus*) of already placed modules. The methods in this class accept as input a list of pins which must be interconnected (the signal sets), the module and backplane dimensions, and pertinent control parameters. They then iteratively operate upon the set of unplaced modules, selecting one of them and positioning it in the partially formed backplane configuration. Once modules are positioned, they are not moved. These methods have the virtue of requiring a comparatively small amount of computation time and yet are sufficient for many applications. The particular rules for *selection* (or *ordering*) and *positioning* of the modules define the specific methods.

Selection Rules. To determine the order of module selection, some measure of the connectivity of the modules is required. That is, the sequencing of modules for placement is a sequential decision rule based upon some notion of how "strongly" unplaced modules are "bound" to modules already placed. We make this notion explicit after we develop some terminology and notation.

p_k is the weight associated with the kth signal set.

J_k is the number of modules in the kth signal set.

$S(X)$ is the set of signal sets of which module X is a member.

S_k is the kth signal set.

$M(X)$ is the set of modules adjacent to X, i.e., all modules in common signal sets with X.

A^j is the set of placed modules in the partial layout configuration after the jth iteration of the selection rule.

B^j is the set of unplaced modules after the jth iteration of the selection rule.

p_k^j is the number of placed modules of the kth signal set in the partial layout after the jth iteration of the selection rule.

Sec. 5.5 Constructive Initial-Placement Methods 227

$$\alpha_k^j = \begin{cases} 1, & \text{for } 0 < p_k^j < J_k \\ 0, & \text{for } p_k^j = 0, J_k \end{cases}$$

($\alpha_k^j = 1$ if and only if, after the jth iteration, there is at least one module of the kth signal set placed and at least one module of that set unplaced.)

$R^j(X) = \{k \mid k \in S(X) \text{ and } \alpha_k^j = 1\}$.

A natural measure of connectivity between two distinct modules, X and Y, is the weighted number of signal sets common to X and Y. We denote this measure by

$$\mathbf{P}(X, Y) = \begin{cases} \sum_{k \in S(X) \cap S(Y)} p_k, & \text{for all } X \neq Y \\ 0, & \text{for all } X = Y \end{cases}$$

Note that $P(X, Y) = c_{ij}$ in the cost matrix of the associated quadratic assignment problem, where the modules X and Y correspond to the vertices i and j. Clearly $P(X, Y)$ is symmetric in X and Y.

An extension of this measure of connectivity $P(X, Y)$ is one which takes into account the signal set size J_k. We accomplish this by defining

$$\mathbf{P}_\lambda(X, Y) = \begin{cases} \sum_{k \in S(X) \cap S(Y)} \left(\dfrac{J_k + \lambda}{J_k}\right) p_k, & \text{for all } X \neq Y \\ 0, & \text{for all } X = Y \end{cases}$$

λ is a parameter determining the relative importance of the size of each signal set. The greater λ is, the larger is the influence of module pairs that belong to small signal sets, whereas a negative λ increases the influence of the large sets. Since the smallest value of J_k is 2, a lower bound on λ is -1.

Note that $P_0(X, Y) = P(X, Y)$. Here, and in the remainder of this chapter, for notational simplicity we frequently use $P(X, Y)$ rather than $P_\lambda(X, Y)$ if it is clear tht either can be employed.

One selection function based on $P(X, Y)$ which can be used in the selection rule is

$$f_1(X) = \max_{Y \in A^j} P(X, Y)$$

where $f_1(X)$ is defined for all $X \in B^j$. Then a selection rule is simply choosing the module X that corresponds to the maximum value of $f_1(X)$. In effect, this amounts to choosing that unplaced module which has the highest connectivity (in the sense defined) to any module already placed. The initial module(s) must be chosen by another selection rule, e.g., select modules X and Y given by max $P(X, Y)$, or be prespecified manually. These selection functions are used in the pair-linking method given in Section 5.5.2.

Another selection function based upon $P(X, Y)$ which incorporates the connectivity from an unplaced module X to all modules already placed is

$$f_2(X) = \sum_{Y \in A^j} P(X, Y), \quad \text{for all} \quad X \in B^j$$

Again, the selection rule is choosing that module X which corresponds to the maximum $f_2(X)$. This amounts to choosing that unplaced module which has the highest connectivity to all modules already placed.

Note that $P(X, Y)$ is independent of the sets A^j and B^j; that is, all $P(X, Y)$ can be computed prior to any iteration of the selection rule. The functions $f_1(X)$ and $f_2(X)$, however, are dependent upon A^j since we sum $P(X, Y)$ over all $Y \in A^j$.

Alternatively, we may define a selection function based upon the number of signal sets to which X belongs and which are partitioned into nonempty subsets of A^j and B^j. For each module in B^j we count how many of its signal sets have at least one module in A^j. We designate this count by $f_3(X)$, and by definition of α_k^j

$$f_3(X) = \sum_{k \in S(X)} \alpha_k^j, \quad X \in B^j$$

We select the module X with the maximum $f_3(X)$ value.

Another selection function based upon the same notions as f_3, but which weights the signal set, is obtained by defining

$$f_4(X) = \sum_{k \in S(X)} p_k^j, \quad X \in B^j$$

This selection function counts the actual number of modules in each signal set that has at least one module in A^j.

It should be noted that f_2 and f_4 are related. Specifically, it can be shown that $f_2(X) = f_4(X)$ if all p_k in $P(X, Y)$ are equal to unity. Alternatively, if we define

$$f_4 = \sum_{k \in S(X)} p_k^j p_k$$

then $f_2(X) = f_4(X)$ for all p_k. We leave the proof of this as an exercise for the reader.

Possibly more appealing than f_4 is an extension which takes into account the size of the signal set under consideration. As an example we define

$$f_5(X) = \sum_{k \in S(X)} \frac{p_k^j}{J_k}, \quad X \in B^j$$

In general, we weight the p_k^j by the associated size, J_k, of the kth signal set, thereby obtaining a relationship between the number of placed modules of the signal set and their size.

We can further refine the relationship between the number of placed modules and the signal set size by considering individual signal sets. To this purpose we define the vectors

$$q(X) = (J_{k_1}, J_{k_2}, \ldots)$$

and

$$q^j(X) = (p^j_{k_1}, p^j_{k_2}, \ldots)$$

enumerating over all $S(X)$, i.e., k_i is taken over all signal sets to which X belongs.

A new type of selection function based on the above vectors is the vector function

$$f_6(X) = q(X) - q^j(X)$$
$$= (J_{k_1} - p^j_{k_1}, J_{k_2} - p^j_{k_2}, \ldots)$$

A typical selection rule defined on f_6 would be to choose that module X which corresponds to the vector that has a minimum nonzero component.

Although other meaningful selection functions and rules can be defined, the selection procedures just discussed generally cover the generic types of initial-placement selection rules. The particular application, characteristics of the signal set, and amount of computer time available dictate the proper choice of selection functions and rules. For example, a selection rule based on the function f_1 uses considerably more computation time than does a selection rule based on the function f_2. A selection rule based on f_1, however, in general yields better results than a selection rule based on f_2 for placement of modules on boards. Such is not necessarily the case for placement of gates on modules. The reason for this is that gates typically belong to three to five signal set†, which implies a high degree of independence amongst the gates. The function f_1 is not designed for this situation. Selection rules based on f_2 and f_3 have been tried for this application [14].

It is clear that we need not search all of B^j when applying a selection rule. In fact, if a module has none of its signal sets partitioned, i.e., all of its signal sets are entirely in B^j, then this module is not a candidate for placement at the next iteration of the selection rule. The set of modules that are candidates is given the name *candidate set* and is designated by C^j. There are several ways to express C^j, one of which is

$$C^j = \{X \in B^j \,|\, f_3(X) > 0\}$$

†Modules on the other hand typically belong on the average to approximately 20 signal sets for a small problem of 100 modules and 530 signal sets. The average value of J_k for module signal sets is between 4 and 5 with the maximum value of J_k going perhaps to 60.

In general, a selection rule does not yield a unique module. In this situation a *tie-breaking rule* is necessary. There are three general approaches to tie-breaking: (a) look-ahead μ steps, (b) optimize on a secondary function and (c) choose arbitrarily.

As an example of a look-ahead rule, consider the selection rule which chooses the module in C^j corresponding to the maximum of f. If there is more than one module corresponding to this maximum value, these modules are tentatively placed—one at a time—and f is computed for the new candidate sets. The module is then chosen which would produce the largest f value in the new candidate sets. If a tie persists, either a choice is made arbitrarily (one-step look-ahead) or else the process is repeated. This method is used in the pair-linking method in Section 5.5.2.

As an example of the second approach to tie-breaking, a different selection function, say $f_k(X)$, is computed for all modules X corresponding to the maximum value of f. The module with the maximum f_k value is then chosen. This method of tie-breaking is used in the cluster-development method.

The selection rules just described are all independent of the positioning rule; hence, the modules can be completely ordered, if desired, before the positioning rule is applied. It is possible to devise selection rules which depend on the relative positions of the modules on the board, but these are more complex.

For future reference we now introduce the function $Q(X)$ which is the sum of the entries in each row of P, i.e.

$$Q(X) = \sum_Y P(X, Y)$$

Equivalently, $Q(X)$ is the sum of the edge weights of all edges incident to module X in the module connection graph (Fig. 5-7).

In some applications the function $Q(X)$ is desired without need for $P(X, Y)$. In this case it is computationally simpler to use the relation

$$Q(X) = \sum_{k \in S(X)} (J_k - 1)p_k$$

This equation follows by substituting the value of $P(X, Y)$ into the preceding one for $Q(X)$ to obtain

$$Q(X) = \sum_{Y \in M(X)} \sum_{k \in S(X) \cap S(Y)} p_k$$

and noting that p_k is added to the expression for $Q(X)$ for every $Y \neq X$ in the kth signal set of $S(X)$. But there are exactly $(J_k - 1)$ modules not equal to X in the kth signal set, and thus the desired equation follows.

We now introduce a measure on the modules which is used in Section 5.5.3, namely

Sec. 5.5   Constructive Initial-Placement Methods   231

$$P_\lambda(X) = \sum_{k \in S(X)} \frac{J_k + \lambda}{J_k} p_k$$

Let $P(X) = P_0(X) = \sum p_k$. If $p_k = 1$ for all k, then $P(X)$ is equal to the number of signal sets of which X is a member, i.e., $P(X) = |S(X)|$. It follows that $P(X)$ and $Q(X)$ are related by the equation

$$P(X) + Q(X) = \sum_{k \in S(X)} J_k p_k$$

For $\lambda \neq 0$ this generalizes to

$$P_\lambda(X) + Q_\lambda(X) = \sum_{k \in S(X)} (J_k + \lambda) p_k$$

where

$$Q_\lambda(X) = \sum_Y P_\lambda(X, Y)$$

We leave the proof of this generalization as an exercise for the reader. If $J_k = 2$ for all k, that is if the placement problem reduces to the quadratic assignment problem, then $P_\lambda(X) = Q_\lambda(X)$.

As a numerical illustration, we compute $Q(X)$ and $P(X)$ for the example of Section 5.2.2 and present the results in Fig. 5-9. In Fig. 5-10 we show the results of computation of all the selection functions on the same eight-module problem. For that computation, assume $\lambda = 0$ and A^j contains A, D and F after the jth iteration of the selection rule. Figure 5-11 presents the redrawn graph of Fig. 5-2, showing only those modules in A^j and C^j. By inspection of Fig. 5-11 and some elementary computation, the values in Fig. 5-10 are obtained.

	$Q(X)$	$P(X)$
A	11	5
B	21	10
C	19	9
D	6	3
E	8	4
F	9	4
G	10	5

Fig. 5-9. P and Q values.

	f_1	f_2	f_3	f_4	f_5
B	5	6	2	3	$\frac{5}{6}$
C	4	8	2	4	$\frac{7}{8}$
E	4	4	1	1	$\frac{1}{3}$
G	0	0	0	0	0

Fig. 5-10. Evaluation of selection functions.

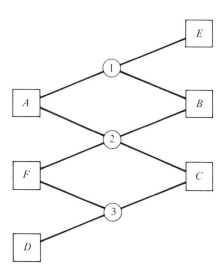

Fig. 5-11. Partial bigraph.

Positioning Rule. In positioning a module we wish to find the best slot for it relative to the modules already positioned. Clearly, it is not necessary to examine all empty slots. Only a comparatively small subset of slots, termed *candidate slots*, can possibly yield the minimum cost. These sets of candidate slots may be accumulated along with the positioning of each new module, or an appropriate set may be regenerated for each positioning.

For each module to be positioned, we compute a cost over the candidate slots which is related to the net distance of all its signal sets. We then position the module in that slot which yields the minimum cost. Ideally the cost function is the actual interconnection rule. In practice, however, this cannot always be done. In general, the computation time would be excessive, and since not all modules are placed, the optimal interconnection cannot be determined at that point. Thus, an approximation rule is frequently adopted. For example, a simple direct linking rule can be used as an approximation for both minimum spanning tree and chaining. In this rule, module X is positioned closest to a placed module Y, where module Y is that module for which $P(X, Y)$ is maximum.

An auxiliary positioning rule can be further used to discriminate a slot among the subset of candidate slots given by the primary approximation positioning rule. For example, any ties among the slots can be resolved by taking the resultant weighted center-of-location of modules which are in common signal sets with module X. These approximation rules are used in the pair-linking and cluster-development methods defined in Sections 5.5.2 and 5.5.3.

To further fix these ideas let us consider another positioning technique for a minimum spanning tree interconnection rule. As each module is positioned, the minimum spanning tree distance for all partially placed signal sets common to that module is computed for every candidate slot. The module is placed in the slot which yields the minimum total distance for all nets. The main disadvantage of this method is that the computing time increases rapidly with large signal sets. The computation time increases by the square of the number of elements in a signal set and linearly with the number of sets.

To decrease the computing time, a simple approximation to the minimum spanning tree can be used. For each candidate slot we compute the distance to the closest module in each partially placed signal set of the module to be placed, and we choose that slot which gives the minimum total distance.†

The candidate slots are generated by adding a predetermined pattern of locations to this subset as each module is positioned. The simplest such pattern is the four adjacent slots of each newly placed module. (As each slot is occupied, it is deleted from the candidate set, and the slots already occupied are never reemployed. Also, unusable slots are never added to the set.) By choosing this simple pattern, the periphery of slots of positioned modules is always in the candidate set. It is not difficult to see that the best location of all unused slots must be on this periphery. Variations of this technique have also been used. As an example see Freitag and Hanan [14].

5.5.2. Pair-Linking Method

The selection rule on which the pair-linking method [32], [35] iterates to choose the next module for positioning in the partial layout is based upon the f_1 selection function of Section 5.5.1. The implications of each choice are taken into account to resolve any ties. The unplaced module is then positioned as close as possible to its already placed mate. Location ties in the positioning are resolved by means of a weighted average of coordinates of those modules already placed and to which the connection pins of the new module must be electrically common. The layout is started by choosing as a nucleus the pair of modules that has the largest sum of common weighted signal sets, i.e., $\max_{x,y} P(X, Y)$. A subset of modules may be manually prepositioned. If such is the case, these modules serve as the nucleus, and the "main body" of the method can then be entered with an appropriate adjustment of various tables and fixing of board boundaries.

†As reported in [14], this approximation and the minimum spanning tree were both employed in an application where gates were placed on modules. For small problems (up to 100 gates) there was an insignificant difference in the final results. For larger problems (550 gates) the total distance increased by about 10% when using the approximation, but the computation time decreased by an order of magnitude. It took about one minute of IBM 7094 time to place 543 gates (largest signal set had 25 gates) using the minimum spanning tree in the positioning rule.

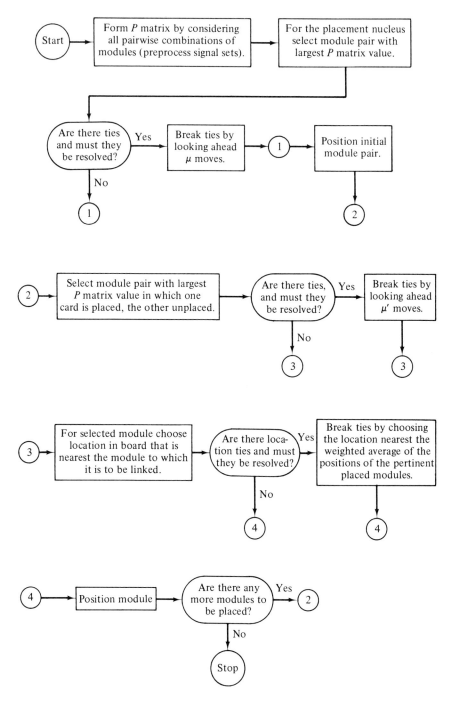

Fig. 5-12. Flow diagram of the pair-linking method.

An outline of the algorithm is given in the flow diagram in Fig. 5-12. The method is defined in detail on an operational level by the following five steps.

Step 1: P matrix formation. The symmetric $n \times n$ P-matrix defined in Section 5.5.1 is constructed where each entry is given by

$$P_\lambda(X, Y) = \begin{cases} \sum_{k \in S(X) \cap S(Y)} \left(\dfrac{J_k + \lambda}{J_k}\right) p_k, & \text{for all } X \neq Y \\ 0, & \text{for all } X = Y \end{cases}$$

J_k is to be interpreted as the number of *distinct* modules in a signal set. Thus, the list of connection pins comprising a signal set must be preprocessed to ensure that only the number of distinct modules determines J_k and that $J_k \geq 2$.

As noted in Section 5.5.1, the choice of λ is a function of the importance of large vs small signal sets. A neutral choice of $\lambda = 0$ simplifies the P matrix computation and is suitable for most applications.

Step 2: Selection of initial module pair. The largest entry in the P matrix is selected. This entry represents a module pair which is then positioned in the backplane. (For details of the positioning, see step 3.)

In the event of ties for the largest entry in the P matrix, the candidate entries are entered in a tie table. The *strips*† X_j and Y_j in the P matrix given by the first tie table entry are searched for the largest entry exclusive of (X_j, Y_j). The search for a larger entry to replace that candidate entry then continues along the strips given by the next tie table entry, and so on, until all the entries in the tie table are exhausted. The tie table entry (X_k, Y_k) is chosen which has, in one of the strips X_k and Y_k, the largest entry thus found. In effect, this is looking a move ahead, that is, the initial pair is selected which tends to maximize the value of the choice of the next module.

If there are ties resulting from the search of strips given by the tie table, the process may continue recursively; that is, a subtie table is formed, and the process just described is repeated. The maximum entry given by the subtie table determines which entry in the tie table is to be chosen. Logically, this process of

---

†The term *strip* is used for brevity. Since only $n(n-1)/2$ elements need be stored, the search proceeds down a column of the P matrix until it is reflected across the row by the boundary of the main diagonal. Thus, the strip X_j corresponds to the row (or column) X_j in a completely stored matrix sans the diagonal elements. In a large sparse matrix only the values and positions of the nonzero elements in a row (column) are stored.

subtables may continue until all ties are resolved, or until we look μ moves ahead and arbitrarily select one of the candidates in the μth tie table. (The condition $\mu = 0$, of course, denotes that no tie table is formed and that an arbitrary choice of one of the largest P matrix entries is to be made.) The degree of independence of the signal sets, i.e., the number of independent modules, largely determines the value of looking ahead.

Figure 5-13 illustrates the selection of the initial module pair

	A	B	C	D	E
A		10	10	4	6
B			1	5	5
C				3	2
D					10
E					

1st Tie Table   2nd Tie Table

(AB)   (AC)

(AC)   (AB)

(DE)

Fig. 5-13. Tie-breaking (pair-linking initial pair selection).

in which two entries in the same strip are tied. Assume that the P matrix shown in Fig. 5-13 has been formed for a five-module problem, and that μ is set equal to two. Since μ is equal to two, the tie-breaking process is truncated with the formation of the second tie table. The first entry in that table is then selected, namely, (A, C). This entry resulted from a search of the strips A and B given by the entry (A, B) in the first tie table. Thus the module pair (A, B) is selected for the nucleus. Future module selections are linked to one of the placed modules (see steps 4 and 5). Module C would be selected next and linked to module A. Intuitively it thus seems preferable in this example to start with a module pair such as (A, C) or (A, B) rather than (D, E).

Step 3: Positioning of initial pair. The pair of modules selected to serve as a nucleus is now positioned in the "center" of an array in computer memory representing the $p \times q$ board grid, one immediately next to another in the closest neighboring location. At this point we note that best results are obtained with the use of a multiplicative factor N ($0 \leq N \leq 1$) in the measurement of distance along the long axis of the board. This parameter compensates for unequal dimensions of the board and biases the placement of modules toward the long axis. For example, if

$p \geq q$, then multiplying the long dimension by $N = q/p$ transforms, loosely speaking, a rectangular board into a square one.

Step 4: Selection of unplaced modules. The selection of the unplaced modules proceeds by selecting a pair of modules, one of which is placed and one unplaced, using the P matrix. This is done using the strips of the placed modules† in the P matrix. On those strips, the largest entry associated with an unplaced module is selected. That is, any entry that represents the connection of a placed module with another placed module is ignored. In terms of Section 5.5.1, we search $[P_\lambda(X, Y)]$, where $Y \in A^j$ and $X \in C^j$, and select the module in C^j corresponding to the maximum $P_\lambda(X, Y)$. In short, we use the selection function f_1.

In the event of ties, a slight modification of the tie-breaking scheme adopted for the selection of the initial pair is used. As before, the strips given by the candidate entries in the tie table are searched. This time, however, each entry (X_1, Y_1) in the tie table represents an unplaced and a placed module. The strips of the placed modules are ignored. Furthermore, in the strips of the unplaced modules, only those entries that denote a connection from an unplaced module to another unplaced module are considered. These entries, if more than one, may in turn be entered into a second tie table. The entries in this tie table are quite different from those in the first tie table in that they represent not one but two unplaced modules. However, in this table, even though both modules are unplaced, only the strip of one module is searched—the strip that represents the module not given by the associated entry in the previous tie table.

In effect, we are examining the implication of treating one of the modules as if it had been positioned. Thus, in a certain sense, one module given by the entry in the second tie table is pseudoplaced, while the other is unplaced in any sense. In this manner of generating couplets of a pseudoplaced module and an unplaced module, new tie tables may be formed. As before, these levels of tie tables may be truncated on the μth level, or upon cessation of ties.

Step 5: Positioning of the selected module. The module selected for placement should be positioned as closely as possible to the

---

†For computational efficiency, as soon as more than half the n modules are placed, the strips of unplaced modules, rather than placed modules, should then be searched for the largest entry representing an unplaced and a placed module.

placed module associated with it. The partial layout is allowed to grow in the horizontal direction until it is p cells wide, and in the vertical direction until it is q cells long. At that time the layout boundaries are then fixed. This floating boundary scheme increases the likelihood for paired modules being placed close to one another.

In the event two or more locations are equidistant from the placed module, the following positioning tie-breaking scheme is adopted. A weighted average is formed of the coordinates (locations) of those modules already placed to which the new module must be connected. This process defines a point $(\bar{u}, \bar{v})$, and the candidate location nearest to this point is chosen. The weighting factors for the appropriate locations are given by the associated entries in the P matrix. Thus to position module Y, it is necessary to evaluate

$$\bar{u} = \frac{\sum_{X \in A^j} P_\lambda(X, Y)(u_X + K)}{\sum_{X \in A^j} P_\lambda(X, Y)} - K$$

$$\bar{v} = \frac{\sum_{X \in A^j} P_\lambda(X, Y)(v_X + K)}{\sum_{X \in A^j} P_\lambda(X, Y)} - K$$

where (u_X, v_X) are the coordinates of the module X. The constant K is used to compensate for the u and v numbering starting from zero.

Timing for the Pair-Linking Method. The amount of processing time required for the pair-linking method is directly a function of the number of modules n, the total number of signal sets, and the frequency distribution of their lengths J_k. The dominating factor, however, is the number of modules, and it is shown in [35] that the order of growth of the pair-linking method is n^2.

Example: We illustrate the pair-linking method with an eight-module problem which is also used for other methods throughout this chapter. We assume a chain interconnection rule with rectilinear distance. Figure 5-14 shows the three signal sets.

$$S_1 = \{A_1, B_1, C_1, E_1, H_1\}, \quad p_1 = 1, \quad J_1 = 5$$
$$S_2 = \{B_2, D_1, E_2, G_1\}, \quad p_2 = 1, \quad J_2 = 4$$
$$S_3 = \{B_3, D_2, F_1\}, \quad p_3 = 1, \quad J_3 = 3$$

Fig. 5-14. Signal sets for an eight-module sample problem.

We take $\mu = 0$, $\lambda = -1$, and $p_k = 1$ for all k. The entries of the P matrix, given in Fig. 5-15, are then defined by

$$P_{-1}(X, Y) = \sum_{k \in S(X) \cap S(Y)} \frac{J_k - 1}{J_k}$$

	A③	B①	C④	D②	E①	F⑦	G⑥	H⑤
A③		0.8③	0.8④		0.8			0.8⑤
B①	0.8		0.8	1.42②	1.55①	0.67⑦	0.75⑥	0.8
C④	0.8	0.8			0.8			0.8
D②		1.42			0.75	0.67	0.75	
E①	0.8	1.55	0.8	0.75			0.75	0.8
F⑦		0.67		0.67				
G⑥		0.75		0.75	0.75			
H⑤	0.8	0.8	0.8		0.8			

Fig. 5-15. The P matrix.

The largest element in the P matrix, (B, E), is selected and the pair of modules is positioned in the board (shown in Fig. 5-16). The next module selected is D, from the entry (B, D), which is positioned (linked) next to B in the board grid. The process continues with the encircled numbers in the P matrix in Fig. 5-15 showing the order of module selection, and the encircled numbers on the modules in the slots in Fig. 5-16 showing the order of pair-linking. The configuration has a cost of ten units. An optimal placement has a cost of nine units (see Fig. 5-21).

An exercise for the reader is to find a placement configuration using the pair-linking method with $\lambda = 0$.

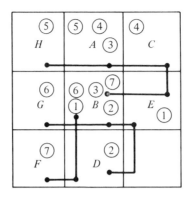

Wire length = 4 + 4 + 2 = 10 units

Fig. 5-16. Placement configuration by pair-linking.

5.5.3. Cluster-Development Method

In the cluster-development method [32], [35], a measure of the expected number of interconnections to the placed modules for each unplaced module is computed as a function of the type of interconnection tree assumed. The unplaced module with the largest value is selected for placement. In the event of ties, the candidate module that is in the most (weighted) signal sets is selected. That module is then positioned in the weighted center-of-gravity of those modules already placed to which its connection pins must be electrically common. New expected interconnection values for the reduced set of unplaced modules are then recomputed and the process re-iterates.

The following five steps implement the selection and positioning rules. Figure 5-17 presents a general flow diagram of the process.

Step 1: P list formation. Each of the n modules to be placed is assigned a number $P_\lambda(X)$, thereby forming the P-list defined in Section 5.5.1. Recall that the number $P_\lambda(X)$ for module X is calculated by

$$P_\lambda(X) = \sum_{k \in S(X)} \frac{J_k + \lambda}{J_k} p_k$$

The list of pin specifications is preprocessed to determine J_k as in the pair-linking method. The neutral choice of $\lambda = 0$ is appropriate for most applications.

Step 2: Selection of the initial module. The module selected for initial positioning is the one with the largest P-number. If two or more P-table entries are tied for largest, one is arbitrarily selected.

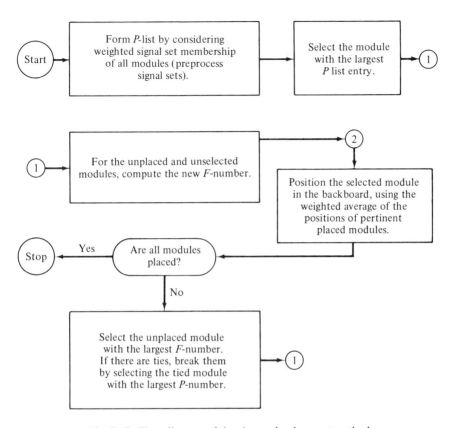

Fig. 5-17. Flow diagram of the cluster-development method.

This module is positioned in step 4. As in the pair-linking method, a subset of modules may be manually prepositioned; the same remarks made for that method also hold here.

Step 3: F list computation. After each new module selection, a number $F(X)$ is computed for each unplaced module X. This is done for the selection of new modules. These numbers $F(X)$ are a measure of the expected number of lines in the interconnection net from the set of unplaced modules to the set of placed modules.

The formula for $F(X)$ is suggested by the following. For every partially placed signal set k, there must be at least one line leading from the unplaced modules to the placed modules. Let T_k be the maximum possible number of such lines in k. Where the signal set is to be interconnected by a chain, T_k is given by

$$T_k = \begin{cases} \min [2p_k, 2(J_k - p_k)], & p_k \neq \dfrac{J_k}{2} \\ 2p_k - 1, & p_k = \dfrac{J_k}{2} \end{cases}$$

Figure 5-18 shows all chain net interconnection crossings from A^j to B^j for the case of $J_k = 7$, $p_k = 2$. (For other types of interconnection nets, T_k would be given by different functions. For example, with minimal spanning trees, $T_k = J_k - 1$.)

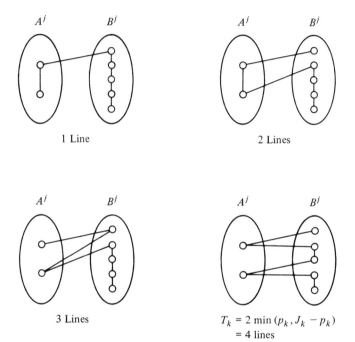

Fig. 5-18. Chain interconnection crossings from A_j to B_j, where $j_k = 7$, $p_k = 2$.

For computational simplicity, the approximation is made that all possible numbers of lines between the two partitions are equally likely. This leads to the expression

$$\frac{1}{T_k} \sum_{j=1}^{T_k} j = \frac{T_k + 1}{2}$$

for the expected number of lines.

Further, we consider the subset of modules in B^j that can be connected to modules in A^j. The maximum number of elements in B^j with this property is T_k since at most that number of lines "cross" the two partitions, and the minimum number is one. Assuming all cases are equally likely, we take $T_k/2$ as the average number of modules in B^j that are connected to modules in A^j. Thus, in a particular signal set k, the expected number of lines per module in B^j that can possibly be connected to modules in A^j is

$$\frac{1/T_k \sum_{j=1}^{T_k} j}{T_k/2} = \frac{T_k + 1}{T_k}$$

For $T_k = 0$ (all the modules in a set are either placed or unplaced, where the newly selected module is treated as if it were actually placed), the contribution is 0. With the this in mind, we define†

$$F(X) = \sum_{R^j(X)} \frac{T_k + 1}{T_k}$$

$R^j(X)$ denotes that the summation is to be taken over all signal sets k in which X is a member and which are partially placed, i.e., $0 < p_k < J_k$. This latter condition ensures that $T_k \neq 0$.

Step 4: Positioning of modules. As in the pair-linking method, the module initially selected is positioned in the "center" of an array in computer memory representing the board grid. For positioning of all other modules, the rule is essentially the same as the pair-linking rule for location tie-breaking. The only difference is that the P list is used to obtain the weighting factors for the locations of those modules already placed and whose pins are electrically common to those on the new module.

Step 5: Selection of modules. The module with the largest F number is selected next. If there is more than one module with the largest value, the candidate module with the largest P number is selected. If there are still ties, an arbitrary choice is made. Steps 3, 4 and 5 are reiterated until all modules are placed.

Timing for Cluster-Development Method. The amount of processing time required for the cluster-development method is a function of the same

†Another candidate for the module base in the expected line averaging is the use of the entire set of $J_k - p_k$ modules in B^j. The alternate expression for F is then $(T_k + 1)/2(J_k - p_k)$. Note that the two expressions for F are identical for $p_k \geq J_k - p_k$. For the special case where $J_k = 2$ for all k (as in the Steinberg 34-module problem reported on in Section 5.8.1), both expressions for F yield precisely the same ordering of modules.

variables as for the pair-linking method. It is shown in [35] that the timing equation is also of the order of n^2. Although this is the same order of magnitude as that for the pair-linking method, the processing time of the steps are much smaller than the corresponding time for the pair-linking subroutines (see Section 5.8.1 for time comparisons).

Example: We illustrate the cluster-development method with the eight-module problem defined by the signal sets shown in Fig. 5-14, with all $\rho_k = 1$. We take $\lambda = -1$, and then the values in the P list are defined by

$$P_{-1}(X) = \sum_{k \in S(X)} \frac{J_k - 1}{J_k}$$

The P list is shown in Fig. 5-19.

A	0.8
B	0.8 + 0.75 + 0.67 = 2.22
C	0.8
D	0.75 + 0.67 = 1.42
E	0.8 + 0.75 = 1.55
F	0.67
G	0.75
H	0.8

Fig. 5-19. P list for eight module sample problem.

Module B is used to initiate the layout, and the F list is computed for the first iteration. The F list is shown in Fig. 5-20, where the columns

	1	2	3	4	5	6	7	8
A	1.5	1.25	1.25	1.25	1.25	X	X	X
B	X	X	X	X	X	X	X	X
C	1.5	1.25	1.25	1.25	1.25	1.25	X	X
D	3.0	2.75	X	X	X	X	X	X
E	3.0	X	X	X	X	X	X	X
F	1.5	1.5	1.5	1.5	X	X	X	X
G	1.5	1.25	1.5	X	X	X	X	X
H	1.5	1.25	1.25	1.25	1.25	1.25	1.5	X

Fig. 5-20. F table for eight module sample problem.

in the table give the F numbers for each iteration. The module with the largest F number, namely E, is then selected for placement and positioned next to B; then the F list is recomputed. The process repeats, and each time a module is selected and positioned, an X is placed in its position in the F list. The placement configuration is given in Fig. 5-21,

Sec. 5.6 Iterative Placement-Improvement   245

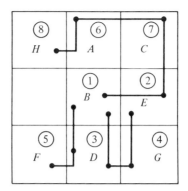

Fig. 5-21. Placement configuration by cluster-development.   Wire length 4 + 3 + 2 = 9 units

with the encircled numbers denoting the order of module selection and positioning. The layout produced is optimum with a cost of nine units.

We leave it as an exercise for the reader to find a placement configuration using the cluster-development method with $\lambda = 0$.

5.6 ITERATIVE PLACEMENT-IMPROVEMENT

In general, iterative placement-improvement methods demand more computer time than the constructive initial-placement methods but are potentially capable of producing better layouts.

The algorithms in this category operate in iterative fashion upon a placement configuration. Typically a subset of modules is selected for change of position along with the associated candidate locations. The modules are then deterministically repositioned so as to reduce the total wire length. Each stage of the operation of the algorithm either results in a new placement with a lower cost or in the retainment of the old configuration.

5.6.1. General Discussion

We can formulate the class of placement-improvement algorithms in a general, abstract sense by the following five steps.

Step 1: Generate initial parameter set.
Step 2: Apply transformation.
Step 3: Compute cost.

Step 4: Modify parameter set on basis of cost.
Step 5: Apply stopping rule with outcomes:
(a) go to step 2;
(b) go to step 1;
(c) stop.

The *parameter set* consists of a placement configuration, a subset of modules, and a subset of slots. The *transformation* is comprised of an *operator* and a parameter set, where the operator specifies the mapping by which the modules of the parameter set are repositioned into the slots given by the parameter set. Hence a layout is mapped into a new configuration by the transformation. The cost of the new layout is then computed and compared to the old. If the cost of the new layout is reduced, then the new configuration is substituted for the old one in the parameter set; otherwise the old placement is retained. A new subset of modules and slots is then generated by some prespecified rule and substituted into the parameter set.

The *stopping rule* specifies conditions for branching to step 2, step 1 or stopping. These steps are equivalent to determining whether to: (a) continue modifying a given placement; (b) generate a new initial placement to restart placement-improvement iterations; or (c) terminate the procedure.

The basis for the decision to terminate the "loop" of step 2 rests upon the rate of improvement in the cost or upon the existence of a predetermined number of transformations without improvement, i.e., without the generation of a new placement. The decision to go to step 1 or stop is a function of the time available. Revisiting step 1 results in a new starting point, i.e., a new initial placement, for subsequent modification. Each initial placement results in an improved placement. The costs of all these placements are compared and the best placement is chosen.

The resultant placement (for each initial placement) is *locally optimum* in the sense that there does not exist a placement with a smaller cost for the given initial parameter set, transformation, parameter set modification rule and stopping rule.

We can generate a large number of locally optimum placements, thereby increasing the probability that at least one of these placements is close to the global optimum. Alternatively, we can find better locally optimum placements, thereby increasing the probability that each initial placement leads to a placement close to the optimum. The formulation of each particular placement-improvement algorithm represents a choice of tradeoffs for these alternatives. These choices range from the simple pairwise interchange to the sophisticated Steinberg assignment technique (defined, respectively, in Sections 5.6.4 and 5.6.2).

To illustrate, a simple transformation such as pairwise interchange yields more locally optimum points (placement configurations) for a given amount of computation time than does a more complicated transformation such as trial interchange of three modules. This triple interchange method in general yields better local optimums than the pairwise interchange method, but at the expense of considerably more computation time.†

We also note that, for a given transformation, there exists a choice of methods to secure initial placements. The simplest is random generation which leads to the greatest number of locally optimum points. However, more elaborate initial placements can lead to better locally optimum points.‡

The methods discussed in this section are a subset of a larger class of methods known as *local optimization techniques* which have been employed to solve other optimization problems. For example Shen Lin [40], Reiter and Sherman [50], and Steiglitz and Weiner [52] have successfully applied such techniques to the traveling salesman problem.

5.6.2. Steinberg Assignment Method

In Section 5.4 the relationship between the linear assignment, the quadratic assignment and the placement problems is discussed. This relationship is exploited by a method due to Steinberg [53] and expanded on by Rutman [51]. Steinberg considered a simplified model of the placement problem, namely the quadratic assignment problem, in which all signal sets contained only two modules. Rutman made the natural extension to the general placement problem and computed actual wire length over the signal sets to determine a placement cost.

Recall that in the linear assignment problem, in which we assign n men to n jobs, each man did not interact with other men in the computation of the individual man-job pairing cost. Such is not the case in the quadratic assignment or placement problems. However, we can still treat certain portions of the total placement problem, i.e., subsets of the modules to be placed, in the same fashion as in the linear assignment problem. The subsets of modules of interest must have the property that no two modules can be members of a common signal set. This property defines a set of *independent modules*§ and forms the basis of the "linearization" of the placement and quadratic assign-

---

†Garside and Nicholson [16] actually investigated the advantages of triple interchange over pairwise (double) interchange and came to the conclusion that the gain in local improvement did not compensate for the comparatively great increase in computation time.

‡Folklore has it that iterative placement-improvement methods are relatively insensitive to the starting point. Experience of others has suggested the opposite belief (which is also the view of the writers). Neither conjecture has been confirmed by a sufficient amount of realistic experiments.

§Steinberg, followed by Rutman, uses the phrase *mutually unconnected modules*. We term these sets *independent* to conform to the current graph theory terminology. (See Section 5.2.2.)

ment problems. The methods of generation of these independent sets are deferred to the end of this subsection.

Iteration Procedure. Assume that n modules are positioned in a board and interconnection nets are established. For the sake of exposition and without loss of generality, we assume that the modules in each signal set are interconnected by means of chains, i.e., no more than two wires on a module pin. We form sets of *maximal independent modules*† and sequentially operate upon them to find new placement configurations.

By definition, every module in an independent set is not adjacent, i.e., not to be connected, to any other member of the set. Consequently, a given module may be sequentially placed in all of the positions occupied by the modules in its independent set (assuming those modules to be removed from the board), and a wiring cost established for each placement. In computing this cost, only the connections of the given module to those in its signal sets need be considered; the other modules cannot possibly have any influence on that wire length.

If there are m modules in the independent set under discussion, occupying m places in the layout, an $m \times m$ matrix results. Clearly, the matrix may be augmented by associating dummy modules with empty slots. Each entry in this $m \times m$ matrix gives the wire length associated with placing the particular module in that particular location; dummy modules are assumed to have infinite cost for any location. Since all of the modules must be placed, the problem is one of assigning the modules such that the sum of m independent entries is minimal—the standard linear assignment problem.

A concise statement of an assignment algorithm for exact solution, the Munkres method [44], and an approximation method [33] is given in the Appendix. For a full discussion of the methods and of the assignment problem, the reader should see standard references such as [8], [13], and [54]. The tradeoff between employing approximation methods for large maximal sets or using an exact method for smaller sets, possibly nonmaximal, has not been explored.

As previously mentioned, an entry in the $m \times m$ matrix represents the wire length of the nets resulting from the associated module being in the given location. For that figure to be meaningful, it is necessary for new nets to be computed and the entry to be based on the new interconnection. Furthermore, it should be stressed that any measure of distance can be employed, and any topology imposed upon the interconnections.

After the set of independent modules is positioned, we consider the next set, and so on. For best results only maximal sets are operated upon, i.e., new sets of modules are not considered if they are proper subsets of other independent

---

†An independent set is *maximal* if the addition of any new module destroys the property of independence.

module sets. By so doing, the greatest choice of locations is made available for the subsets of modules. We emphasize, however, that the technique does not logically require maximal sets.

Considerable repositioning of modules could result after one pass through the list of maximal independent sets of modules. This repositioning could make available new candidate locations for any given set. In particular, the effect of tie solutions, which are arbitrarily broken, could offer alternate candidate locations. However, the previous placement is maintained unless there is an improvement. Thus, the process terminates if there is no reduction in the cost for $p - 1$ steps through p maximal independent subsets. The flow diagram for the Steinberg assignment method is shown in Fig. 5-22.

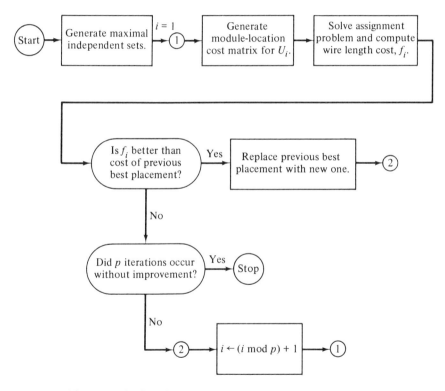

Fig. 5-22. The flow diagram of the Steinberg assignment method.

Generation of Maximal Independent Sets. We generate all the independent sets in recursive fashion. First, we generate all independent two-element sets. An efficient way of obtaining these sets, with the typical distribution of signal set sizes found in realistic placement problems (see first footnote in Section 5.5.1), is to generate all pairs of modules within each signal set and delete these from the set of all possible pairs of distinct modules.

We assume a list of module pairs to exist, say, in upper triangular form; hence, the deletions and accesses occur rapidly with a table lookup technique. Observe that the module pair (X, Y) is independent if and only if $S(X) \cap S(Y) = \phi$.

For $k > 2$, the $(k + 1)$-element independent sets are generated from k-element sets. We select a k-element independent set and seek another set of size k such that the intersection contains $k - 1$ elements. If such a set is found and if the pair of noncommon elements exists in the set of two-element independent sets, then we take the union of these two k-element sets to form the $(k + 1)$-element set. All k-element sets $(k > 2)$ that are subsets of the $(k + 1)$-element sets should be deleted. We leave it to the reader to prove that the cited conditions for such a $(k + 1)$-element set to exist are necessary and sufficient.

It is only realistic, though, to generate the set U of all maximal independent sets for small problems. For large problems we adopt the policy of generating a subset of U of size p and then treating the subset as the complete set. The maximal independent sets in the subset should possess a certain structure. For example, there should be a certain percentage of overlap from one set to another, and for computational efficiency, there should be a lower bound on the number of elements in the set. (See Rutman [51].)

These goals may be satisfied by the following generative procedure. For the first maximal independent set, U_1, consider the full set of modules, M, and randomly select an element in it. Put that module in the set U_1 and form the set M' by deleting all modules in M that are members of common signal sets (adjacent) with the module in U_1. Randomly select an element out of M', place it in U_1, and delete all modules in M' that are members of common signal sets with the newly selected element to form a reduced M'. The set U_1 is completely generated when the reduced M' is empty.

The $(k + 1)$ maximal independent set, U_{k+1}, is generated with the use of U_k by selecting at random β elements from U_k, where β is determined by the degree of overlap desired. These modules are placed in U_{k+1} and a new M' is formed as before by deleting the modules in common signal sets with the set of β elements. The remaining elements in U_{k+1} are selected from the reduced M' as is done for U_1.

If the number of modules in U_{k+1} is below the required number, then U_{k+1} must be re-formed and the process repeated with another random choice in U_k to initiate U_{k+1}. Further, the set U_{k+1} must be tested to ensure that it is not a previous U_i set. This "Monte Carlo" process continues until p maximal independent sets are generated.

Example: To illustrate the method, we show it operating upon our usual eight-module sample problem given in Fig. 5-14, assuming an initial layout configuration given in Fig. 5-23.

Sec. 5.6   Iterative Placement-Improvement   251

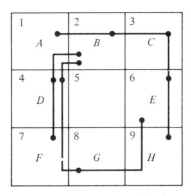

Fig. 5-23. Initial layout for the Steinberg assignment method.

Wire length = 4 + 6 + 3 = 13 units

We generate all the two-element independent sets (Fig. 5-24) by examining each module pair to determine whether they are common members of any signal set. We list those that are not. Next, three-element independent sets are constructed from the two-element sets. It is expedient to tag with asterisks all the two-element sets that are subsets of three-element sets. Since we cannot construct any four-element sets we are finished. The set of maximal independent sets (Fig. 5-25) consists of all the three-element sets and the nontagged two-element sets.

We form the cost matrix for our assignment problem by taking for

Two-element Sets

1. $\{A, D\}$
2. $\{A, F\}$*
3. $\{A, G\}$*
4. $\{C, D\}$
5. $\{C, F\}$*
6. $\{C, G\}$*
7. $\{D, H\}$
8. $\{E, F\}$
9. $\{F, G\}$*
10. $\{F, H\}$*
11. $\{G, H\}$*

Three-element Sets

1. $\{A, F, G\}$
2. $\{C, F, G\}$
3. $\{F, G, H\}$

Fig. 5-24. Set of independent sets.

1. $\{A, F, G\}$
2. $\{C, F, G\}$
3. $\{F, G, H\}$
4. $\{A, D\}$
5. $\{C, D\}$
6. $\{D, H\}$
7. $\{E, F\}$

Fig. 5-25. Set of maximal independent sets.

	1	7	8	5
A	4	5	4	4
F	2	3	4	2
G	4	5	6	4
Z	L	L	L	L

Cost Matrix—Iteration No. 1

(a)

Assignment	Cost
A:8	4
F: 1	2
G: 5	4

total cost = 10 units

Module-Location Assignment

(b)

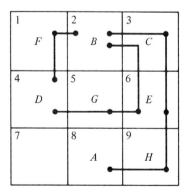

Wire length = 4 + 4 + 2 = 10 units

Layout from Placement Improvement—Iteration No. 1

(c)

Fig. 5-26.

our rows the modules in the first maximal independent set, and for the columns, their board locations in the initial placement. A dummy module, Z, associated with the empty location 5 with very large costs, L, is included for all the assignments of that module. The necessary wire-length cost computations are shown in Fig. 5-26a. Solving the resulting cost matrix for a minimal cost by an assignment algorithm yields the assignment shown in Fig. 5-26b. The total cost is ten units, and the resulting placement is shown in Fig. 5-26c.

Similarly, we treat the second maximal independent set using the new placement, and obtain the following cost matrix (iteration No. 2):

	1	3	5	7
C	5	4	4	5
F	2	3	2	3
G	4	4	4	5
Z	L	L	L	L

Iteration No. 2

and the following module-location assignment:

Assignment	Cost
C: 3	4
F: 1	2
G: 5	4
total cost = 10 units	

Once again we obtain exactly the same configuration with a final wiring cost of ten units—or no improvement.

Interations 3, 4, 5, 6, and 7 lead to precisely the same wiring cost; thus we stay with the same placement configuration:

Cost Matrix

	1	5	7	9
F	2	2	3	5
G	4	4	5	5
H	5	4	5	4
Z	L	L	L	L

Module-Location Assignment

Assignment	Cost
F: 1	2
G: 5	4
H: 9	4
total cost = 10 units	

Iteration No. 3

	4	7	8
A	5	5	4
D	6	8	7
Z	L	L	L

Assignment	Cost
A: 8	4
D: 4	6
total cost = 10 units	

Iteration No. 4

	3	4	7
C	4	6	5
D	5	6	8
Z	L	L	L

Assignment	Cost
C: 3	4
D: 4	6
total cost = 10 units	

Iteration No. 5

	4	9	7
D	6	7	8
H	6	4	5
Z	L	L	L

Assignment	Cost
D: 4	6
H: 9	4
total cost = 10 units	

Iteration No. 6

	1	6	7
E	8	8	8
F	2	4	3
Z	L	L	L

Assignment	Cost
E: 6	8
F: 1	2
total cost = 10 units	

Iteration No. 7

We note, however, that other placement configurations also could have been chosen. We adopt, though, the rule that we maintain the previous placement unless we can improve on it. At this point the procedure is terminated since there has not been any improvement for $p = 7$ iterations. The best obtained is ten units, even though the cost of an optimal placement (Fig. 5-21) is nine units.

In this example, the total cost is identical to the wiring cost of the placement configuration since the modules in each independent set cover all the signal sets. In general, for computational efficiency, we compare the incremental change in cost of only the signal sets effected by the modules in the independent set under consideration.

5.6.3. Relaxation

Relaxation methods are applicable to the quadratic assignment problem. Since, as we observed in Section 5.4.2, the placement problem can be transformed to a quadratic assignment problem, a method for handling that problem is useful for an approximate solution for the placement problem. (Recall that an optimum solution for the associated quadratic assignment problem is not necessarily optimum for the placement problem.) In this section we define the class of methods known as relaxation techniques. A representative algorithm was first reported in the open literature by Fisk, Caskey, and West [11], who termed their method "force-placement."

Conceptually, we can regard the modules in a placement to be joined together by lines of force, similar to springs or rubber bands, that attract the modules to one another. With this analogy in mind, we can invoke Hooke's law and define the mutual force of attraction, **F**, on each pair of modules by $\mathbf{F} = k\mathbf{s}$, where **s** is the distance vector separating them and k is the spring constant given by the edge weights, c_{ij}, of the derived edge-weighted graph of the associated quadratic assignment problem. (Recall, in that graph the modules are linked together on the basis of a weighted signal set membership.) The total force on each module is then the sum of forces of all module pairs to which it belongs. This produces a resultant force vector, $\mathbf{F}_M$, for each module M in the layout configuration

$$\mathbf{F}_M = \sum_i c_{Mi} \mathbf{s}_{Mi}$$

See Fig. 5-27 for an example.

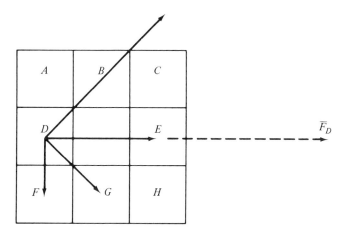

Fig. 5-27. The forces on module D.

If all modules are simultaneously allowed to move freely, then stability is reached upon the system of modules setting in its state of minimum tension, i.e., its lowest energy state. If only one module M is allowed to move, then the point of stability which it reaches is the *target point* $t(M)$, where the sum of forces upon it is equal to zero. A *target location* $l(M)$ is then that slot location nearest to the target point. The force vector, $\mathbf{F}_{t(M)}$, with its origin at the module gives the direction and distance to the zero-tension target location and is obtained from the resultant force vector $\mathbf{F}_M$. That is

$$\mathbf{F}_{t(M)} = \frac{1}{\sum_i c_{Mi}} \mathbf{F}_M$$

256 *Placement Techniques*   Chap. 5

Equivalently, the coordinates $(\bar{x}_M, \bar{y}_M)$ of the target point for module M are given by

$$\bar{x}_M = \frac{\sum\limits_i c_{Mi} x_i}{\sum\limits_i c_{Mi}}$$

$$\bar{y}_M = \frac{\sum\limits_i c_{Mi} y_i}{\sum\limits_i c_{Mi}}$$

where the (x_i, y_i) are the locations of the modules to which M is adjacent in the associated quadratic assignment problem.

An interesting interpretation of the point $(\bar{x}_M, \bar{y}_M)$ can be made. If the module M_i is considered to be a mass with weight equal to the weight of the edge joining M and M_i, then $(\bar{x}_M, \bar{y}_M)$ is the center of gravity of the masses M_i. (See Fig. 5-28 showing the computation of $\bar{x}_M$ and $\bar{y}_M$ for all modules in the sample problem of Fig. 5-14 with initial layout of Fig. 5-23. Figure 5-29 shows the target point vectors.) We leave it as an exercise to the reader to verify these formulas for the zero-tension target point.

M	x	y	$\bar{x}_M$	$\bar{y}_M$	$(\mathbf{F}_M)_x$	$(\mathbf{F}_M)_y$	$\mathbf{F}_{t(M)}$
A	0	2	$\frac{7}{4}$	$\frac{5}{4}$	$\frac{7}{4}$	$-\frac{3}{4}$	$(\frac{7}{4}, -\frac{3}{4})$
B	1	2	1	$\frac{8}{9}$	0	$-\frac{10}{9}$	$(0, -\frac{10}{9})$
C	2	2	$\frac{5}{4}$	$\frac{5}{4}$	$-\frac{3}{4}$	$-\frac{3}{4}$	$(-\frac{3}{4}, -\frac{3}{4})$
D	0	1	1	1	1	0	$(1, 0)$
E	2	1	1	$\frac{9}{7}$	-1	$\frac{2}{7}$	$(-1, \frac{2}{7})$
F	0	0	$\frac{1}{2}$	$\frac{3}{2}$	$\frac{1}{2}$	$\frac{3}{2}$	$(\frac{1}{2}, \frac{3}{2})$
G	1	0	1	$\frac{4}{3}$	0	$\frac{4}{3}$	$(0, \frac{4}{3})$
H	2	0	$\frac{5}{4}$	$\frac{7}{4}$	$-\frac{3}{4}$	$\frac{7}{4}$	$(-\frac{3}{4}, \frac{7}{4})$
Z	1	1	0	0	0	0	$(0, 0)$

Fig. 5-28. Target-point computations.

The relaxation methods take advantage of the zero-tension target location concept to modify a layout by sequential relaxation of tension on single

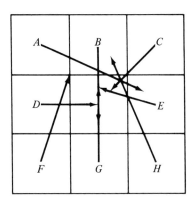

Fig. 5-29. Target-point vectors.

modules. A module is selected (either stochastically or deterministically) and the target location is computed. If this target location is available, i.e., unoccupied by another module, then we are guaranteed that the overall forces of the associated quadratic assignment problem are reduced. If it is occupied, then an alternative location is defined. Parenthetically, we note that the probability of finding the target location available is low. Thus, there is no guarantee that the total tension of even the associated quadratic assignment problem is reduced after each module relaxation, let alone the cost of the layout configuration. After a complete cycle of relaxation throughout all the modules, the resulting placement is accepted only if there is improvement. The process terminates otherwise.

An implementation for this process developed by the authors [22], and termed *force-directed relaxation*, is the following. $Q(X)$ is computed for all modules X. Recall that $Q(X) = \sum_y P(X, Y) = \sum_i c_{X_i}$. The module with the largest Q-value is selected and its zero-tension vector computed. It is then positioned in its target location. If no other module had occupied that position, then the module with the next largest Q-value is chosen for relaxation. If the target location had been occupied by a module, then that one is chosen and, with a sole pathological exception, treated exactly like the initial module; the exception being that the nearest location to its target location is used if the previously placed module now occupies that equilibrium point. An open location is preferred in the event of choice of two or more candidate locations of equal distance from the target position.

A new module for relaxation is chosen on the same basis as before. A displaced module is relocated next, and if none is displaced, then priority passes to the nonrelaxed module with the largest Q-value. This new partial arrangement of modules is accepted only if it results in a cost improvement. Once a module is relaxed and repositioned, it is not moved until all modules have been relaxed. The cycle of relaxation of all modules is repeated for a desired number of iterations or is terminated upon no further reduction in the wire length of the configuration.

This process can be implemented by the simultaneous relaxation of two or more modules. Naturally this adds a great deal of computational complexity. In fact, we note that if the n modules selected for simultaneous relaxation are not members of independent sets, then the problem of finding the proper set of target points is equivalent to the n-body problem in physics.

5.6.4. Pairwise Interchange

Pairwise interchange methods are extremely simple in concept. Two modules in a layout are interchanged and the wire length of the new layout is computed. If there is an improvement, the new layout replaces the old one; otherwise it does not. Another pair of modules is selected and the process is repeated. The wire cost is derived by modification of the old cost, namely, the incremental change is computed by consideration of only those signal sets that contain a module undergoing interchange, i.e., a location change.

The choice of modules (or locations) can be purely stochastic akin to a Monte Carlo method, or, more efficiently, by a systematic sequential interchange of the $n(n-1)/2$ pairs of n modules. For example, the pairs of modules (or locations) can be interchanged in a prearranged order independent of both the placement configuration and the particular signal sets.

As a more refined approach, the order of trial interchanging is based on the incidence of modules in the signal sets. For example, Glaser [19] ordered the modules by use of a "connection table" which is precisely the Q-list defined in Section 5.5.1. The position of the module that is "most connected," i.e., the largest Q-value, is then sequentially interchanged with those of all the other modules. With each trial interchange, a total wire length cost is secured and the configuration with minimal wire length is actually retained. The module that is next heavily connected is then selected for candidate interchanges with those of lower value on the Q-list, and so on. Interchanging all of the modules constitutes a cycle. The method terminates upon successive improvement for a cycle falling below a given bound or percentage improvement.

In Garside and Nicholson [16] an interesting variant is adopted. With an initial layout, they compute the cost contribution of module i, $C(i)$, to the total cost

$$C(i) = \sum_j c_{ij} d_{p(i)p(j)}$$

where c_{ij} is the weight of the edge joining modules i and j in the derived edge-weighted graph, and $d_{p(i)p(j)}$ is the distance between slots $p(i)$ and $p(j)$ containing modules i and j, respectively.

The module which has the maximum $C(i)$ value, say M, is then chosen for trial interchange with all other modules. The trial interchange that results in the maximum reduction in total cost is the only one actually made. (If the maximum reduction is zero then no interchange is made.)

The process is repeated with the recomputation of $C(i)$, for all $i \neq M$, and the iteration cycle continues until all modules have been treated. The process terminates just as in the usual pairwise interchange methods. It is interesting to note that the policy of choosing the maximum cost-reduction interchange, rather than accepting the first cost-reduction interchange, is analogous to the approach taken in gradient vector optimization methods.

Most of the computation time in the interchange methods is spent on developing interconnections for the modules in signal sets that have been effected by the trial interchanges. In fact it has been estimated that 90% of the computation time is spent developing interconnections for the trial layouts.

In a recent note of the authors [22], an attempt has been made by means of force-directed interchanges to delimit the complete set of the $n(n-1)/2$ pairs of trial interchanges for one cycle. In this manner the proportion of favorable interchanges to trial interchanges can be expected to increase.

Two methods suggest themselves. Adopting notions from the class of relaxation methods, we derive the edge-weighted graph of the associated quadratic assignment problem. We then consider the edge weights and the distance vectors separating the module to define lines of forces that operate upon the modules in a layout, attracting them to one another. In effect, we combine features of the relaxation methods and pairwise interchange. By this means, we develop a resultant force vector, $\mathbf{F}_M$, and a zero-tension target vector, $\mathbf{F}_{t(M)}$.

In the first method, which we term *force-directed pairwise relaxation*, we compute the zero-tension target location of all modules in a layout (see Figs. 5-28 and 5-29) and choose that pair of modules (X, Y) such that the target location of module X is within a given neighborhood of the location of Y and, alternately, the target location of module Y is within the given neighborhood of the location of X. We make the interchange subject to the condition that the placement configuration is thereby improved. We sequentially exhaust all module pairs with the above properties, treating at each time modules which have not been interchanged during the current cycle.

The zero-tension target locations of the modules in the new placement are then computed and another cycle commences. The procedure terminates upon the unavailability of any new module pairs with the mutual location-target neighborhood property.

In the second method, which we term *force-directed interchange*, the Q-list is computed and the modules ordered inversely on their Q-values. Instead of exhaustively interchanging all pairs, as in the standard pairwise interchange, we compute a resultant force vector $\mathbf{F}_M$ for the first module on the ordered Q-list and thus obtain its target point $t(M)$ and target location $l(M)$. The location of the module and the target location point can be considered to form a rectangle within the grid.

The module is trial interchanged with its adjacent modules in the direction of the target location point (equivalently, the vector $\mathbf{F}_M$ can supply the direction). First, the horizontal or vertical interchange which lies along the largest dimension of the rectangle is tried. If no cost improvement is secured, the other direction given by the smaller dimension of the rectangle is then tried. Finally, if there is still no cost improvement, the diagonal module is tried.

A new rectangle, which determines the order of trial interchanges for the given module, is defined after each successful interchange. The trial interchanges for the module cease upon lack of improvement or upon the target location being reached. As a minor sophistication in an attempt to minimize the expected number of unsuccessful trial interchanges, the target location point may be recomputed after each successful interchange. Naturally, the target point can at most change one unit after each interchange.

The next module on the ordered Q-list is then selected and in turn is

260   *Placement Techniques*   Chap. 5

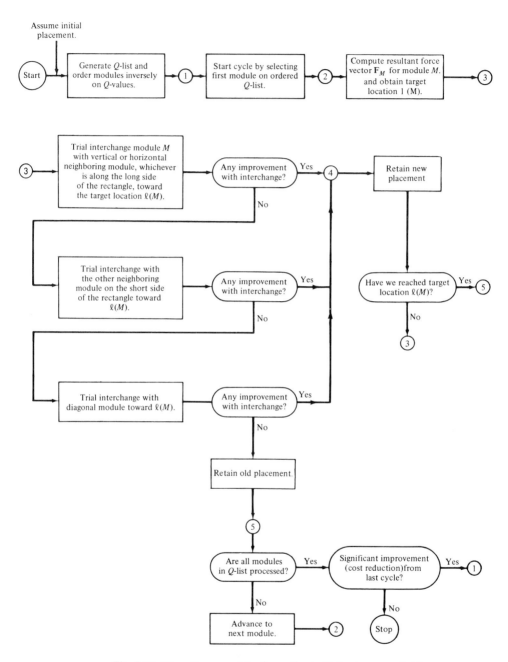

Fig. 5-30. Flow diagram of the force-directed interchange method.

processed in precisely the same manner. This process repeats until all modules on the Q-list have been treated, thereby determining a cycle. The method terminates upon successive improvement from one cycle to another falling below a given bound. (See flow diagram, Fig. 5-30.)

Example: The force-directed interchange method is now illustrated with our sample problem of eight modules (Fig. 5-14), using the initial layout shown in Fig. 5-23. We derive the matrix of the associated quadratic assignment problem which is shown in Fig. 5-31. The modules are inversely ordered on the basis of their Q-values, as shown in Fig. 5-32.

	A	B	C	D	E	F	G	H	Z	$Q(X)$
A	0	1	1	0	1	0	0	1	0	4
B	1	0	1	2	2	1	1	1	0	9
C	1	1	0	0	1	0	0	1	0	4
D	0	2	0	0	1	1	1	0	0	5
E	1	2	1	1	0	0	1	1	0	7
F	0	1	0	1	0	0	0	0	0	2
G	0	1	0	1	1	0	0	0	0	3
H	1	1	1	0	1	0	0	0	0	4
Z	0	0	0	0	0	0	0	0	0	0

Fig. 5-31. Connection matrix of the associated quadratic assignment problem.

	X	$Q(X)$
1	B	9
2	E	7
3	D	5
4	A	4
5	C	4
6	H	4
7	G	3
8	F	2
9	Z	0

Fig. 5-32. Modules inversely ordered on $Q(X)$.

The resultant force vector $\mathbf{F}_B$ is computed to determine the path of horizontal and vertical trial interchanges. We compute this vector by the formula

$$\mathbf{F}_B = (\mathbf{F}_B)_x \Delta x + (\mathbf{F}_B)_y \Delta y$$

Substituting numeric values, we obtain

$$(\mathbf{F}_B)_x = -1 + 1 + 2(-1) + 2(1) - 1 + 0 + 1 = 0$$
$$(\mathbf{F}_B)_y = 0 + 0 + 2(-1) + 2(-1) - 2 - 2 - 2 = -10$$

The interchange $B \longleftrightarrow Z$ is performed as specified by the direction of $\mathbf{F}_B$, resulting in the configuration shown in Fig. 5-33. The length of the first chain increases to six units, but the lengths of the second and third decrease to four and two units, respectively. This leads to a reduced cost of twelve units. Thus, the modified configuration is retained as the new standard.

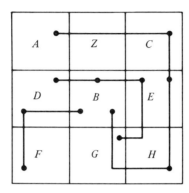

Wire length = 6 + 4 + 2 = 12 units

Fig. 5-33. Placement after first interchange.

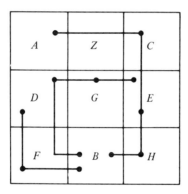

Wire length = 5 + 4 + 2 = 11 units

Fig. 5-34. Placement after second interchange.

The next interchange, $B \longleftrightarrow G$, results in a further decrease in cost to eleven units in the placement shown in Fig. 5-34. Since the boundary of the board is reached, the interchanges for module B terminate for this cycle.

We leave it as an exercise to the reader to compute the resultant force vector for module E and to perform the indicated trial interchanges with it.

5.6.5. Stochastic Methods

Monte Carlo approaches have been independently experimented with by the authors [23], [32], and by others [7]. Stated very briefly, the Monte Carlo method maps modules randomly into board slots on the basis of a density distribution, typically a uniform (rectangular) one. The cost of the resulting placement configuration is computed and compared with the previous best configuration. (Initially, to start the procedure, an infinite cost is assumed.) The better layout and its cost figure are then retained. This process is repeated until a preassigned time has elapsed or until a preassigned number of configurations have been generated and considered.

The method is motivated by the hope that there are a comparatively large number of good placements in the space of all feasible placements, i.e., that the probability of randomly selecting a good placement is reasonably high. Unfortunately this is not the case. The results of experiments are not favorable for that hypothesis. The ratio of good placements to the enormous space of possible layout configuration is miniscule; thus this method is not useful as an effective device for module placement.

The Monte Carlo method, however, is of value in providing distributions from which one can estimate the value of a layout produced by other methods. More specifically, a plot of the frequency of occurrence of costs produced by these methods versus the cost values can be closely approximated by a normal distribution. By use of this phenomenon, we can state, in a statistical sense, that the layout we have generated by the use of some placement technique is so many percentiles out on the tail of the distribution of all layouts. This estimation can provide a means for evaluating a placement algorithm.

A variant of the Monte Carlo method, the *adaptive Monte Carlo*, was developed by Kurtzberg [32] to accelerate the convergence rate toward a good placement. In this method, the distribution from which modules are mapped into slots is modified according to the results of the past layouts so as to bias the placement of modules into favorable locations, which leads to low cost configurations.

The probability of module M being mapped into slot location l is therefore no longer a constant $1/pq$, where pq is the number of slots in the $p \times q$ board, as it is in the standard Monte Carlo approach. Instead of the one uniform density function, there is a set of dynamic density functions, one for each module, giving the probability of a module being mapped into the various locations. The modification of the density functions takes place with each perturbation and cost calculation.

To illustrate, assume that module M is in location l in the ith placement configuration which has cost c_i, and define a density function associated with M over slots j as $d_M(j)$, $1 \leq j \leq pq$. The value $d_M(l)$ is augmented by a value which is inversely proportional to the placement cost c_i. The remaining $d_M(j), j \neq l$, are diminished by an amount such that the distribution is once again normalized.

Further, once a value in a density function exceeds a given bound, then that value is set to unity and all other values in that density function are set to zero. This adjustment results in modules becoming locked into locations, thereby reducing the order of the necessary permutations for each iteration and eliminating any danger of oscillations because of layout symmetry considerations.†

Unfortunately, the computation time for a single iteration (perturbation, cost calculation and distribution modification) is extremely high. Also, an excessively large number of iterations still appears necessary, thereby rendering this technique to be strictly of academic interest for the placement problem.

5.7 BRANCH-AND-BOUND

Branch-and-bound methods are applicable to the quadratic assignment problem and can be used to find an optimum solution to that problem if the order is less than fifteen modules. It is the only class of methods discussed in this chapter which has this feature. These methods have been employed in a variety of problems (see Lawler and Wood [39] for an excellent account). Gilmore [18] and Lawler [38] independently applied these methods to the quadratic assignment problem. Although the method dates back to 1958, the earliest reference to the name "branch-and-bound" is Little, et al. [41], 1963.

The branch-and-bound technique is treated in Section 5.7.1, and Gilmore's approximation of the exact method is presented in Section 5.7.2. The material in the next two sections is based on the two Gilmore papers [17] and [18]. In particular, the example is taken from [17].

5.7.1. Exact Branch-and-Bound

Briefly stated, the branch-and-bound method proceeds as follows. The set of all feasible solutions is partitioned, and a search for the optimum is made in each partition. A lower bound is computed for the solutions in each partition, and the search in any partition is terminated upon the lower bound exceeding the cost of some previously found feasible solution. (The feasible

---

†It is interesting to note that a similar "fixing" technique has been tried successfully by Shen Lin [40] in his algorithm for the traveling salesman problem. Shen Lin starts with a given tour and tries all triple interchanges of branches (links in the tour). He selects the best resulting tour (which is termed *3-opt*) and then randomly selects a new starting tour. Every ten iterations he examines the set of links that remain invariant over the set of 3-opt tours which he has constructed. Those he fixes for all future initial tour generation so that the order of the problem is monotonically decreasing.

solution may be obtained by any method, e.g., by a constructive initial-placement technique or by the branch-and-bound algorithm itself.) The partitioning is continued until either the search in every partition is terminated or a solution is reached. All solutions arrived at in this way must be optimum.

It is evident that methods must be defined for computing lower bounds and partitioning sets of feasible solutions. We explore these ideas and present specific methods by treating the quadratic assignment problem. Recall that in the quadratic assignment problem there is a connection matrix $C = [c_{ij}]$ and a distance matrix $D = [d_{ij}]$, and we seek a permutation p such that

$$\sum_{i,j} c_{ij} d_{p(i)p(j)}$$

is a minimum.

As a preliminary, consider the branch-and-bound method from the viewpoint of an $(n-1)$-stage decision process. At each stage k, a module i is chosen from the remaining $(n-k)$ modules and assigned to all the remaining $(n-k)$ slots. A decision tree is used to represent this process in Fig. 5-35. (The use of the term "branch-and-bound" stems from viewing the method as a search in a decision tree.)

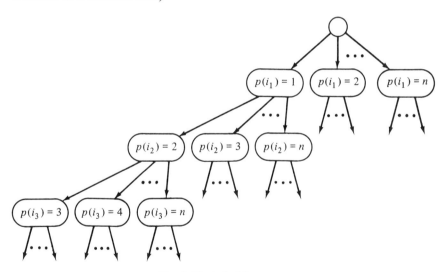

Fig. 5-35. A decision tree.

On the first level of the tree in Fig. 5-35, module i_1 has been chosen, and each of the n first-level vertices represents a possible slot for this module. On the second level, module i_2 ($\neq i_1$) has been chosen and assigned to any of the remaining slots on that level. In Fig. 5-36, the complete decision tree for the special case of $n = 3$ is illustrated. The notation† $\Delta 1 \Delta$ denotes module 1 is

†In this section for notational simplicity we designate modules by numbers rather than by letters.

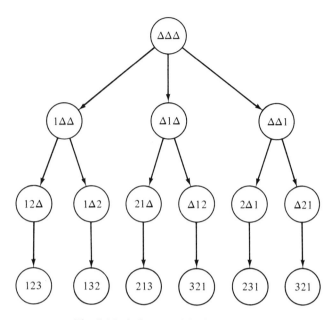

Fig. 5-36. A three-module decision tree.

assigned to slot 2, and slots 1 and 3 are still empty. We call this a *partial permutation* and denote it by q.

For each vertex in the decision tree, a lower bound is computed. If this lower bound exceeds the cost of some known feasible solution, the search along that branch of the tree is terminated. That branch cannot possibly lead to a solution with a lower cost than the cost of the feasible solution.

The efficiency of the method depends on both the bounding and the branching techniques. For example, if no lower bound is computed then the branch-and-bound method reduces to a pure enumerative technique. The function of the lower bound is to prune the decision tree; the better the bound, the sharper the pruning. However, a trade-off does exist. In general, the better the bound, the more computation time is necessary to compute this bound.

Gilmore developed two lower bounds for $\sum c_{ij} d_{p(i)p(j)}$ based on the following observation. If

$$\mathbf{c} = (c_1, c_2, \ldots, c_m) \quad \text{and} \quad \mathbf{d} = (d_1, d_2, \ldots, d_m)$$

are two vectors, then the minimum dot product of $\mathbf{c}$ and $\mathbf{d}$, i.e., the minimum of $\sum c_i d_{p(i)}$ over all permutations, p, is obtained by arranging $\mathbf{c}$ in increasing order and $\mathbf{d}$ in decreasing order, and then multiplying term by term.

A simple lower bound for the quadratic assignment problem is obtained by treating the nondiagonal elements of the matrices C and D as vectors of length $n(n-1)$ and performing the proper ordering and multiplication. The diagonal elements are eliminated (thereby obtaining a better bound) under the realization that no pin is connected to itself and two modules cannot occupy the same slot. If some modules are already placed, then the contribution of these placed modules to the total distance is computed, the corresponding entries from the C and D matrices are deleted, and a new lower bound is computed.

This lower bound for the partial permutation q can be obtained by the following four steps (again deleting all diagonal elements):

1. Compute the contribution of all placed modules.
2. For each row i of C corresponding to a placed module i, form a vector of the entries not used in step 1, and for each row $q(i)$ of D form a similar vector. Compute the lower bound of the dot product of these two vectors. Repeat this operation for each column i corresponding to a placed module i. Do this for all placed modules, and sum the lower bounds.
3. Form the vectors of the remaining entries in C and D, and find the minimum dot product.
4. Add the results of the first three steps.

We illustrate this method by computing the lower bound for the partial permutation $5\Delta\Delta3\Delta$ corresponding to the C and D matrices given in Fig. 5-37a and b. Note that C and D are symmetric as is the usual case, that is, $c_{ij}d_{p(i)p(j)} = c_{ji}d_{p(j)p(i)}$. We exploit this fact by computing in the following four steps only $c_{ij}d_{p(i)p(j)}$ instead of $c_{ij}d_{p(i)p(j)} + c_{ji}d_{p(j)p(i)}$. We thus obtain one-half the lower bound.

	1	2	3	4	5
1	—	7	4	5	8
2	7	—	4	3	0
3	4	4	—	3	7
4	5	3	3	—	4
5	8	0	7	4	—

(a) Connection matrix

	1	2	3	4	5
1	—	8	6	6	5
2	8	—	1	2	4
3	6	1	—	3	0
4	6	2	3	—	4
5	5	4	0	4	—

(b) Distance matrix

Fig. 5-37.

1. Since module 5 is placed in slot 1 and module 3 is placed in slot 4, the contribution of this term is

$$c_{53} \cdot d_{14} = (7)(6) = 42$$

2. The entries in row 5 of C not already used are c_{51}, c_{52} and c_{54}, whose values are 8, 0 and 4, respectively. The entries of row $q(5) = 1$ of D not already used are d_{12}, d_{13}, and d_{15}, whose values are 8, 6 and 5, respectively. The minimum dot product of these two vectors is

$$(0, 4, 8) \cdot (8, 6, 5) = (0)(8) + (4)(6) + (8)(5) = 64$$

The corresponding minimum dot product for module 3 is

$$(3, 4, 4) \cdot (4, 3, 2) = (3)(4) + (4)(3) + (4)(2) = 32$$

This corresponds to $c_{34} \cdot d_{45} + c_{31} \cdot d_{43} + c_{32} \cdot d_{42}$. Hence the total contribution of step 2 is 96.

3. The remaining entries in C are c_{12}, c_{14} and c_{24}, which have values of 7, 5 and 3, respectively. The remaining entries in D are d_{23}, d_{25} and d_{35}, which have values of 1, 4, and 0, respectively. The minimum dot product of these two vectors is

$$(3, 5, 7) \cdot (4, 1, 0) = (3)(4) + (5)(1) + (7)(0) = 17$$

4. The lower bound for the partial permutation $5\Delta\Delta3\Delta$ is then given by twice the sum of numbers obtain in the first three steps

$$2(42 + 96 + 17) = 310$$

We now give another method of computing a lower bound for the empty permutation (i.e., no modules placed).

1. Delete the c_{ii} entry from the ith row of C and the d_{jj} entry from the jth row of D, and form the corresponding vectors $\mathbf{c}_i$ and $\mathbf{d}_j$.
2. Find the minimum dot product of $\mathbf{c}_i$ and $\mathbf{d}_j$ and call this minimum value a_{ij}.
3. Form the assignment matrix $A = [a_{ij}]$.
4. Solve the assignment problem defined by A.

Since the total cost of placing module i into slot j is $\sum c_{ik} d_{jp(k)}$, it is not difficult to show that a_{ij} is the lower bound on the cost of placing module i into slot j. A permutation Π for which $\sum a_{i\Pi(i)}$ is minimum then yields a lower bound for the original quadratic assignment problem. Finding the minimum of $\sum a_{i\Pi(i)}$ is the assignment problem defined in Section 5.4.1 for which algorithms are given in the Appendix. The matrix A of the above example is given in Fig. 5-38.

		\multicolumn{5}{c}{Slot}
		1	2	3	4	5
	1	144	74	46	81	68
	2	77	27	13	38	28
Module	3	107	55	34	60	47
	4	91	49	31	52	43
	5	106	38	19	53	44

Fig. 5-38. Module-slot assignment matrix.

To extend this method to the kth stage of the decision tree, i.e., to a partial permutation where k modules are already placed, we must find the corresponding $(n - k) \times (n - k)$ assignment matrix A_k. We leave the derivation of A_k as an exercise to the reader. The derivation may be compared with Gilmore [18].

This latter method of computing the lower bound yields a sharper (i.e., larger) lower bound than the first method, but it requires considerably more computations. It also has the added feature that a solution to the assignment problem represented by matrix A supplies a feasible solution to the quadratic assignment problem.

Finally, the branch-and-bound methods have an important feature not shared by the other methods presented in this chapter. If we seek a suboptimum solution, that is one within a multiple r, $0 < r < 1$, of the optimum, then branches are terminated more rapidly. Hence, a feasible solution with cost C_f causes the termination of any branch where the lower bound exceeds rC_f. It is interesting to note that a solution to within any prespecified accuracy of the optimum is guaranteed without ever finding the optimum.

5.7.2. Gilmore's Approximation Methods

We note that if we seek a suboptimum solution, the method in the previous subsection even with $r < 1$ is still not computationally feasible for $n > 15$. Two suboptimum methods which are based on the second lower bound are now presented. The methods are heuristic techniques for determining which branch of the decision tree to follow. These approximations to the branch-and-bound method examine only a small subset of the branches of the decision tree. In particular, we examine only one edge leaving any vertex of the decision tree.

Recall that at the kth step in the decision process, module i is chosen and placed in all remaining $n - k$ slots. In the exact branch-and-bound method the choice of the module i is immaterial as regards convergence to the optimum solution (although it might affect rate of convergence) since branches are deleted only on the basis of the lower bound. In an approximate method where not all branches are examined, the choice of the module and the slot is crucial.

We now give two heuristic approaches to determine the module and slot chosen at any stage in the decision process. Both are based on the assignment matrix A_k derived in the kth step of this process. Rationales for them are given at the end of this section. The two methods are:

1. Find the minimum entry in A_k and delete the corresponding row and column to form A'_k; then find the minimum entry in A'_k and similarly form a new A'_k. Continue until A'_k is empty. Choose the maximum of the $n - k$ elements found and branch on that element.
2. Solve the assignment problem for A_k and branch on the maximum of the $n - k$ elements appearing in the assignment problem solution.

In method 1 we secure an approximate solution for the assignment problem and base our decision branching on it, and in method 2 we obtain an optimum solution for our branching.

It can be shown that method 1 takes the order of $(n - k)^3$ elementary operations, where an elementary operation is defined to be any one of addition, multiplication or comparison of two numbers. Hence an algorithm based on this method takes on the order of $\sum_{k=1}^{n-1}(n - k)^3$ or n^4 elementary operations. The Munkres' algorithm to solve the assignment problem A_k takes on the order of $(n - k)^4$ elementary operations, so that an algorithm based on method 2 takes on the order of n^5 elementary operations.†

Method 1 is based on an approximate method for the assignment problem, termed *matrix-scan*. Any approximate method can serve as the basis of a heuristic algorithm. The *row-scan* method given in the Appendix operates on a row-by-row basis to solve the assignment problem. A branch-and-bound algorithm based on this method takes only on the order of $\sum_{k=1}^{n-1}(n - k)^2$ or n^3 elementary operations. Further, the expected solution values of the row-scan and the matrix-scan methods differ only by a constant independent of the size of the assignment problem [33].

Hillier and Connors [26] propose still another variant. The rows of A are scanned for the smallest and second-smallest entries s_1 and s_2. The arithmetic difference $s_2 - s_1$ between these two numbers is computed, and the row with the largest of these differences is noted. The same calculations are made for the columns. The minimum entry in the row or column with the maximum difference is then chosen for branching.

†Although we conform to Gilmore's derivation, empirical results [33] indicate that the expected order of growth for solving the assignment problem using Munkres' algorithm is only n^3, which then implies that Gilmore's "n^5-algorithm" is really of order n^4. In a private communication R. Karp informed the authors that he and J. Edmonds developed an assignment algorithm based upon a "shortest-path method," which appears superior to other assignment methods. Their algorithm is of order n^3.

The rationale behind these heuristics is that the assignment problem is an approximation to the quadratic assignment problem in which the interconnections of the modules are ignored, i.e., all modules are treated as being independent of one another. Thus, the solution of the assignment problem produces a reasonable start for the quadratic assignment problem. The maximum element (module-location pair) is chosen under the philosophy of "doing the most difficult job first." The arithmetic difference supplies a notion of the penalty one pays for not being able to use the minimum value. Thus, Hillier and Connors follow the policy of positioning the element first for which this penalty is highest.

The results of the n^4, n^5 and Hillier-Connors heuristics compare favorably to the results of Steinberg for a 34-module problem (see Section 5.8.1). The n^3 heuristic has never been programmed.

5.8 CONCLUSIONS

We now present known experimental and theoretical results for various comparisons of methods for the placement and quadratic assignment problems. Remarks and recommendations for future research are also given.

5.8.1. Comparison of Methods

In comparing algorithms it is necessary to consider the interacting factors of goodness of solution and computation time required. In addition, a variety of problems should be used as test cases. Ideally, these test problems should be drawn from real applications or, at the least, should be similar in structure to them.

So far, there is only one standard problem for comparison purposes. Steinberg [53] in 1961 presented a 34-module associated quadratic assignment problem, derived from an existing computer, to which Gilmore [18] and Kurtzberg [35] compared their methods. Also, Hillier and Connors [26] compared their modification of Gilmore's algorithm and Hillier's pairwise interchange method to the same problem. The results† are presented in Fig. 5-39. While the computation time has been normalized to the IBM 7090, it

---

†It should be noted that the full power of Steinberg's algorithm was not employed in [53] since Steinberg was limited to independent sets with no more than ten modules. His program for the assignment problem could accept only matrices of order 12 or less; the two empty locations in his 36-slot board preempted two rows (columns). Further he used only twenty-five randomly selected independent sets.

Also in [18] it is stated that the n^4 algorithm yields smaller costs than does the n^5 algorithm. However, in a private communication Gilmore notes that there was a programming error in the implementation of the n^5 algorithm.

should be noted that direct comparison of computational results is difficult because of the different machines used and the varying relative efficiency of the different programming languages. Also, no strong implication can be drawn about the comparative values of the algorithms on the basis of this one problem.

Method	Actual Computation Time	Computation Time Normalized to IBM 7090	Wire Length
Steinberg's algorithm	Roughly 8 hours	5.2 minutes	4894.54
Pair-linking + interconnection program	7.3 minutes + 2.5 minutes = 9.8 minutes	0.2 minute (total time)	4873.09
Cluster-development + interconnection program	0.6 minutes + 2.5 minutes = 3.1 minutes	0.06 minute (total time)	5264.74
Gilmore's n^4 algorithm	1 minute	1 minute	4547.54
Gilmore's n^5 algorithm	3 minutes	3 minutes	4680.36
Hillier-Connors' modification of Gilmore's	5 minutes	5 minutes	4821.78
Hillier's interchange	3 minutes	3 minutes	4475.28

Fig. 5-39. Comparison summary for a 34-module problem.

Hillier and Connors [26] also tested their algorithms against Gilmore's on sixteen randomly generated† cost matrices for twelve-module quadratic assignment problems. They found an average difference of approximately 5% in favor of their interchange method over Gilmore's n^4 method and with the aid of some statistical analysis concluded that their method was better.

Nugent, Vollman, and Ruml [45] carry these comparisons further. In addition to the Hillier's pairwise interchange method, they tested three other pairwise interchange schemes and concluded their method was best. Eight test problems were used, ranging in size from five to thirty modules. For the thirty-module problem they found that their version of the pairwise interchange method is approximately 2% better than an interchange method based on examining all pairs and interchanging only the pair which yields the maximum improvement. However the computation time for their method is fourteen times greater on this thirty-module problem (210 minutes vs 15 minutes, normalized to the IBM 7090).

Finally, the exact order of computation growth of the various algorithms is usually a complicated function of the specific characteristics of the problem being processed and frequently the order of growth is not known. In general,

---

†That is, the entries c_{ij} in the cost matrix are chosen randomly from a uniform distribution.

it ranges from n^2 for the pair-linking and cluster-development constructive methods, and some of the interchange techniques, to n^4 for the approximations to the branch-and-bound method.

Naturally, the order of growth does not tell us the amount of processing time a specific problem requires or the size of problems the various methods can conveniently handle. For example, the constructive methods can handle far larger problems than the complete interchange techniques even though both are of order n^2. The order of computation growth is useful strictly for purposes of extrapolation of necessary solution time.

While the cited papers represent the most complete comparisons of methods published to date, their usefulness to the placement problem is severely limited for the following reasons. First, the problem size is extremely small, and second, random graphs are not typical logic structures. Methods which are best for small random graphs are not necessarily best for large structured graphs.

An alternate approach taken by some investigators [23], [36] to evaluate placement algorithms is to generate structured problems for which the optimum placement is known. The simplest means of doing so is to consider a rectangular array of modules and then generate signal sets, following the usual distribution of signal set sizes, J_k, and module membership, $S(X)$, based upon realistic problems. Each signal set would be constructed of neighboring modules in the array such that a typical chaining or minimum spanning tree interconnection results in the array possessing the minimum possible distance of $\sum_k (J_k - 1)$.

The methods of testing algorithms discussed so far are all of an experimental nature. Donath [10] has done some interesting work along a theoretical vein. He employed some methods of combinatorial analysis to predict the lower bound for the minimum cost of a class of randomly generated cost matrices for quadratic assignment problems. He then used the complete pairwise interchange method to assign the modules. While the obtained cost is as much as 40% above his theoretically predicted minimum, Donath conjectures that this is due to the inability of the pairwise interchange method to obtain optimum placements.

5.8.2. Remarks and Recommendations

In this chapter we have explored a variety of placement techniques. The relative value of each method depends upon the particular application at hand and the amount of computer time available. For example, in non-stringent and repetitive applications, and where computer time is a factor, the use of a fast initial-placement method is recommended. With highly critical problems, more time should be invested on placement. The high speed technology of today, with promise of even faster circuitry in the future, suggests that more and more placement problems are going to be classified as critical.

With this in mind, we recommend a fast initial-placement method coupled with an iterative placement-improvement method. The bottlenecks of the configuration can be eliminated by such a combination of methods to make the layout conform to the circuitry requirements.

For critical problems, it may be necessary to employ a more realistic model for the cost function. In this case, a constructive method can be used to secure an initial placement with total wire length as a measure of cost. The placement-improvement method then can use a more sophisticated, refined measure to take into account the additional constraints, e.g., no wire can be more than t units in length.

The strongest candidates for securing the initial placement are the pair-linking and cluster-development methods (or methods similar to them using typical selection rules discussed in Section 5.5.1). The pair-linking method, although requiring more processing time, produces a better layout than does the cluster-development method. The exact relative value of each initial-placement method is, however, dependent upon the *sensitivity* of the problem. The relative value of the cluster-development method, for example, increases with a decrease in problem sensitivity. By sensitivity we mean the relative effect of randomly interchanging a few modules. The degree of independence of the signal sets determines the sensitivity.

The strongest candidates for an iterative placement-improvement method are the pairwise interchange and the Steinberg assignment methods. A combination of the two may be used as is done in Rutuman's variant of Steinberg's method. Also variants of pairwise interchange may be combined to achieve improvements in the layouts.

It is clear, however, that there must be a great deal more experimental and theoretical investigations of the placement techniques. For instance, it has been conjectured that the time to obtain an optimum solution for the quadratic assignment problem must grow exponentially with the order of of the problem. In experimental studies, networks with structures identical (or highly similar) to realistic problems must be used. Misleading impressions can be gained from the use of unstructured, random graphs as vehicles for experimentation. For example, for the class of unrealistic random graphs, it appears that the pairwise interchange methods are superior and relatively insensitive to the initial starting configuration.

Another area for future investigation is the determination of choice of parameters (e.g., the selection functions in the initial-placement methods) from the given signal-set distributions and estimation of problem sensitivity. Also, more refined cost functions, i.e., better problem models, must be defined, particularly for the highly critical circuitry. These cost functions would more directly take into account the primary purpose of a good layout, namely increasing the probability of the unit being easily wired.

Finally, research into the natural combination of partitioning, placement and wiring (interconnecting and routing) should provide a major improvement in board and backplane formation techniques. This last area of investigation is particularly difficult and may produce overly time-consuming algorithms. Nevertheless it potentially offers an extremely fruitful area for future work.

ACKNOWLEDGMENTS

The authors wish to express their appreciation to P. C. Gilmore and I. Pohl for discussion on the branch-and-bound technique and to W. E. Donath for a conversation on his work. We wish to thank R. L. Russo for critically reading and commenting on the manuscript. We also wish to thank H. A. Hochstein for her extreme care and infinite patience in the typing of the manuscript.

PROBLEMS

1. Construct the signal-set bigraph, the module connection matrix, and the module adjacency graph of the following set of signal sets.

$$S_1 = \{A_1, B_1, C_1, D_1\}$$
$$S_2 = \{C_2, D_2, E_1\}$$
$$S_3 = \{A_2, B_2, F_1\}$$
$$S_4 = \{A_3, B_3\}$$

2. Verify that the configurations shown in Fig 5-8b and c are optimum, respectively, for the associated quadratic assignment problem and the placement problem defined by the signal sets given in Section 5.4.3. Recall that we assume a chain interconnection rule with a rectilinear metric.

3. Find an example, using a minimum spanning tree interconnection rule, where the optimum layout for the associated quadratic assignment problem is not the optimum layout for the placement problem from which it was derived.

4. Verify the entries in the tables shown in Figs. 5-9 and 5-10.

5. Find an ordering for the modules in the sample problem in Fig. 5-14 using the selection rule $\max_{X \in B^j} f_1(X)$. Find four other orderings by using the selection rule $\max_{X \in B^j} f_i(X)$ for $i = 2, 3, 4, 5$. Choose module B as the first module, and break ties by choosing lexicographically, i.e., A before B, etc.

6. Prove that if we define

$$f_4 = \sum p_k^j p_k$$

then $f_2(X) = f_4(X)$ for all $X \in B^j$.

7. Show that

$$Q_\lambda(X) = \sum (J_k - 1)\frac{J_k + \lambda}{J_k}p_k$$
$$= \sum (J_k + \lambda)p_k - P_\lambda(X)$$

Note that this implies the $Q_\lambda(X)$ can be found without first computing $P_\lambda(X, Y)$.

For problems 8 and 9, use the sample problem of Fig. 5-14 with the same interconnection rule (chain), metric (rectilinear distance) and array of slots as in Section 5.5.2. Break ties by choosing lexicographically.

8. Find a placement configuration using the pair-linking method with $\lambda = 0$.
9. Find a placement configuration using the cluster-development method with $\lambda = 0$. Use either

$$\sum_{R^j(X)} \frac{T_k + 1}{T_k} \quad \text{or} \quad \sum_{R^j(X)} \frac{T_k + 1}{2(J_k - p_k)}$$

as the expression for $F(X)$.

10. Find all independent sets of modules for the set of signal sets given in Section 5.2.2.
11. Show that if α and β are two independent sets of modules with k elements, then $\alpha \cup \beta$ is an independent set with $k + 1$ elements if and only if α and β have $k - 1$ elements in common and the pair of noncommon elements is independent.
12. Prove that if the tail of the vector

$$\mathbf{F}_{t(M)} = \frac{1}{\sum_i c_{Mi}} \mathbf{F}_M$$

is placed at the module M then the head will be at the zero-tension (equilibrium) point $(\bar{x}_M, \bar{y}_M)$, where

$$\bar{x}_M = \frac{\sum_i c_{Mi} x_i}{\sum_i c_{Mi}}$$

$$\bar{y}_M = \frac{\sum_i c_{Mi} y_i}{\sum_i c_{Mi}}$$

(x_i, y_i) and c_{Mi} are as defined in Section 5.6.3.

13. In Section 5.6.4, the resultant force vector $\mathbf{F}_B$ is computed, and the interchanges for module B are performed. Compute the resultant force vector $\mathbf{F}_E$ for this new layout (Fig. 5-34), and perform the appropriate interchanges.

14. (a) Using Gilmore's first method, find a lower bound for the associated quadratic assignment problem for the sample problem of Fig. 5-14. Can you find a (weaker) lower bound by a simpler method?
    (b) Use Gilmore's first method to find a lower bound for the partial permutation $\Delta E \Delta \Delta BD \Delta \Delta \Delta$ for this problem.
    (c) Compute the cost of this associated quadratic assignment problem for the configuration given in Fig. 5-21.

15. (a) Extend Gilmore's second method of computing a lower bound for the quadratic assignment problem to the partial (nonempty) permutation.
    (b) Using the example in Section 5.7 and the results of part (a), find the 3×3 matrix for the partial permutation $5\Delta\Delta3\Delta$.
    (c) Solve the assignment problem from part (b) to find a lower bound for the partial permutation $5\Delta\Delta3\Delta$.

16. Verify some of the entries in the module-slot assignment matrix A of Fig. 5-38.

17. Solve the assignment problem represented by matrix A of Fig. 5-38 to obtain a lower bound for the quadratic assignment problem of Section 5.7. Compare your result to the lower bound obtained in the text. Also compute the cost for the feasible solution obtained by solving the assignment problem.

18. Give expressions for (weak) lower and upper bounds for any placement problem.

REFERENCES

[1] M. Bellmore and G. L. Nemhauser, "The Traveling Salesman Problem: A Survey." *J. Oper. Res.*, Vol. 16 (1968), pp. 538–558.

[2] C. Berge, *The Theory of Graphs and Its Applications*. New York, John Wiley & Sons, Inc., 1962.

[3] M. A. Breuer, "General Survey of Design Automation of Digital Computers." *Proc. IEEE*, Vol. 54, No. 12 (December, 1966), pp. 1708–1721.

[4] M. A. Breuer, "The Formulation of Some Allocation and Connection Problems as Integer Programs." *Naval Res. Logist. Quart.*, Vol. 13 (March, 1966), pp. 83–95.

[5] R. G. Busacker and T. L. Saaty, *Finite Graphs and Networks*. New York, McGraw-Hill Book Company, 1965.

[6] R. L. Clark, *A Technique For Improving Wirability in Automated Circuit Card Placement*. Rand Corporation Report R-4049, August, 1969.

[7] J. F. Cooper, *Monte Carlo Positioning of Connected Elements on a Carrier*. IBM Report 64-520-005, June 29, 1964.

[8] G. Dantzig, *Linear Programming and Extensions*. Princeton, N.J., Princeton University Press, 1963.

[9] G. Dantzig, D. Fulkerson, and S. Johnson, "Solutions of a Large Scale Traveling Salesman Problem." *J. Oper. Res.*, Vol. 2 (1954), pp. 393–410.

[10] W. E. Donath, "Statistical Properties of the Placement of a Graph." *J. SIAM Appl. Math.*, Vol. 16, No. 2 (March, 1968), pp. 376–387.

[11] C. J. Fisk, D. L. Caskey, and L. L. West, "ACCEL: Automated Circuit Card Etching Layout." *Proc. IEEE*, Vol. 55, No. 11 (1967), pp. 1971–1982.

[12] M. Flood, "The Traveling Salesman Problem." *J. Oper. Res.*, Vol. 4, No. 1 (February, 1956), pp. 61–75.

[13] L. Ford and D. Fulkerson, *Flows in Networks*. Princeton, N.J., Princeton University Press, 1962.

[14] H. Freitag and M. Hanan, *A Placement Algorithm*. IBM internal report, November, 1966.

[15] R. L. Gamblin, M. Q. Jacobs, and C. J. Tunis, "Automatic Packaging of Miniaturized Circuits," in G. A. Walker, ed., *Advances in Electronic Circuit Packaging*, Vol. 2 (New York, Plenum Press, 1962), pp. 219–232.

[16] R. G. Garside and T. A. Nicholson, "Permutation Procedure for the Backboard Wiring Problem." *Proc. IEE* (January, 1968), pp. 27–30.

[17] P. C. Gilmore, *A Solution to the Module Placement Problem*. IBM Report RC430, April 26, 1961.

[18] P. C. Gilmore, "Optimal and Suboptimal Algorithms for the Quadratic Assignment Problem." *J. SIAM*, Vol. 10, No. 2 (June, 1962), pp. 305–313.

[19] R.H. Glaser, "A Quasi-Simplex Method for Designing Suboptimal Packages for Electronic Building Blocks." *Proc. 1959 Computer Appl. Symp.*, at Armour Research Foundation, Illinois Institute of Technology, pp. 100–111.

[20] M. Hanan, *Net Wiring For Large Scale Integrated Circuits*. IBM Report RC1375, February, 1965.

[21] M. Hanan, "On Steiner's Problem with Rectilinear Distance." *J. SIAM*, Vol. 14 (1966), pp. 255–265.

[22] M. Hanan and J. M. Kurtzberg, *Force-Vector Placement Techniques*. IBM Report RC2843, April, 1970.

[23] M. Hanan and P. K. Wolff, Sr., Private communication.

[24] F. Harary, *Graph Theory*. Reading, Mass., Addison-Wesley Publishing Co., Inc., 1969.

[25] M. Held and R. Karp, "A Dynamic Programming Approach to Sequencing Problems." *J. SIAM*, Vol. 10 (1962), pp. 196–210.

[26] F. S. Hillier and M. M. Connors, "Quadratic Assignment Problem Algorithms and the Location of Indivisible Facilities." *Management Science*, Vol. 13, No. 1 (September, 1966), pp. 42–57.

[27] J. S. Hughes and K. W. Lallier, *Automated Component Placement and Pin Assignment*. IBM Report 65-825-1422, February, 1965.

[28] U. R. Kodres, *Geometrical Positioning of Circuit Elements in a Computer.* Conference Paper No. CP 59-1172, AIEE Fall General Meeting, October, 1959.

[29] U. R. Kodres, "Formulation and Solution of Circuit Card Design Problems Through Use of Graph Methods," in G. A. Walker, ed., *Advances in Electronics Circuit Packaging*, Vol. 2 (New York, Plenum Press, 1962), pp. 121-142.

[30] J. B. Kruskal, Jr., "On the Shortest Spanning Subtree of a Graph and the Traveling Salesman Problem." *Proc. Amer. Math. Soc.*, Vol. 7 (1956), pp. 48-50.

[31] H. W. Kuhn, "The Hungarian Method for the Assignment Problem." *Naval Res. Logist. Quart.*, Vol. 11 (March-June, 1955), pp. 83-97.

[32] J. M. Kurtzberg, *Backboard Wiring Algorithms for the Placement and Connection Order Problems.* Burroughs Report TR 60-40, June 28, 1960.

[33] J. M. Kurtzberg, "On Approximation Methods for the Assignment Problem." *J. ACM*, Vol. 9, No. 4 (October, 1962), pp. 419-439.

[34] J. M. Kurtzberg, "Computer Mechanization of Design Procedures." *Proc. Sixth Annual AIIE Conference*, Detroit, Michigan, October, 1964.

[35] J. M. Kurtzberg, "Algorithms for Backplane Formation," in *Microelectronics in Large Systems* (Spartan Books, 1965), pp. 51-76.

[36] J. M. Kurtzberg and B. Estes, *Initial Card Placement Algorithms: An evalustion.* Burroughs Report TR 61-44, August 21, 1961.

[37] J. M. Kurtzberg and J. Seward, *Program For Star Cluster Wiring of Backplane.* Burroughs Internal Report, January, 1964.

[38] E. L. Lawler, "The Quadratic Assignment Problem." *Management Science*, Vol. 9 (1963), pp. 586-599.

[39] E. L. Lawler and D. E. Wood, "Branch and Bound Methods: A Survey." *J. Oper. Res.*, Vol. 14 (1966), pp. 699-719.

[40] S. Lin, "Computer Solutions of the Traveling Salesman Problem." *Bell Sys. Tech. J.*, Vol. 44 (1965), pp. 2245-2269.

[41] J. D. C. Little, K. G. Murty, D. W. Sweeney, and C. Karel, "An Algorithm For the Traveling Salesman Problem." *J. Oper. Res.*, Vol. 11 (November, 1963), pp. 972-989.

[42] H. Loberman and A. Weinberger, "Formal Procedures for Connecting Terminals With a Minimum Total Wire Length." *J. ACM*, Vol. 4 (October, 1957), pp. 428-437.

[43] J. S. Mamelak, "The Placement of Computer Logic Modules." *J. ACM*, Vol. 13 (October, 1966), pp. 615-629.

[44] J. Munkres, "Algorithms for the Assignment and Transportation Problems." *J. SIAM*, Vol. 5 (1957), pp. 32-38.

[45] C. E. Nugent, T. E. Vollman, and J. Ruml, "An Experimental Comparison of Techniques for the Assignment of Facilities to Locations." *J. Oper. Res.*, Vol. 16 (1968), pp. 150-173.

[46] O. Ore, *Theory of Graphs*. Providence, R. I., American Mathematical Society, 1962.

[47] T. Pomentale, "An Algorithm For Minimizing Backboard Wiring Functions." *Comm. ACM*, Vol. 8, No. 11 (November, 1965), pp. 699–703.

[48] T. Pomentale, "The Minimization of Backboard Wiring Functions." *SIAM Rev.*, Vol. 9, No. 3 (July, 1967), pp. 564–568.

[49] R. C. Prim, "Shortest Connection Networks and Some Generalizations." *Bell Sys. Tech. J.*, Vol. 36 (November, 1957), pp. 1389–1401.

[50] S. Reiter and G. Sherman, "Discrete Optimizing." *J. SIAM*, Vol. 13 (September, 1965), pp. 864–889.

[51] R. A. Rutman, "An Algorithm For Placement of Interconnected Elements Based on Minimum Wire Length." *Proc. SJCC* (1964), pp. 477–491.

[52] K. Steiglitz and P. Weiner, "Some Improved Algorithms For Computer Solution of the Traveling Salesman Problem." *Proc. of 6th Annual Allerton Conf.* (October, 1968), pp. 814–821.

[53] L. Steinberg, "The Backboard Wiring Problem: A Placement Algorithm." *SIAM Rev.*, Vol. 3, No. 1 (January, 1961), pp. 37–50.

[54] S. Vajda, *Mathematical Programming*. Reading, Mass., Addison-Wesley Publishing Company, Inc., 1961.

[55] M. N. Weindling, "A Method For Best Placement of Units on a Plane." *Proc. 1964 Share Design Automation Workshop*, and Douglas Paper 3109, Douglas Aircraft Co., Santa Monica, California.

APPENDIX

Assignment Algorithms

A.1 MUNKRES' ALGORITHM FOR THE ASSIGNMENT PROBLEM

There are a number of algorithms specifically directed at the solution of the assignment problem. Many of these are variants of Kuhn's method [31] (also known as the Hungarian method). We assume a cost matrix, A, with nonnegative elements. The following steps, as outlined by M. Flood [12], must be implemented to find a minimal cost assignment.†

---

†A maximal cost assignment can be found by operating with the given algorithm upon a transformed cost matrix. Hence, if the elements of the cost matrix (A_{ij}) lie in $[a, b]$ then the transformed matrix (A'_{ij}) is defined by $(A'_{ij}) = (b - a_{ij})$.

Step 1: Subtract the smallest element in matrix A from each element of A, obtain a matrix A_1 with nonnegative elements and at least one zero.

Step 2: Find a minimal set S_1 of lines, n_1 in number, which contain all the zeros of A_1. If $n_1 = n$, there is a set of n independent zeros,† and the elements of A in these n positions constitute the required solution.

Step 3: If $n_1 < n$, let h_1 denote the smallest element of A_1 which is not in any line of S_1. Then $h_1 > 0$. For each line in S_1, add h_1 to every element of that line; then subtract h_1 from every element of A_1. Call the new matrix A_2.

Step 4: Repeat steps 2 and 3 using A_2 in place of A_1. The sum of the elements of the matrix is decreased by $n(n - n_k)h_k$ in each application of step 3. Thus, after a finite number of steps, there must be a zero which is not on any line of S_1. Therefore the process must terminate after a finite number of steps.

These steps must be operationally specified by some algorithm. An example of one that does so in an efficient manner is the Munkres' assignment method. That algorithm, as given in [44], is now stated in a paraphrased and condensed form (the proof of the correctness of the algorithm is omitted).

In the description of the algorithm, certain rows and/or columns of the matrix are referred to as "covered," i.e., certain rows and/or columns are tagged. A zero element is referred to as *zero star* ($z*$) if coverd twice, a *zero bar* ($\bar{z}$) if covered once, and a *zero* (z) if not covered.

1. On each row, subtract the smallest element from all elements on the row. Next do the same for the columns of the resulting matrix.

2. Select a zero, z. If there is no $z*$ in its row or column, star z. Repeat, considering each zero in the matrix. Next cover every column containing a starred zero.

3. Choose a noncovered zero (a zero in a noncovered row or column) and bar it. If there is no starred zero in the row of the newly created $\bar{z}$, go directly to step 4. If there is a starred zero in the row, cover the row and uncover the column of the $z*$ in the row of the new $\bar{z}$. Repeat until all zeros are covered, and then go to step 6.

4. Form a sequence of alternating $z*$ and $\bar{z}$ that starts with an uncovered $\bar{z}$; and next has a $z*$ in the $\bar{z}$ column (if no $z*$ exists in $\bar{z}$ column, terminate the sequence); and next has a $\bar{z}$ in the $z*$ row; and next has a $z*$ in the previous $\bar{z}$ column (if no $z*$ exists in the $\bar{z}$ column, terminate the sequence); etc.

5. In the sequence just formed, unstar each $z*$ and star each $\bar{z}$. Now, erase all bars, uncover every row, and cover every column containing a $z*$. If all columns are covered, the set of $z*$'s forms the desired assignment solution; otherwise, return to step 3.

†Zeros whose rows and columns are all distinct.

6. Let h denote the smallest noncovered element of the matrix. Add h to each covered row; then subtract h from each uncovered column. Return to step 3 without altering any stars, bars, or covered lines.

A.2 APPROXIMATION METHODS FOR THE ASSIGNMENT PROBLEM

For large assignment problems it is frequently necessary to employ approximation techniques. Kurtzberg [33] has investigated empirically and analytically the error bounds of some approximation techniques and has determined the average number of operations required in terms of the order of the assignment problem.

One of these methods, the row-scan method, is recommended for consideration as an alternative to the exact assignment alrogithm in the event that many assignment sets must be found and the expense of computation time is a problem.

In this method, all the rows are sequentially examined, and the minimal uncovered element in each row is selected. The row is then assigned to the column of the selected element, and the column covered, i.e., all the elements in that column are covered. Each row and column are thus uniquely assigned to one another, thereby constituting an assignment solution. The cost is computed for the assignment solution by summing the values of all the selected elements.

The method is quite easy to mechanize (indeed, it is simple enough to accomplish using a card sorter or even by unaided hand), and surprisingly good solutions are obtained with this straightforward technique.

The bound on the expected relative error for the row-scan method finding a maximal solution (see first footnote in Section A.1) where the entries in the cost matrix are uniformly distributed between 0 and 1 is $[ln(n + 1)]/n$, i.e., the error tends toward zero as n becomes large. The order of computation time growth is n^2.

chapter six

ROUTING

SHELDON B. AKERS

General Electric Company
Syracuse, New York

6.1 INTRODUCTION

One of the most difficult and yet most intriguing steps in the overall computer design process is that of *routing*—the process of formally defining the precise conductor paths necessary to properly interconnect the elements of the computer. The basic problem is easily stated: Given an interconnection diagram (or the information therefrom), such as in Fig. 6-1, and a circuit board on which the elements in the diagram have been previously placed (Fig. 6-2), lay out the necessary conductor paths on the board to achieve the indicated electrical connections *subject to the imposed constraints*. Needless to say, it is the many and varied constraints that cause the trouble: No conductor can be wider than 25 mils; at most two conductors per pin;

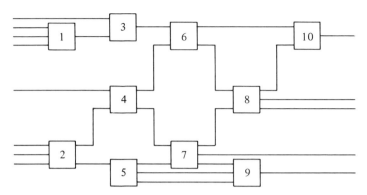

Fig. 6-1. Interconnection diagram.

don't allow conductors to cross; minimize feedthroughs; maintain at least a 25-mil spacing between conductors; keep all lengths under 5 inches; locate all input conductors on layer 2; don't use more than 5 layers; etc. Each new problem seems to add another constraint.

In the discussion which follows, a step-by-step approach to the general problem of routing will be described. Its purpose is twofold: to present a series of algorithms—some formal but many heuristic—which collectively have been proven to constitute an efficient interconnection program; and to attempt to identify specific problems and alternate approaches whose investigation and application can lead to new and better techniques.

The process of algorithm development for a design automation system is usually one of compromise. A detailed study of a particular problem typically shows that a "poor" solution can be obtained in about 5 seconds of computer time, and the "best" solution, in about 5 hours. Thus, a compromise between running time and optimality of solution must be effected. Accordingly, since the process which will be described consists of quickly and systematically solving a series of small problems, the reader should anticipate few ventures into the realm of, say, quadratic programming. On the other hand, as the critical stages in the process are reached, neither will he be asked to simply cut and try.

6.1.1. The Multilayer Board and Feedthroughs

Figure 6-2 shows a typical multilayer board on which the various interconnections are to be made. As a result of an automated (or manual) placement procedure, the various elements to be interconnected have been assigned specific locations on this board. The pins of these elements are assumed to extend through all of the several layers. Thus, if pin A is to be connected

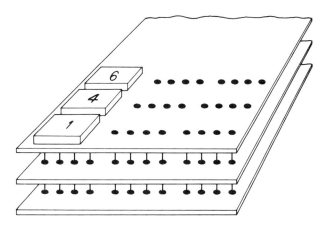

Fig. 6-2. Multi-layer board.

to B, such an interconnection may be laid out on any one of the layers. Moreover, if it is not possible to achieve such a connection on any of the layers, then a *feedthrough pin* (or *via*) may normally be employed. This is essentially a dummy pin, also extending through all layers, which may be employed to route an interconnection path from one layer to another. Thus, another way to interconnect A and B would be to lay out a path on layer 1 from A to a feedthrough pin and then on layer 2 to lay out a path from this same feedthrough pin to B. Feedthroughs may take the form of additional pins which are inserted as needed (often in one of a set of prespecified locations) or they may be actual element pins which are not otherwise used. For both electrical and mechanical reasons, it is often desirable to hold the number of feedthroughs to a minimum, although their complete elimination is seldom possible. For this reason, many of the procedures to be described are aimed at maximizing the number of wires that can be laid out completely on a single layer, thereby minimizing the need for feedthroughs.

6.1.2. The Rectangular Grid

In any routing problem the first constraints that must be considered are those imposed on the dimensions of the conductors which are to be laid out. These constraints are typically of two types: a minimum conductor width, w, and a minimum spacing, s, which must be maintained between conductors. It is often convenient to combine these two requirements into a single center-to-center constraint, Δ, where $\Delta = w + s$. Now if a uniform rectangular grid is defined having an incremental spacing of Δ, then by requiring that all conductors follow the lines of this grid, one automatically ensures that the width and spacing requirements will be satisfied. The most common value for Δ on present-day circuit boards is 50 mils, with the pins of the circuit elements usually located on 100-mil centers.

The use of such a grid automatically precludes any curved or diagonal runs for the conductors such as are typically seen in manual layouts. However, this slight loss in generality is usually far outweighted by the ease with which conductor runs may be represented and by the avoidance of repeated geometrical considerations.

6.1.3. Manhattan Distance

An immediate consequence of this grid approach is that the conventional Euclidean distance formula no longer applies. Thus, in Fig. 6-3 the minimum conductor distance between points A and B is simply $x + y$ since diagonal runs are not permitted. This distance measure has been called *Manhattan distance*, owing to a reputed tendency of Manhattan cab drivers to give distances as, "6 blocks west and 11 blocks north—17 blocks." Throughout this discussion, all distances will be Manhattan unless otherwise specified.

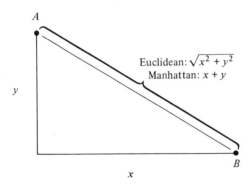

Fig. 6-3. Euclidean / Manhattan distance.

Manhattan geometry has a number of interesting properties: circles become diamonds; ellipses become hexagons; there are, in general, many shortest paths between two points; etc. One practical advantage is that the calculation of Manhattan distance is typically an order of magnitude faster than that required in the Euclidean case.

6.1.4. The Basic Approach

Many persons who have not followed through the steps of an operational wire layout program tend to dismiss the program as "just an application of Lee's algorithm." Certainly Lee's algorithm is a valuable tool in the final layout process. In practice, however, a successful layout results from a

series of closely interlocked steps, each designed towards the ultimate goal of making the final layout procedure as simple and as efficient as possible. The four basic steps which are to be followed are:
1. wire-list determination;
2. layering;
3. ordering;
4. wire layout.

Roughly speaking, these four steps supply, respectively, the same information as that of the traditional well-written news story: what, where, when, and how. In the first step, wire list determination, a listing of precisely *what* wires are to be laid out is compiled. Since the given interconnection data normally allow considerable freedom in exactly which wires are chosen, a judicious selection at this stage can greatly simplify the subsequent procedures.

In the second layering step, an attempt is made to decide exactly *where* each wire should be located. This consists of tentatively assigning each wire to a particular layer of the board so that within each layer the number of "conflicting" wires is minimized. The next step of ordering decides for each layer precisely *when* each wire assigned to that layer is to be laid out, in other words, the *order* in which the wires will be processed. Finally, the wire-layout step answers the ultimate question of exactly *how* each wire is to be routed.

6.2 WIRE-LIST DETERMINATION

The input information for this first step consists of an *interconnection list* which tells for each module pin those other pins to which it must be connected. Figure 6-4 shows such a list. Here module-1-pin-1 must connect to module-3-pin-6; module-1-pin-2 must connect to module-4-pin-1; etc. These single-pin to single-pin connections are entered directly onto the wire list as wires to be laid out by the program. However, when a pin must be connected to more than one other pin, complications arise. Pin 1-7, for example, must connect to pins 2-1, 2-6, and 4-9. One possibility, of course, would be to simply require that three wires be laid out from 1-7, one to each of the three load pins. On the other hand, perhaps a better solution would be to connect 1-7 only to 2-1 and to 4-9 and then to connect 2-1 to 2-6. The exact answer must depend on the interconnection constraints imposed by the designer and the actual pin locations.

In general, to interconnect n pins, $(n - 1)$ wires are required. If no constraints are imposed on how the interconnections are to be made, there exists a very neat algorithm for finding the interconnecting set of wires

288 Routing   Chap. 6

Source	Loads
1-1	3-6
1-2	4-1
1-4	4-3
1-7	2-1, 2-6, 4-9
1-11	7-6
2-2	C
2-4	6-6
2-5	C, 4-7
3-1	4-2, 5-8, 5-11, 6-2
.	.
.	.
.	.

Fig. 6-4. Interconnection list.

having the *smallest total length*. This is known as the minimum tree algorithm [35], [39].

6.2.1. The Minimum Tree Algorithm

A simple example will illustrate its operation. Consider the six points (pins) shown on their Manhattan grid in Fig 6-5a. How should these six points be interconnected so that the total wire length is a minimum? The algorithm

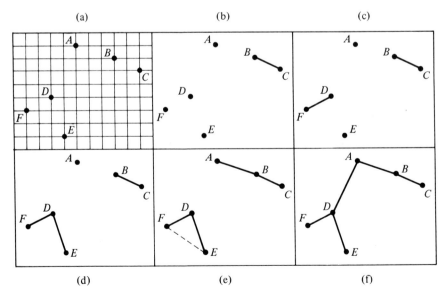

Fig. 6-5. Minimum tree example.

begins by simply interconnecting the pair of points which are closest together. In this case, pairs *B-C* and *D-F* are both a Manhattan distance of 3 apart. Either may be selected. Say pair *B-C* is selected, as shown in Fig. 6-5b. (It is convenient to show the selected interconnections as straight lines even though they ultimately must follow the rectangular grid lines.) Now the next closest pair (*D-F*) is interconnected, etc. However, *whenever the next closest pair involves two pins for which an interconnection path already exists, this connection is skipped.* Here, for example, after four interconnections have been made, the next closest pair is *E-F* (Fig. 6-5e). Since there already exists a path (via *D*) between these points, this connection would be skipped. The next closest pair (*A-D*) would then be selected, yielding the "tree" shown in Fig. 6-5f. The resulting five wires would now be added to the wire list. (Note that the computer is not actually making connections at this step. It is merely deciding what wires will eventually be laid out.)

For the designer who applies this algorithm manually to points plotted on graph paper, the tedious part is the determination of the distances between the various pairs of points. The checking as to which pairs of pins are connected is easily done by observation. Ironically, with a computer the situation is reversed. At any step, the computer quickly finds the next closest pair of pins, but the problem of deciding whether or not there *already* exists a path between these pins is something else again.

6.2.2. The Interconnection Matrix

A method of resolving this difficulty involves the use of an *interconnection matrix*, [*I*]. This is an $n \times n$ symmetric binary matrix where $I_{ij} = 1$ if a single branch exists between point *i* and point *j*, and 0 otherwise. (I_{ii} is initially taken as 1.) With such a matrix, a record can be kept of each interconnection as it is made. Figure 6-6 shows this matrix after the first four interconnections in the previous example have been made. Now it turns out that if we square this matrix in the usual manner, except that we use logical instead of arithmetic summation, the resulting binary matrix, [I^2], will have 1's wherever two points are connected by a path having at most *two* branches. In general, [I^k] will indicate all points connected by paths with at most *k* branches. Clearly, to discover all such paths, it is only necessary to continue raising the power of the matrix until the number of 1's stops increasing.

Here, for example, [I^2] introduced 1's for pairs *A-C* and *E-F*, indicating that they are each already interconnected by a path with two branches. [I^3], however, produced no new 1's, so the search for additional paths would end. Thus, the next interconnection to be made would be between the closest pair having a 0 in [I^2].

A number of variations on this interconnection matrix approach are possible, depending on the degree of sophistication required. Actual interconnec-

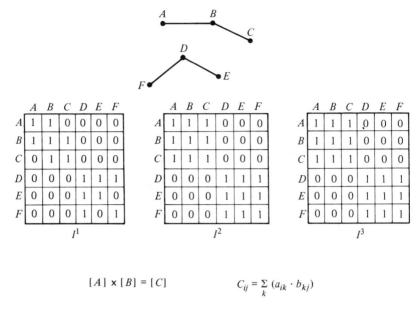

Fig. 6-6. The interconnection matrix.

tion lengths may be entered, for example, and by suitably altering the basic matrix operations, a tabulation of the minimum path lengths between every pair of points may be obtained. Various list-structure approaches [46] are also available for keeping track of the connected points.

Unfortunately, the designer is often not free to interconnect a set of pins in any manner he chooses. Constraints such as maximum number of wires per pin or maximum allowable distance between any pair of pins are quite common. Typical is the requirement that the pins be interconnected serially by one continuous chain. The problem of finding the particular chain which has the minimum length turns out to be equivalent to the classical traveling salesman problem [13], [28].

6.2.3. The Traveling Salesman Problem

In its most common formulation, this problem involves a traveling salesman who desires to start at, say, Albany, New York, visit the other 49 state capitals, and return to Albany by the shortest path (such a path will be called a *loop*). Since there are 49!/2 possible loops, an exhaustive search is clearly impossible. This problem has received a great deal of attention, particularly in the field of Operations Research; as yet, however, no neat algorithm has been discovered. (Another unsolved problem in graph theory which is closely related is that of finding a simple test to determine whether or not a graph has a Hamiltonian circuit.)

Sec. 6.2   Wire-List Determination   291

It is easy to show the equivalence of this problem to that of finding the shortest chain (rather than loop) through a set of points. (A *chain* results when the salesman halts at the 50th capital rather than returning to the starting capital.) Assume that a 51st state capital was introduced which was the same distance, D, from each of the other 50. Then clearly the minimum loop through all 51 state capitals would be of length $C + 2D$, where C is the length of the minimum chain through the original 50. This means that any algorithm for finding the shortest loop will also allow the shortest chain to be found. (To show the reverse, that an algorithm for the shortest chain also solves the shortest-loop problem, is slightly more involved, but also straightforward.)

An intuitively appealing method of attacking the shortest-chain problem is to choose a starting point and then at each step simple *go to the closest point not already in the chain*. Although this method will not always give the shortest chain, in practice, it has the virtue of being quite fast and, more often than not, of yielding a near minimum chain. Because of its simplicity and speed, there is usually time enough to try each of the n points as a starting point and then to select the minimum of the n chains so obtained. Figure 6-7 shows the result of applying this approach to the six points examined previously. The C-chain, *C-B-A-D-F-E*, would be selected.

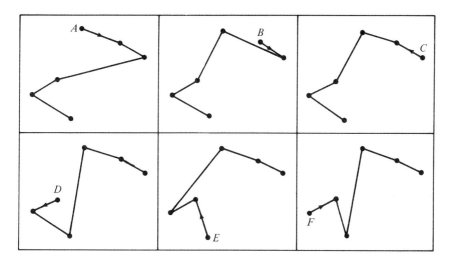

Fig. 6-7. Traveling salesman example.

6.2.4. Alternate Approaches

Other interconnection constraints can often be accommodated by modifying one of these algorithms. The minimum tree algorithm is particularly useful in this regard since, as each branch is selected, a check can be made that the inclusion of this branch will not violate any additional constraints which may be involved. (If a violation is indicated, the next-best branch is selected.)

There is another useful version of the minimum tree algorithm which has the virtue of avoiding the need for an interconnection matrix. This version is not as easily modified to accommodate additional constraints, but when such constraints are absent, it is quite easy to apply. We begin by putting a check on any point. Thereafter, we simply find the shortest branch between a checked and an unchecked point. This branch is added as a connection and its unchecked point is checked. The process halts when all points are checked. If, for example, point E was initially checked in Fig. 6-5a, then the branches which would be successively added would be: *D-E*, *F-D*, *A-D*, *B-A*, and *C-B*, again yielding the minimum tree of Fig. 6-5f.

Another approach to this problem of interconnecting a set of points involves the introduction of intermediate points so that an even shorter interconnection tree may be generated. This can be done in several ways. One possibility is to postpone the problem until the actual layout step and then let the routing procedures dictate the best choices. This approach is described in Section 6.5.10. A disadvantage of this approach is that, without a definition of the exact wires which will be laid out, it becomes quite difficult to determine the interrelationships between these interconnections and others during the layering and ordering processes. Hence, when the time comes for the final layout, a much longer, rather than shorter, set of interconnections may be required.

6.2.5. Steiner's Problem

Another possibility leads directly to the Manhattan version of Steiner's problem [12], [21]. This is the problem of finding, for a fixed set of n points in the plane, the shortest network for completely interconnecting them (i.e., the minimum network which contains a path between any pair of points). Another way of stating the problem is: Add k additional points (k may be 0) so that the minimum tree for the $n + k$ points is a minimum. This problem has never been solved in general for either the Euclidean or the Manhattan case. However, results are available for small numbers of points.

Particularly with a rectangular grid, the known results [23], plus obvious rules of thumb, should make near optimal solutions readily obtainable. Again, a possibly offsetting disadvantage of this approach is that new "pins" now appear on the wire list which may or may not cause data-handling complications.

Sec. 6.2   Wire-List Determination   293

6.2.6. Pin-To-Connector Assignment

A second task presented by the interconnection list (Fig. 6-4) involves those pins which must be tied to some connector pin, C. A board will normally contain a number of such pins, usually located along its edges, which serve to route signals to and from other boards. Ordinarily, when an element pin must be tied to a connector pin, any such pin may be chosen. Hence, the problem arises of just how to most efficiently make these pin-to-connector assignments.

Since the ultimate goal is to make the wire layout process as simple as possible, an obvious criterion for such an assignment is to attempt to ensure that, when straight lines are drawn between the element pins and their corresponding connectors, *no crossovers occur*. A simple way of accomplishing this is shown in Fig. 6-8. Given a row of eight connector pins and eight element

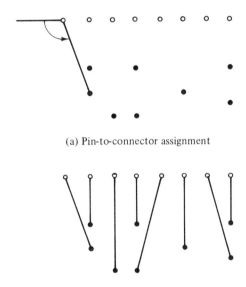

(a) Pin-to-connector assignment

(b) Final assignment

Fig. 6-8. Pin-to-connector assignment.

pins which must be connected to them, the leftmost connector pin is selected and a vector is rotated counterclockwise (as shown) until the first of the element pins is reached. This pin is then assigned to that connector. If two or more pins are reached simultaneously (as occurs in the fourth step in this example), the *closest* pin is chosen.

When the connector pins are in a straight line, this process ensures that no intersecting lines will result. However, if there are considerably more connector pins than needed, then a skewed assignment may result. (See Fig. 6-9.)

294   Routing   Chap. 6

One way of avoiding this effect is to calculate at each step the "slack" available, i.e., the ratio of the extra connectors remaining to the number of as yet unassigned pins. If this ratio is greater than or equal to one *and* if the indicated element pin is closer to some other available connector pin, then the current connector pin is skipped. The lower portion of Fig. 6-9 shows this rule in operation.

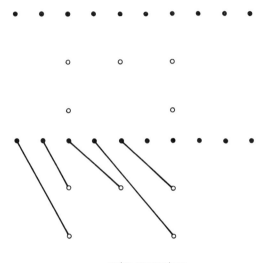

$$\text{Slack} = \frac{\text{extra connectors}}{\text{unassigned pins}}$$

Rule: Skip connector if (a) indicated pin is closer to some other connector or (b) slack ≥ 1.

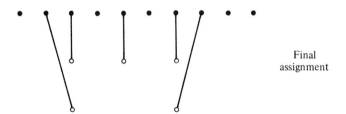

Final assignment

Fig. 6-9. Pin-to-connector assignment (excess connectors).

6.2.7. Input-Pin Permutation

A final technique, which may be employed at this step in the process in order to simplify the later wire-routing problems, involves an inspection of the input wires to the various elements. Usually many of the pins to which these wires connect are *logically equivalent*. In other words, the wires to such pins may be interchanged without affecting the logical operation of the circuit. (A typical case is the set of input pins for a symmetric logic gate such as an AND, OR, or NOR gate.) Again, a good rule for interchanging wires is to avoid straight-line intersections. Figure 6-10 shows how a series of potential wiring conflicts can be avoided by simply noting the equivalence of various pins and permuting the wires accordingly.

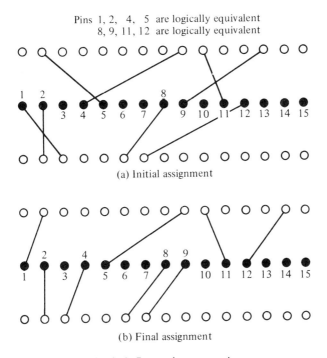

Fig. 6-10. Input pin permutation.

Incidentally, this is a good example of a case which could be rephrased as an intellectually challenging assignment problem. However, since in practice the number, p, of pins in an equivalent set seldom, if ever, exceeds five or six, it is quite feasible to simply search through all $p!$ permutations and pick the best one.

6.3 LAYERING

Once the list of wires to be laid out has been precisely defined, the problem of deciding exactly where and when each wire should be laid out is considered. The "where" part of this process is known as *layering* and involves tentatively assigning each wire to one of the layers of the given multilayer board in such a way as to minimize the wiring difficulties on each layer. One way to visualize this problem is to assume that, on a single layer, every wire is drawn as a straight line between the two pins involved. (See Fig. 6-11.) Now, if there are, say, three layers available, then the layering

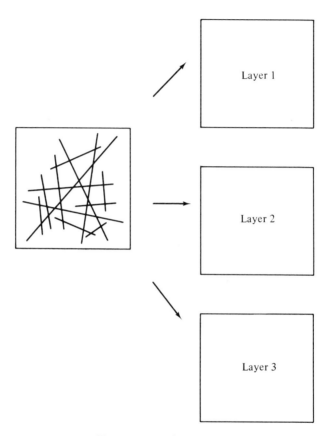

Fig. 6-11. Layering problem.

problem becomes one of locating the wires on the three layers so that on each layer the number of crossovers is minimized. Since the final layout of any wire must consist of only horizontal and vertical segments, the fact that the straight (Euclidean) lines between two pin-pairs cross does not mean that their minimum Manhattan paths must necessarily cross. (See Fig. 6-12.) On the other hand, the fact that the straight lines do cross does imply that there is certainly a potential conflict between this pair. (It is easily shown that whenever the minimum Manhattan paths *must* intersect, then the Euclidean paths will also intersect.)

Sec. 6.3 Layering 297

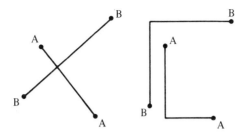

Euclidean intersection   No Manhattan intersection

Fig. 6-12. Euclidean/Manhattan intersection.

There are a number of possible approaches to the layering problem, depending on the desires of the user. It can, in fact, be rephrased as a classical problem in graph theory (Section 6.3.2). There is, however, an initial observation that can be made which can lead to considerable simplification of the general problem:

If a set of wires is to be partitioned between k layers, any wire which intersects less than k others may be ignored.

Consider a wire X which intersects just two others, A and B. Assuming that three layers are involved, the rule says that X may be ignored. The reason is that, if X is ignored and the other wires are partitioned between the three layers, there will always be one layer on which neither A nor B appears. Hence, X can always be located on this layer without intersections. Likewise, with k layers, and $k - 1$ wires intersecting X, there will always be at least one layer where none of the $k - 1$ wires appear and, hence, where X may be located.

In practice this rule can be quite useful. Consider again the set of wires in Fig. 6-11. With three layers, the rule permits the removal of the short wire in the right-center since it intersects only two others. (See wire 1 in Fig. 6-13.) With this wire gone, the lower of the two wires it intersected now has itself only two remaining crossovers. Hence, it also can be removed (wire 2 in Fig. 6-13). Proceeding in this way, we can successively remove five wires from further consideration. Note, however, that *the order in which these wires were removed must be remembered, and they must be relocated in the opposite order.* Wire 2, for example, could be removed only *after* wire 1 was gone; hence, it must be replaced *before* wire 1.

An immediate consequence of using this rule is that, in the remaining set of wires, each wire will intersect at least k others. At this point, assume all of the remaining wires are on layer 1. The general strategy is to now begin removing wires from this set and placing them on layer 2 until no inter-

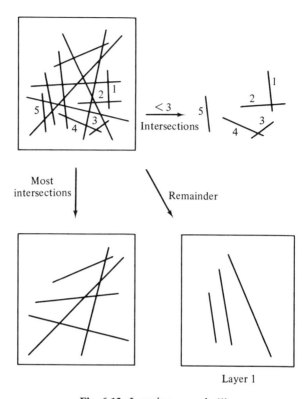

Fig. 6-13. Layering example (1).

sections remain on layer 1. A reasonable way to accomplish this is by a second rule: *At each step remove that wire having the most intersections.* In this manner, five wires are removed to layer 2, and three (nonintersecting) wires remain on layer 1 (bottom of Fig. 6-13).

Now the entire process may be repeated. (See Fig. 6-14.) Since only two layers remain, only wires having one intersection may be removed by the first rule (wire 6 in Fig. 6-14.) Again the order of removal must be retained. Two wires then go onto layer 3, and two remain on layer 2. It is interesting to note that the remaining wires on layer 2 are roughly orthogonal to those on layer 1 (Fig. 6-13). This is a typical result of this process. Although no explicit calculation of direction is made, each layer will normally end up with a clearly discernible polarization.

When the last layer (in this case, layer 3) is reached, wires with the most intersections are again removed—this time into a *residue* class—until again no intersections remain (Fig. 6-15). Each remaining wire in the residue class is now placed on the layer where it "does the least damage," i.e., where it will intersect the fewest other wires. (In this example, the one wire in the res-

Sec. 6.3   *Layering*   299

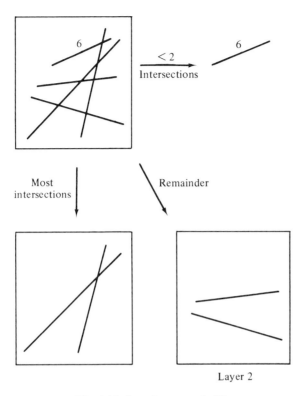

Fig. 6-14. Layering example (2).

idue class would go back to layer 3.) Finally, all of the wires removed by the application of the first rule are placed in the opposite order from that in which they were removed. (Note that all of these will go on with no intersections—it is simply a question of finding the available layer.) When this is done in the example, the final layering shown in Fig. 6-16 results.

A final touch-up procedure can now be applied if desired. This consists of simply examining each wire individually and checking to see if there is some other layer to which it can be moved with a resulting decrease in the number of wires which it intersects. If such a move is possible, then it is made. This process continues until every wire is on its best layer.

6.3.1. Intersection Calculation

The application of the foregoing procedure requires continual reference to the question of whether or not a given pair of wires intersect. The ideal way to resolve this problem is to generate, initially, a binary *intersection matrix* where a 1 in cell ij means that wires i and j intersect, and 0 means that

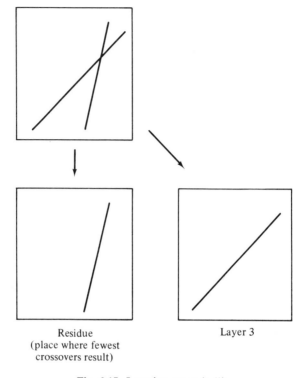

Fig. 6-15. Layering example (3).

they do not. All of the foregoing rules can then be quickly carried out by simply referring to this matrix. Unfortunately, with typical problems, the number of wires involved may be up in the hundreds, if not in the thousands; thus the size of such a matrix may become prohibitively large. On the other hand, the alternative of continually rechecking to see whether or not a given pair intersects is equally unattractive from the standpoint of running time. Clearly, some sort of trade-off between memory size and running time is needed. Since this type of problem occurs quite often in the process of programming the various design automation algorithms, it is worthwhile to describe how it can be resolved for this case.

The basic approach is to calculate and store for each wire—not the labels of the wires which it intersects, but rather just the total number of such wires. With W wires, this requires a list W long instead of a matrix of W^2 elements. Now such questions as, "Does wire A intersect less than three others?", or, "Which wire intersects the most others?", are quickly answered. The decrease in storage requirements is achieved at the cost of having to update the intersection list each time a wire is removed. This requires that the complete list be processed to find the remaining wires which intersect the removed

Sec. 6.3 Layering 301

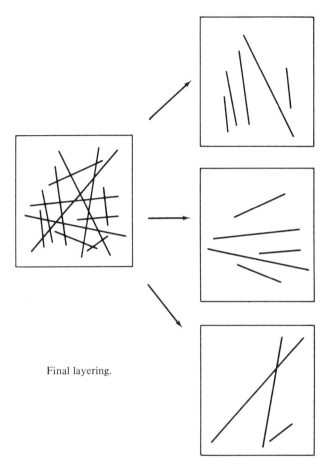

Final layering.

Fig. 6-16. Final layering.

wire so that the intersection counts of these wires can be lowered by one. However, this increase in processing time is normally not prohibitive and is more than offset by the resulting decrease in memory size.

The problem of deciding whether or not a given pair of wires intersect is a simple exercise in analytic geometry. Knowing the coordinates of the two pins to which a wire must connect, we can calculate the equation of the straight line through these two pins. If the two resulting lines are parallel but not collinear, then no intersection exists. If they are collinear, then it is a simple task to determine if the two wires share a common point on the line. When the two lines are not parallel, their equations may be solved simultaneously to find their point of intersection. A check is then made to see if this point lies on both of the wires. Figure 6-17 shows these various cases.

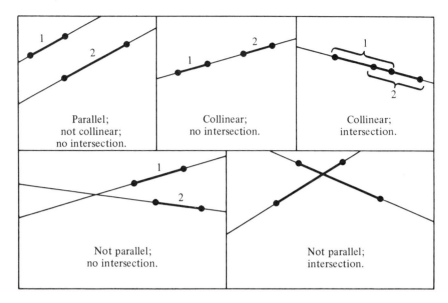

Fig. 6-17. Possible intersections.

6.3.2. The Intersection Graph and Chromatic Numbers

As with many of the problems associated with Design Automation, the layering problem may be viewed as a classical problem in graph theory. To do this we present the initial input data, i.e., the information regarding which pairs of wires intersect, in the form of a linear graph known as the *intersection graph*. Consider the five intersecting wires in Fig. 6-18. Their

Nodes = wires

Node A connected to node B
if wire A intersects wire B

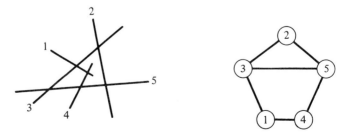

Fig. 6-18. The intersection graph.

Sec. 6.3 *Layering* 303

intersection graph, shown at the right of the figure, has one node corresponding to each wire; a branch between a pair of nodes indicates that the corresponding wires intersect. Clearly such a graph gives a much better picture of the various intersection relationships.

To restate the layering problem in terms of this graph, it is useful to employ a well-known concept from graph theory, namely, the chromatic number of a graph:

The *chromatic number* of a graph, G, is the smallest number of colors with which the nodes of G may be colored such that no branch connects nodes of the same color.

With the intersection graph, coloring the nodes corresponds to assigning each wire to a (colored) layer. Moreover, if each branch has its two nodes colored differently, then no layer will have any intersecting wires. Therefore, the minimum number of layers required for an intersection-free partitioning of the wires is precisely equal to the chromatic number of the intersection graph.

Even a modest survey of the many results concerning chromatic numbers is beyond the scope of this chapter. However, some examples of the applicability of these results will be given.

6.3.3. Koenig's Theorem and a Two-Color Problem

One of the best-known results concerning chromatic numbers is Koenig's theorem:

A graph can be two-colored if and only if it has no cycles of odd length.

In the present context, this says that a set of wires can be partitioned between two layers without intersections if and only if their intersection graph has no cycles of odd length. Consider the set of wires shown in Fig. 6-19 together with its intersection graph. Since this graph has only even cycles, according to Koenig's theorem, it should be possible to two-color it. To accomplish this, an arbitrary node (say 3) is selected and colored (say red). Now all nodes adjacent to 3 (in this case, 4 and 12) are colored black; all uncolored nodes adjacent to 4 or 12 are now colored red; etc. Note that as this process continues, the presence of an odd-length cycle would eventually result in a conflict, i.e., a branch with nodes of the same color. However, since none are present, the required two-coloring will always result. Figure 6-20 shows the layering induced by this coloring.

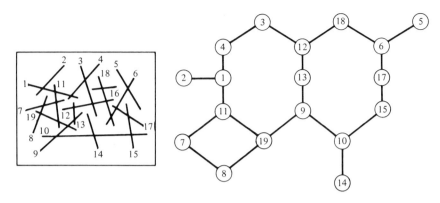

Fig. 6-19. Two-color example.

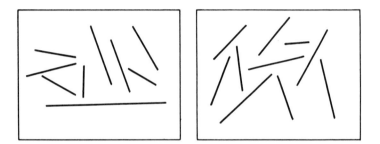

Fig. 6-20. A two-layer partition.

Unfortunately, in actual cases the probability that a set of wires will lead to an intersection graph without odd cycles is virtually zero. Thus, with two-layer boards, a more practical problem is how to two-color the intersection graph so that the number of *improper* branches, i.e., branches with nodes of the same color, is minimized. This problem occurs in a variety of practical applications and can, in fact, be restated in a number of different ways. (For example, find the minimum number of branches whose removal eliminates all odd cycles; find the bipartite subgraph having the maximum number of branches; divide the nodes into "transmitters" and "receivers" so that the maximum number of messages may be sent.)

This problem can, of course, be attacked using various heuristic techniques such as those described in Section 6.3. However, there is a formal result [2] which can be usefully employed:

A two-coloring is minimal if and only if in every cut at least half of the branches are proper.

A branch is said to be *proper* if its nodes are of different colors; otherwise, it is *improper*. A *cut* is any set of branches whose removal leaves two disconnected sets of points.

The proof is straightforward. If a cut does not meet the foregoing condition, then by complementing the colors of all nodes on one side of the cut, the propriety of all branches in the cut (and only those branches) will be reversed, thus giving fewer improper branches. Conversely, if a given two-coloring is not minimal, then a cut violating the condition may be found if we consider a better two-coloring and then divide the nodes into two classes: those receiving the same color in both colorings, and those receiving different colors. To apply this result to a given two-coloring, a search is made for cuts having more improper branches than proper. Whenever such a cut is found, a better two-coloring automatically results. In Section 6.3.5, another approach to this problem for two-layer boards is discussed.

6.3.4. The Four Color Conjecture

The most famous unresolved conjecture in graph theory and perhaps in all of mathematics is directly concerned with chromatic numbers. It states simply that:

Any planar graph can be four-colored.

Despite the efforts of many of the foremost mathematicians over the last century, this conjecture remains unresolved [42]. Five colors have been shown to be sufficient to color any planar graph, but no one has yet found a planar graph which *needs* five colors. In terms of the layering problem, this conjecture becomes: A set of wires may be partitioned between four layers without intersections whenever their intersection graph is planar. If by some chance the reader has occasion to use this result, he is strongly urged to do so despite the fact that it is merely a conjecture. If and when his procedure breaks down, it will be because he has found a counter example to the four-color conjecture—a small price to pay for a permanent niche in mathematical history!

6.3.5. Layering for Two-Layer Boards

As mentioned earlier, the primary goal of the layering process is to partition the various wires between the available layers so that hopefully each wire can then be routed solely on the layer to which it is assigned. This, in turn, has the effect of minimizing the number of feedthroughs which will ultimately be required. For boards of three or more layers, the process

306   *Routing*   *Chap. 6*

described in Section 6.3 has proven quite successful. There is, however, an important exception, namely, the two-layer board.

Here the emphasis is typically much different. Years of manual experience with the routing problem for such boards have shown that by far the most efficient approach is to route virtually all horizontal runs on one layer and all vertical runs on the other. This implies the use of a much greater number of feedthroughs, since many connections will require runs on both layers.

The use of a procedure such as that described in Section 6.3 will, in practice, normally result in clearly discernible horizontal and vertical layers, i.e., the average slope of wires assigned to one layer will be roughly horizontal, while that on the other layer will be nearly vertical. The shortcoming of such a procedure, when applied to a two-layer board, is that it assigns each wire solely to one layer or the other and, also, that it ignores the various feedthrough locations which may well be needed during the actual layout of that wire.

An attractive alternative is to consider the possibility of assigning a wire not to a single layer but to both layers after first breaking it into two segments, one as nearly horizontal as possible, and the other nearly vertical. (See Fig. 6-21). Since the break may occur at any available feed through location, these may be systematically examined in order to select the one which most nearly results in the desired vertical and horizontal segments.

Of course not all wires need be "broken." In particular, if the wire is

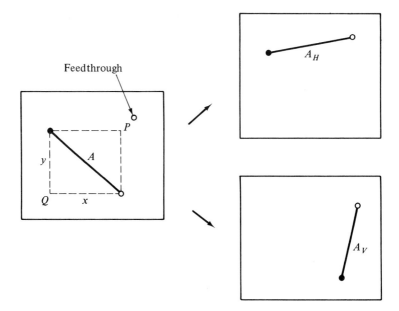

Fig. 6-21. "Breaking" a wire.

Sec. 6.3   Layering   307

already nearly horizontal (vertical), it is probably better to assign it completely to the indicated layer. A simple rule for making this decision is to define a small incremental distance, δ, and to assign to the horizontal layer all wires having a y-distance (see Fig. 6-21) less than or equal to δ. Likewise, those with an x-distance less than or equal to δ can go on the vertical layer. Other wires are "broken" as described before.

As an example of this process, consider the five wires shown as connected pin-pairs in Fig. 6-22. Assume that eight feedthrough locations are available,

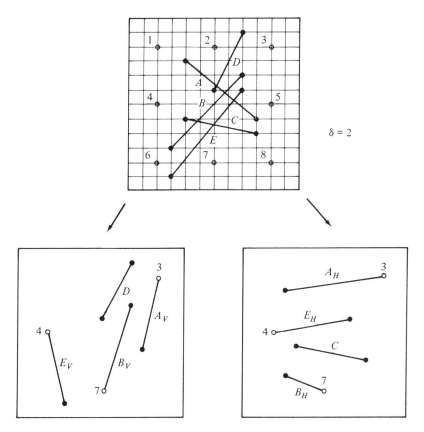

Fig. 6-22. Two-layer layering example.

as indicated by the hollow circles. If δ is chosen equal to 2 units, then wire C would be assigned to the horizontal layer, and wire D to the vertical layer. Wires A, B, and E would become candidates for the breaking process. Consider wire A. Its "rectangle" defines two ideal feedthrough locations—points P and Q in Fig. 6-21. A simple rule for selecting the feedthrough for A is to select that one whose distance to P or Q is a minimum. Thus in Fig. 6-22 feedthrough 3 would be selected for A. Likewise for wires B and E, feedthroughs 7 and 4, respectively, would be chosen. The lower portion of Fig. 6-22 shows the resulting two-layer partition. As with the other layering schemes, a final partitioning can often be enhanced if we reexamine each wire, in turn, to see if some more attractive alternative is available.

6.4 ORDERING

Once each wire or wire segment has been tentatively assigned to an individual layer of the board, each layer is then examined, in turn, to determine exactly *when* each wire on that layer should be laid out; in other words, in what *order* should the wires on each layer be processed.

Figure 6-23 illustrates the importance of this question. Assume that two

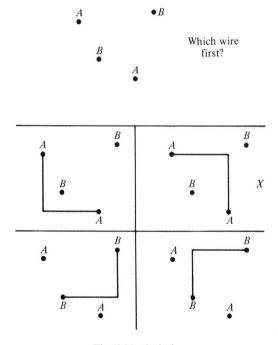

Fig. 6-23. Ordering.

wires are to be laid out—one between the pair of A-pins and one between the pair of B-pins. If each wire is laid out with just one turn, then for each wire there are only two choices—start vertically, then turn horizontally, or start horizontally and turn vertically. The two choices for each wire are shown in Fig. 6-23. Which choices cause difficulty in laying out the other wire? Inspection of the four cases shows that in just one case (when A is laid out first, with the turn at the upper right) will the other wire not go in optimally.

Sec. 6.4 Ordering 309

Thus, in this situation it would be desirable to lay out B first, since A will then go in without complication regardless of the (minimal) path followed by B.

A closer study of this two-wire case yields the following guideline: If a B-pin appears in the A-rectangle (i.e., the rectangle having the A-pins on opposite corners), then the B-wire should be laid out first. (See Fig. 6-24). Unfortunately, this rule is, at best, difficult to apply. Consider the three pin-pairs at the top of Fig. 6-25. When their rectangles are drawn, a squared

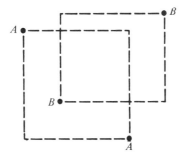

Fig. 6-24. B in A-rectangle.

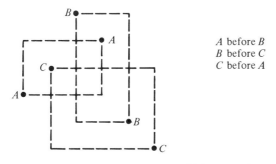

A before B
B before C
C before A

Fig. 6-25. Non-transitivity example.

version of a famous beer trademark is obtained which leads directly to the three requirements: A before B, B before C, and C before A. Thus, the above rule lacks the basic property of *transitivity* which is necessary in any ordering process.

6.4.1. An Ordering Rule

The rule does, however, have a message: When pin Y is in rectangle X, it is desirable to lay out wire Y before wire X. Likewise, if both Y-pins are in the X-rectangle, it is even more desirable that Y precede X. And, in general, the greater the number of pins in the X-rectangle, the greater the number of wires which should be laid out before X. This suggests the following simple rule for ordering the wires. Let

Priority number of wire X = number of pins in the X-rectangle

Wires are then laid out in the order of their priority numbers.

As an application of this rule, consider the four pin-pairs (A, B, C, D) at the top of Fig. 6-26. When the four rectangles are drawn, their priority numbers are seen to be 5, 1, 0, and 2, respectively; thus for this set the indicated layout order is C, B, D, A. Figure 6-27 shows the layout which would result when this order is followed.

As can be seen from the example, this rule normally leads to a nesting of the wires. Short horizontal and vertical runs tend to be laid out first; then almost vertical and horizontal wires surrounding them are laid out. Finally, the long runs are laid out and tend to be on the outside of the others. As with all of these algorithms, there is nothing sacred about this particular rule. It has the virtue of being strictly quantitative—only a simple pin-count is required—and, as such, leads to short running times and contented programmers. On the other hand, if a closer examination of the actual location of the various pin-pairs is made, a more efficient ordering procedure can surely be derived. Other nesting techniques are described in Sections 6.5.2 and 6.5.7.

6.4.2. Critical-Length Wires

Another ordering consideration that can occur involves critical-length wires. Particularly when high speed logic is involved, the designer may be told that the length of all wires must be less than a certain critical value. This constraint becomes of primary importance in the placement process and, of course, in the actual layout routines where the wire length is formally determined. However, it is usually advisable to also check during the ordering process for those wires whose pin-pairs are almost too far apart. Clearly, these wires should be laid out very early in the process to ensure that their actual lengths will, in fact, meet the required constraint.

Wires to be ordered

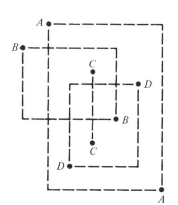

Ordering

A: 5 4th
B: 1 2nd
C: 0 1st
D: 2 3rd
C-B-D-A

Fig. 6-26. Ordering example.

Order: C-B-D-A

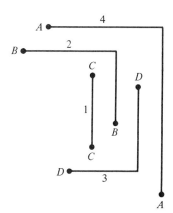

Fig. 6-27. Final layout.

6.5 WIRE LAYOUT

With all of the wires assigned to specific layers and the processing order defined for each layer, the moment of truth finally arrives. How are the individual wires to be laid out? Practically every routine since early in the partitioning and placement processes and on through wire-list determination, layering, and ordering has been specifically oriented towards this particular problem. Yet many complications still remain. Fortunately, there is a fundamental tool available which can greatly simplify this process.

6.5.1. Lee's Algorithm

This technique [33] is actually an application of the shortest-path algorithm widely used in Operations Research and graph theory [5], [41]. Here the paths under consideration must follow the cells of a rectangular grid; hence the algorithm becomes even simpler to apply. Consider the two pins, A and B, in Fig. 6-28. Is there a path between them and, moreover, if

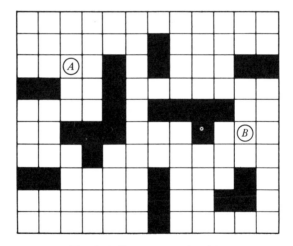

Fig. 6-28. Interconnect A and B.

there is more than one such path, which one is shortest? (At this point, it is convenient to assume that the basic rectangular grid is displaced slightly so that paths go from cell to cell rather than from intersection to intersection. Also, all nonempty cells will be so labeled.)

The process begins with the selection of either pin as a starting point. (As will be seen later, one pin may be more desirable than the other to minimize running time. In theory, though, either will work.) Say A is selected. Initially, a 1 is entered in each empty cell immediately adjacent to the cell containing

Sec. 6.5   Wire Layout   313

pin *A*. (See upper-left diagram in Fig. 6-29.) Next, 2's are entered in all empty cells adjacent to those containing 1's (upper-right diagram in Fig. 6-29). Next, 3's are entered adjacent to 2's; and so on. The process continues in this manner until one of two results occurs. Either all paths become blocked, i.e., on the kth step there are no empty cells adjacent to those containing $(k - 1)$, or, as in this case, the target cell, *B*, is reached. Here this occurs on the 13th step (lower-left diagram, Fig. 6-29). Immediately, two facts are known: There is a path from *A* and *B*, and the shortest such path is of length 13. All that remains is to actually find this path and to label the cells involved.

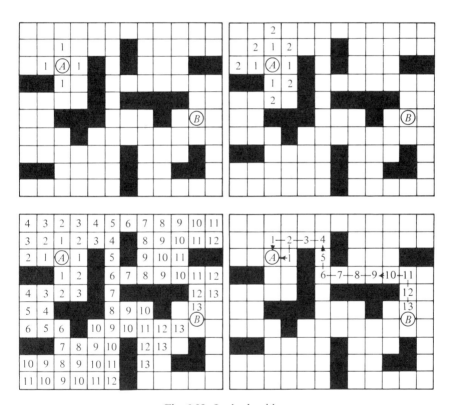

Fig. 6-29. Lee's algorithm.

6.5.2. Retracing

Since *B* was reached on the 13th step, it follows that there must be a 12 immediately adjacent to *B*. Likewise, this 12 must be adjacent to an 11, etc. Proceeding in this manner, it is an easy task to find a path leading back to *A* (lower right, Fig. 6-29).

A complication occurs when there is a choice of cells, i.e., when there are two or more cells with $(k - 1)$ immediately adjacent to k. (This happens with 2 in the example.) In theory any of these cells may be chosen and a path will still be found. However, in practice, there are several useful guidelines for making such choices. The first is: Don't change direction unless you have to. Thus, in the example, the segment from 3 to 2 is horizontal; therefore the 1 to the left of 2 would be chosen. This rule tends to minimize the number of turns in a given path. As the number of wires already laid out increases, cases are soon reached where in a given path a turn *must* be made, and there are two possible choices for the turn. Here a useful technique is to store a list of "compass" priorities, e.g., *N-E-W-S*. Using this rule, north (i.e., up) is chosen whenever possible; if not, east is tried; etc. The particular permutation chosen is relatively unimportant. What is significant is that whenever possible, the *same* direction is consistently selected. This leads to a uniform nesting effect which might not otherwise be achieved. (For a more sophisticated nesting technique, see Section 6.5.7.) This technique is also useful at the start of the retrace process when a choice of directions exists.

6.5.3. Three-Dimensional Paths

There is an almost limitless number of variations possible on this basic algorithm. One of the most useful involves its extension to three-dimensional grids—in particular, to multilayer boards with feedthroughs. As an example of this case, consider a popular three-dimensional maze which was widely distributed several years ago [19]. See Fig. 6-30. The maze consists

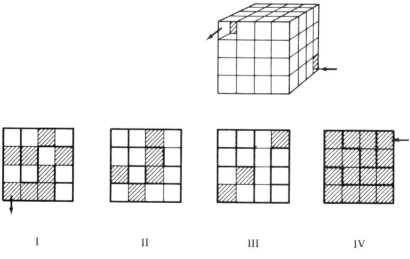

Fig. 6-30. 3-D example (1).

of 64 unit cubes arragned in a 4 × 4 × 4 cellular array, with each cell having one or more openings to adjacent cells. The problem is to find a path for a marble to follow from the starting cell at the bottom to the opposite opening at the top. Figure 6-30 shows the precise construction of each of the four layers, with heavy lines indicating "walls" and shaded areas indicating "floors." To find the desired path using Lee's algorithm, we again enter a 1 in the starting cell and follow precisely the same procedure as in the two-dimentional case, except that *all* adjacent empty cells must be labeled at each step, particularly those which are accessible from above and below, as well as those on the same layer. Figure 6-31 shows the resulting diagram when the target cell is reached on the 20th step. As before, a simple retrace procedure yields the path shown in Fig. 6-32. Note that between 16 and 13, two alternate paths are possible.

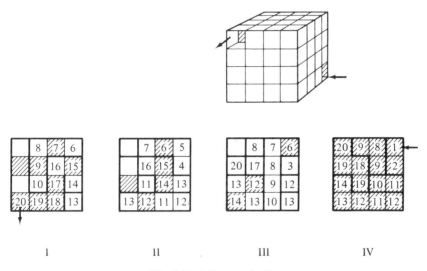

Fig. 6-31. 3-D example (2).

6.5.4. Some Speed-Up Techniques

In a typical wire layout program, 70 to 80% of the actual running time is spent on Lee's algorithm. Accordingly, any techniques that can be employed to speed-up its execution have immediately observable payoffs. Figure 6-33 shows three commonly used techniques. The first is the previously mentioned problem of starting-pin selection. Consider the two pins, X and Y, at the top of Fig. 6-33. If X is chosen as the starting point, less than half of the total cells will have to be labeled before Y is reached. On the other hand, starting at Y results in practically all of the cells being labeled before X is reached. The conclusion is obvious: Start on the pin farthest from the

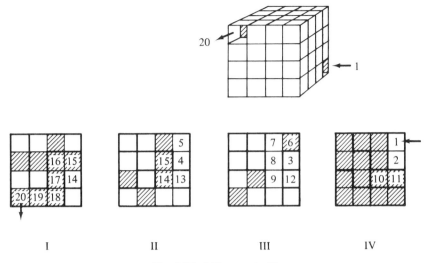

Fig. 6-32. 3-D example (3).

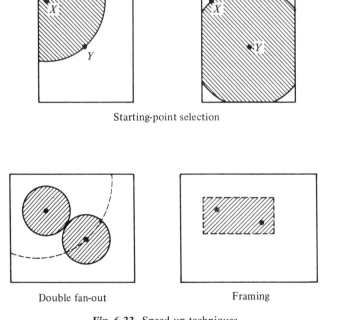

Fig. 6-33. Speed-up techniques.

center of the board. (This rule becomes increasingly less important as the number of laid-out wires increases, but initially it can prove quite helpful.)

A second *double fan-out* technique is shown at the lower left of Fig. 6-33. Here the trick is to begin at *both* pins and continue labeling until a point of contact is reached. Obviously the programming is more complicated, but the resulting saving in time may warrant it.

A technique that has proved even more beneficial in practice is that of *framing* (lower right in Fig. 6-33). Here an artificial rectangular boundary is imposed about the pin-pair being processed, and no labeling is allowed beyond this boundary. Typically this frame is 10 to 20% larger than the rectangle defined by the pins. Since most of the wires which are processed first go in quite easily within the frame, this approach considerably speeds up their layout. Of course, if no path is found, the frame can be removed and the process repeated. Likewise, when this begins to occur frequently, the frame can be eliminated entirely.

Another standard time-saving device involves saving a list of the locations of the "latest cells." If the fan-out process is being performed on, say, a 100 × 100 cell grid, it is certainly not desirable to have to scan all 10,000 cells at each new step. We avoid this by keeping a list of the current latest cells. Now when the next step begins, only those cells adjacent to cells on this list need be examined. A new latest-cell list can now be generated which, in turn, replaces the old list.

6.5.5. Storage Problems and a Coding Procedure

The primary *disadvantage* of using Lee's algorithm is the amount of computer memory required. For every cell in the grid being processed, there must be a corresponding memory location within the computer. With large-size boards, this can become a serious problem. Assume, for example, that a (not unusual) 20 × 20 inch, 6-layer backpanel is to be wired. With a grid spacing of 50 mils, this leads to 160,000 cells per layer. If wire lengths up to, say, 20 inches are expected, then any cell may receive a number as large as 400 (20 units/inch) during the labeling part of the algorithm. This means that at least 9 bits must be allocated to each memory location, or something like a million and a half bits of storage per layer! Clearly, a more efficient storage scheme is needed. To see how this might be done, consider again the configuration of cell labels which results at the final step of the labeling process. (See Fig. 6-29 or 6-31.) A close examination of these labels reveals that for any label k, all adjacent labels are either $k-1$ or $k+1$. Thus, when this label is reached during the retrace process, it is only necessary that the "predecessor" labels be distinguishable from the "successor" labels. This suggests that a simpler labeling sequence might suffice *if each label has its preceding label different from its succeeding label.* The simplest such sequence is 1-1-2-2-1-1-2-2-1-.... Figure 6-34 shows how the previous two-dimensional problem could be solved using this sequence. Again 1's are entered in the empty cells adjacent to *A*. Next 1's are entered again; then 2's, then 2's again; etc. Once more, pin *B* is reached to the 13th step. The retrace is now accomplished as before by retracing on the given sequence (which happens to be the same when reversed). As before, a choice of paths occurs just before *A* is reached [1]. With this coding scheme, it follows that each cell will be in one of just four states: empty, filled, or labeled with a 1 or a 2. Hence, two bits will suffice to specify the state of any cell.

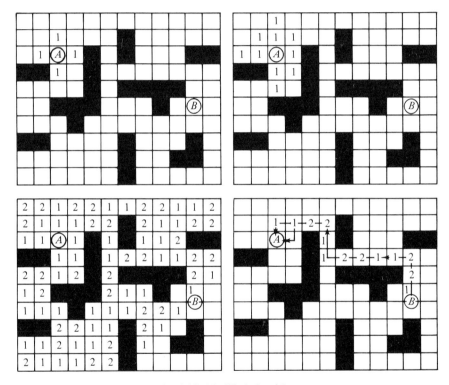

Fig. 6-34. Modified algorithm.

6.5.6. Some Useful Variations

Perhaps the main reason for the widespread use of Lee's algorithm for the layout process is the ease with which it can be modified to accommodate various specialized situations and constraints. A number of papers have appeared describing its use in such cases and many of these are included in the references. Typical modifications involve varying the incremental weights by which the labels are increased at each step in order to give certain paths priority over other less desirable ones. Paths that cross feedthrough locations or tend to block other pins, for example, can be minimized by a suitable modification of the labeling process.

Nearly all such modifications automatically preclude the simple 1-1-2-2 coding scheme described before, but usually a similar code can be found. (The basic requirement on the labeling sequence is that the retrace path be unambiguously defined at each cell.) Likewise, multilayer coding techniques are available [20]. The next two sections describe two patticularly useful variations of Lee's algorithm.

6.5.7. Routing to Preserve Grid Segments

Most operating versions of Lee's algorithm are designed to route a wire so as to minimize its total "length." The exact definition of its length is often a weighted sum of such parameters as the number of grid cells it traverses, the number of feedthroughs it uses, the number of bends, etc. Yet, it may be argued that the actual length of the wire is really not the important factor. Once it is established that feasible paths exist, the significant problem might be said to be not finding the shortest such path, but rather *finding that path which will cause the least difficulties for subsequent path layouts*. Certainly many short paths tend to have this effect, but they explicitly offer no firm guarantee that this will always be the case.

What path then *does* minimize future wiring problems? There is no pat answer. However, a reasonable choice seems to be that path which uses up the *fewest grid segments* (as opposed to grid intersections). Figure 6-35 will clarify this distinction.

Assume two wires, X and Y, have already been routed on a grid as shown in Fig. 6-35a, and that A is now to be connected to B. The normal procedure is to assume that all grid intersection points are available for routing, except those explicitly occupied by wires X and Y. The algorithm then proceeds to find a shortest path through these intersection points. However, a clearer picture of the wiring situation is obtained if we look not just at the available grid intersections, but also at the available grid segments (Fig. 6-35b). When this is done, we see that 21 segments are not really available at all since they connect directly to wire X or to wire Y.

Moreover, once an A-B path is laid out, a number of other segments will no longer be available. If, for example, the short one-turn path in Fig. 6-35c is chosen, then 22 segments are eliminated, in addition to the 9 segments occupied by the path itself. On the other hand, the two-turn path of Fig. 6-35d

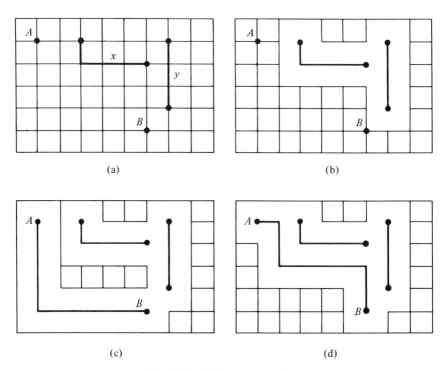

Fig. 6-35. Grid segment routing.

(also of length 9) eliminates only 14 other segments. Certainly future layout problems look much rosier in this second case. (In particular, the 16-segment dead-end of Fig. 6-35c is avoided.)

Fortunately, it is a relatively simple matter to modify Lee's algorithm to find the particular path which minimizes the number of eliminated segments. For each unoccupied cell (grid intersection), it is necessary to know the number of available grid segments adjacent to that grid intersection (from 0 to 4). Using this cell in the path will cause all of these segments to be eliminated. If, however, for a given path, the segments eliminated by each cell are added up, we obtain not the total, E, of eliminated segments, but rather $E + L$, where L is the number of segments in the path itself. This happens because each segment in the path will have been counted twice. The trick, therefore, is to label each cell with the number of available segments adjacent at that point *minus one*. Now the sum of the cell labels will equal not E, but $E - 1$. (A path of length L involves $L + 1$ cells; hence $E + L$ will have been reduced by $L + 1$.) However, since we are only interested in the relative value of the E's, this is of little consequence. Once the cells have been so labeled, a minimum path search can be made. Figure 6-37a shows the resulting array when the cells are labeled in this fashion. Note that, in practice, it is not necessary to store all of this information simultaneously. As each cell is reached during the fan-out process, its weight can be calculated by simply noting the contents of its immediate neighbors. In Section 6.5.9, we shall return to this array and demonstrate the precise method by which the minimum path can be found.

6.5.8. Routing on a Two-Layer Board

As mentioned in Section 6.3.5, the key to success when routing on a two-layer board is to keep most of the horizontal runs on one layer and the vertical runs on the other. A layering process as described in that section can provide an important step in this direction. However, during the actual routing process, it is even more important that this polarization be maintained. Even one modestly long vertical run on the horizontal layer, for example, can cause subsequent wiring difficulties for literally dozens of wires. What is needed then is a way of constraining the routing algorithm so that vertical runs *want* to go on the vertical layer, and horizontal runs, on the horizontal layer.

A simple method of accomplishing this is to vary the weights of the incremental steps during the fan-out process so that the desired directions are given priority on the two layers. Figure 6-36 shows how this may be done. If, for example, the neighbors of a cell on the horizontal layer are about to be labeled, those to the left and to the right (i.e., in the horizontal direction) would receive labels increased by 1; while those above or below (vertical

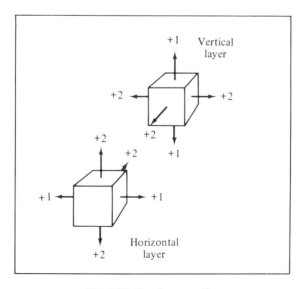

Fig. 6-36. Two-layer routing.

322   *Routing*   Chap. 6

direction) would receive labels incremented by 2. Likewise, if a feedthrough is permitted from the cell, the label assigned to the corresponding cell on the vertical layer would also be 2 greater. A similar incrementing procedure is used for cells on the vertical layer with only those labels in the vertical direction receiving increments of 1.

This scheme does not preclude vertical runs on the horizontal layer (and vice versa), but it does ensure that they will, in general, be quite short and will occur only when the alternative paths are even less desirable. The particular increments employed here—1 for the preferred direction and 2 for the other direction and for feedthroughs—may, of course, be varied as the situation dictates. In particular, when one layer is likely to become heavily congested, the incrementing weights can be adjusted accordingly. Likewise, the feedthrough weights can be chosen to reflect their degree of desirability.

6.5.9. Lee's Algorithm with Arbitrary Weights

Both of the routing procedures of the previous two sections, and, in fact, most modified versions of Lee's algorithm involve a more complicated fan-out and retrace procedure then those illustrated in Sections 6.5.1 and 6.5.3. This is due to the fact that the incrementing weights can vary considerably, depending on both the cell involved and the rules being followed. However, since even the most general case is easily handled, one final example should suffice. Consider the array of Fig. 6-37a which results from the problem of finding the path which eliminates the fewest grid segments. (It will be convenient to refer to this array during the description which follows, but, as already noted, it need not be stored at all since the required weight information for each cell can be quickly calculated when needed.)

If pin A (with a weight of 3) is chosen as a starting point, the process begins by assigning to each empty cell adjacent to A a value equal to its weight, W, plus the weight of A. This value will be called the *score*, S, for that cell. Each such cell is then tagged (circled) as a latest cell. (See Fig. 6-37b.) From here on, the process continues in successive steps. At each step, each latest cell C^*, is examined. (Let its score be S^*.) Any empty cell (with weight W) adjacent to C^* receives a score equal to $S^* + W$ and itself becomes a latest cell for the next step. If a cell adjacent to C^* already has a score, S, and $S^* + W$ is less than S, then again the cell is assigned a score of $S^* + W$ and becomes a new latest cell. Otherwise, the labeling of the adjacent cell is unchanged. Note that some old latest cells may also become new latest cells—if their scores are decreased.

Figure 6-37c shows the result of this process at the end of the second step. At the end of the sixth step, the situation is that shown in Fig. 6-37d. Now an interesting (but not unusual) effect is about to occur. Notice the circled

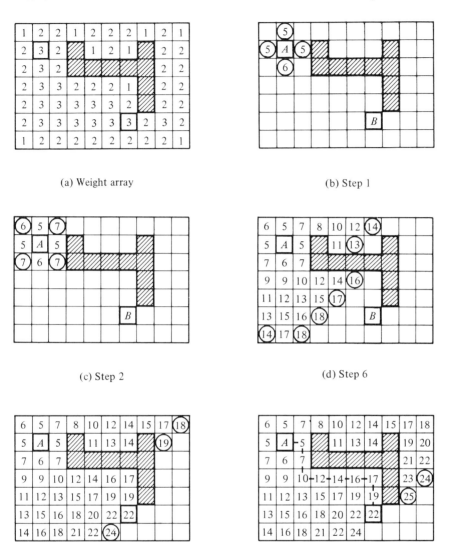

Fig. 6-37. Lee's algorithm with arbitrary weights.

cell in the lower left corner (with a score of 14). As the next step proceeds, this cell will cause the score of the cell to its right (presently at 17) to be reduced to 16; hence this cell will now again become a latest cell. (What is happening is that a shorter path, one which follows the left edge of the grid, is about to *override* an older path.)

Figure 6-37e occurs after the ninth step when the target cell, *B*, is reached for the first time (by a path with score 22). However, since the "first player to reach the clubhouse" is not necessarily the winner, a check must be made to see if any *possible* contenders are still "on the course." It turns out that two are (with lower scores of 18 and 19); so the process continues, terminating after three more steps when the remaining scores all go past 22.

The retrace procedure should be obvious. Consider the final array in Fig. 6-37f. Beginning with *B*, a search is made for the adjacent cell which led to the score of 22, i.e., that cell whose own score, plus its weight, equals 22. Clearly, it is the cell immediately above B $(19 + 3 = 22)$. Now the cell which gave 19 is sought; etc. As with the previous retrace procedures, there will be cases where more than one cell may be chosen, and, as before, any will work. Figure 6-37f shows the path selected in this example. It is precisely the path previously examined in Fig. 6-35, where its value of 22 $(E - 1)$ was predicted.

6.5.10. Layout of Complete Nets

When a number of pins must be electrically tied together, the approach which has been described (Section 6.2) was to initially use the minimum tree algorithm to define a set of pin-to-pin connections (wires) which collectively interconnect all of the given pins. These wires were then individually processed through the various layering, ordering, and layout steps without any further reference to the fact that, collectively, they constituted a single *net*. (Researchers seem to have great difficulty in agreeing on what to call the interconnecting network for a set of electrically equivalent pins. Other terms include *signal*, *node*, *logic string*, *tree*, and *network*.)

This approach of initially reducing each net to a set of pin-to-pin connections had the advantage of greatly simplifying the data-handling problems during the subsequent steps since all connections were now of the same type, i.e., a single pin-pair. Also, when there are precise constraints on exactly how a net can be laid out, this approach can verify that such constraints will be observed.

On the other hand, particularly with the larger unconstrained nets, this approach will seldom, if ever, lead to the optimum layout for that net. For this reason, some designers prefer to lay out an entire net at one time, using a suitable modification of the routing techniques. One popular approach corresponds closely to the second version of the minimum tree algorithm (Section 6.2.4). Consider the eight pins shown in Fig. 6-38a. Initially one pin (say *E*) is chosen. The usual Lee's algorithm fan-out process is now begun with *E* as a starting point and the other seven pins as targets. It continues until the first target pin (in this case, *D*) is reached. This path is then laid out (wire 1 in Fig. 6-38b). Now *all cells* in this wire are tagged as starting points,

Sec. 6.5   Wire Layout   325

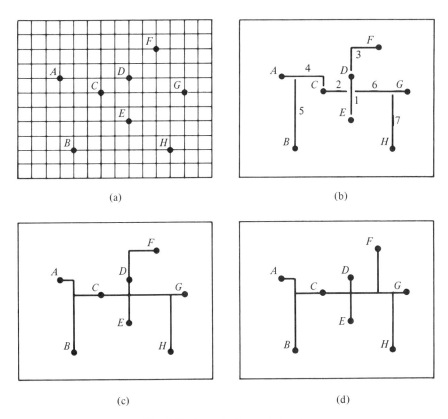

Fig. 6-38. Layout of complete nets.

with the other six pins remaining as targets. Again the fan-out process continues until the first target pin (C) is reached. Note that now when the corresponding path is laid out, it goes not to D or E, but to a point on the wire connecting them. Every cell in this C-D-E net now becomes a starting point, and the process continues in this manner. Figure 6-38b shows the final layout with the numbers indicating the order in which the paths were laid out.

It would be nice to be able to state that the nets obtained by this process are of minimum length; but such (even in this example) is not the case. There is, however, a fairly simple technique which will often allow even shorter nets to be found. It is well known that removing any branch from a net will result in two disjoint subnets (one may be just a single point). Now it is a simple matter to find the shortest path between these two subnets. (All cells in one serve as starting points for the fan-out process, and all cells in the other serve as targets. The process halts when the first target cell is reached.) If the length of this shortest path is less than the length of the deleted branch, then by inserting this path between the two subnets, a new shorter length net is obtained. Applying this technique to the net of Fig. 6-38b, we find that when the branch from the intermediate point on wire A-C to pin C is deleted, a shorter path can be found. (See Fig. 6-38c.) Likewise, when branch D-F is deleted, another shorter path exists (Fig. 6-38d).

6.5.11. Other Layout Techniques

Aside from the large memory requirements, the most common objection to Lee's algorithm is the fact that it essentially looks at only one wire at a time. Unless the algorithm is constrained in some way, it will go to any lengths to lay out the particular wire under consideration, even at the expense of permanently blocking dozens of other paths. In a sense it is sometime almost too clever. (One practical means of avoiding this effect is to check on the indicated wire length *before* the retrace process begins. If the indicated length is excessively long, it may be more desirable to move on to other wires.)

This single-minded feature has caused a number of researchers to look for alternative approaches to the layout problem and several techniques have been proposed [18], [31]. While none appear to have the generality that Lee's algorithm affords, some do seem attractive for specialized situations.

6.5.12. Discrete-Wire Routing

All of the routing techniques which have been described have been concerned with the etched circuit technology where the conductor paths are deposited directly on the surface of the board or chip. Practically all automated routing programs currently under development are directed towards this case. There are, however, a number of wire routing programs still in active use which were developed for *discrete-wire routing*, i.e., for the case where an interconnection is made by simply running a wire directly between the two pins involved [4], [8]. Some large computers still use discrete wiring for interconnecting the various circuit cards on the backpanel.

The advent of the automatic wirewrap machine provided the initial impetus for the development of these early wire routing programs. Since the input to such a machine was simply a tape telling for each wire the pins to which it should connect and the path it should follow, it was a natural-next step to let a computer generate such an input tape. Many of the techniques presently employed for etched circuit routing were originally developed for the dicrete-wire case.

In a typical discrete-wire routing problem, the pins themselves define the channels which the wires must follow. Paths in a variety of directions are normally permitted, provided that they do not overlap the pins. Figure 6-39 shows some typical paths which the wires may follow. The pins are commonly located on 100- or 125-mil centers.

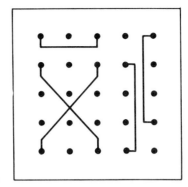

Fig. 6-39. Discrete wire routing.

Crossovers such as the one shown are normally permitted, provided the resulting buildup, i.e., the stacking of wires on top of each other, does not become excessive. On the other hand, the case illustrated by the two vertical wires can cause problems. Here both wires occupy a significant portion of the same channel, and as a result interwire noise can result. A great portion of the effort spent on developing the discrete-wire routing algorithms has been directed towards avoiding this situation.

Other constraints and specifications which these programs can accommodate include:

1. a maximum number of wraps per pin (typically two);
2. step-down constraints, i.e., if a wire is wrapped at different heights on the two pins, then it cannot be shorter than a specified length;
3. preassigned wires;
4. detection of "bridges," i.e., crossing a wire at the wrong height during step-down;
5. generation of regular layouts to ease troubleshooting.

Virtually all of these programs use some version of the minimum tree algorithm for initially routing the various wires. Needless to say, a great number of geometric calculations have to be made to specify bend locations, avoid bridges, define channels, etc. Precise formulas are usually available for determining whether or not the interwire noise generated by an overlapping pair of wires is excessive. Normally, considerable rerouting of the wires is undertaken to find the most efficient layout.

6.6 FINAL COMMENTS

In this chapter, the various steps of the routing process have been followed, starting with the initial interconnection list and on through to the final layout of the necessary interconnections. A number of algorithms, both formal and heuristic, have been described, nearly all of which have at one time or another been successfully employed on actual routing problems. Many of these algorithms were initially developed because of a new circuit board constraint or specification which could not be accommodated by the then existing techniques. Others were developed, not to satisfy some particular constraint, but rather to simplify the final layout process. Accordingly, their applicability and usefulness will vary considerably with the problem under consideration.

For this reason, many of these algorithms are incorporated into a routing program as *optional* procedures which may or may not be employed, depending on the desires of the designer. As operational experience increases, the precise role of each of these various optional procedures becomes more clearly defined. Some, for example, are found to be of great help with boards of many layers, but of little or no value with two-layer boards. Others prove to be quite essential for dense problems, but unnecessary for the easier cases.

With regard to the question of the degree of difficulty of a particular layout problem, there is a simple formula which can be quite helpful. Assuming that the wire-list determination procedure of Section 6.2 has been employed, the total Manhattan length, L, of all pin-to-pin wires on this list can then be determined. Likewise, the total number of *available* grid segments on *all* layers may be calculated. (With an $a \times b$ grid in which all segments are available, this value would be $a(b + 1) + b(a + 1) = 2ab + a + b$.) Let this value be S. The ratio $r = S/L$ turns out to be an accurate measure of the difficulty of the given problem. A typical critical value for r is 3. If r is greater than 3, the layout problem is relatively easy, and many of the layout-enhancement routines can often be omitted. When r is less than 3, a difficult routing problem is present and an all-out attack is probably called for. (Incidentally, the routing procedure of Section 6.5.7 is specifically designed to keep this ratio as large as possible.) Of course, each routing program will have its own critical ratio, depending on its efficiency and the constraints which must be handled. However, after a relatively small number of problems, its value can usually be accurately estimated.

No routing program, no matter how sophisticated, can guarantee to lay out all of the necessary interconnections on a given board. The minimum-length paths, for example, may require more grid segments than the total available. For this reason, the possibility of "can't connects" must always be allowed for. Normally, when these occur, the designer is provided with a list of precisely which connections were not made, along with a printed image of the actual layout. Hopefully, he can then manually rework the layout so that the missing connections are completed. These changes are then inserted into the program through a *manual-update* routine which systematically deletes and inserts the indicated paths. Fortunately, with most large operational routing programs, the occurrence of "can't connects" is the exception rather than the rule. On the average, such programs will complete better than 99% of the required connections.

The running times and memory requirements for these programs are influenced by a variety of factors. Typically, however, a small (100 × 100 unit grid), 2-layer board with 200 connections can, at present, be processed on a large general purpose computer in less than five minutes using a 32K memory. A large (400 × 400 unit grid), 6-layer board with 2500 connections will require an 128K memory and roughly three hours of running time.

PROBLEMS

1. Does the triangle inequality hold for Manhattan distance?
2. Assume that a straight line between points A and B is approximated by a "staircase" of short horizontal and vertical segments. What happens to the length of this approximation as the segments become shorter and shorter?
3. Find the minimum tree for the eight points in Fig. 6-38a using first the method in Section 6.2.1 and then the alternate method in Section 6.2.4.
4. Show a simple example where the technique of Section 6.2.3 will not yield the shortest loop.
5. Prove that an algorithm for solving the shortest-chain problem also solves the traveling salesman problem. (*Hint*: First show that, in the latter problem, reducing all distances to a given city by the same amount does not alter the final solution.)
6. Assume that a connector pin is located at each of the fifteen grid points on the top edge of Fig. 6-38a. Make a pin-to-connector assignment for the eight pins.
7. Number the thirteen wires in Fig. 6-11 and draw the corresponding intersection graph.
8. What is the simplest graph that cannot be two-colored? that cannot be three-colored?
9. Can the layering of Fig. 6-16 be improved, i.e., can another layering be found where all crossovers are eliminated? (*Hint*: Apply the result of problem 8 to the intersection graph of Fig. 6-11.)
10. The intersection graph of Fig. 6-11 can be shown to be nonplanar, yet the layering of Fig. 6-16 shows that it is easily four-colored. Does this contradict the four-color conjecture?
11. In the layering process, the wires between two pairs of pins were said to intersect if their shortest Euclidean paths intersected. An alternative definition would be to require that their shortest Manhattan paths intersect. Under what conditions will this occur? (Note that there will usually be many shortest Manhattan paths.)
12. Show a two pin-pair counter-example to the rule of Fig. 6-24.
13. If the five wires in Fig. 6-22 were to be laid out on a single layer, in what order should they be processed?

14. Apply Lee's algorithm to the problem of Fig. 6-28 assuming that the cell labeled 5 in Fig. 6-29d is filled.
15. Select starting pins for the five wires of Fig. 6-22.
16. Repeat problem 14 using the 1-1-2-2 procedure.
17. Using the grid of Fig. 6-28 as modified in problem 14, find the path from A to B which uses the fewest grid segments.
18. Using the procedures of Section 6.5.10, find a net which interconnects the six pins in Fig. 6-5a.

REFERENCES

[1] S. B. Akers, "A Modification of Lee's Path Connection Algorithm." *IEEE Trans. on Elec. Comp.*, Vol. EC-16, No. 1 (February, 1967), pp. 97–98.

[2] S. B. Akers and F. O. Hadlock, "Graph Theory Models of Electrical Networks and Their Minimum Crossover Layouts." *University of the West Indies Conference on Graph Theory and Computing*, January, 1969.

[3] S. B. Akers, "Some Problems and Techniques of Automatic Wire Layout." *Digest of First Annual IEEE Computer Conference* (September, 1967), pp. 135–136.

[4] G. W. Altman, L. A. Decampo, and C. R. Warburton, "Automation of Computer Panel Wiring." *Trans. AIEE*, Vol. 79, Part. 1 (May, 1960), pp. 118–125.

[5] C. B. Berge, *The Theory of Graphs and Its Applications*. New York, John Wiley & Sons, Inc., 1962.

[6] B. Bittner and U. Ulrich, *A Method for Improving the Location of Connecting Paths for Circuit Card Wring*. Report from Thomas Bede Foundation.

[7] M. A. Breuer, "The Formulation of Some Allocation and Connection Problems as Integer Programs." *Naval Res. Logist. Quart.* Vol. 13, No. 1, March, 1966.

[8] R. R. Brown and G. F. Putnam, "The Automation of Backwiring Design and Topological Layout." *AIEE Trans.* (1962), pp. 136–139.

[9] F. K. Buelow, "A Circuit Packaging Model for High-Speed Computer Technology," *IBM J. Res. and Dev.*, Vol. 7 (July, 1963), pp. 182–189.

[10] P. W. Case, H. H. Graff, L. E. Griffith, A. R. Leclercq, W. B. Murley, and T. M. Spence, "Solid Logic Design Automation for IBM System/360." *IBM J. Res. and Dev.*, Vol. 8 (April, 1964), pp. 127–140.

[11] L. Cooper, "Heuristic Methods for Location-Allocation Problems." *SIAM Rev.*, Vol. 6 (January, 1964) pp. 37–53.

[12] R. Courant, *What is Mathematics?* London and New York, Oxford University Press, 1941.

[13]   G. B. Dantzig, R. Fulkerson, and S. Johnson, "Solution of a Large-Scale Traveling Salesman Problem." *J. Oper. Res.*, Vol. 2, (1954), pp. 394–410.

[14]   G. V. Dunn, "The Design of Printed Circuit Layouts by Computer," from *Proc. Third Australian Computer Conf.*, Canberra, Australia (1967), pp. 419–423.

[15]   G. F. Fielding, "Automatically Processed Wire Lists." *Western Electronic News* (June, 1962), p. 22.

[16]   C. J. Fisk, D. L. Caskey, and L. E. West, "ACCEL: Automated Circuit Card Etching Layout." *Proc IEEE*, Vol. 55 (November, 1967), pp. 1971–1982.

[17]   B. Franqui, *Automation of MLB Interconnection Layout*. Technical Memorandum 343-04-04, Data Systems Division, Autonetics, Anaheim, Calif.

[18]   H. Freitag, *Design Automation for Large Scale Integration*. IBM Watson Research Center, Yorktown Heights, N. Y., Presented at the Western Electronic Show and Convention, August 22–26, 1966.

[19]   M. Gardner, "Mathematical Games." *Scientific American*, November, 1963.

[20]   J. M. Geyer, "Connection Routing Algorithm for Printed Circuit Boards." *IEEETCT*, Vol. C-20, January, 1971.

[21]   E. N. Gilbert and H. O. Pollak, "Steiner Minimal Trees." *SIAM J. Math.*, Vol. 16 (1968), pp. 1–29.

[22]   R. K. Grim and D. P. Brouwer, "Wiring Terminal Panels by Machine." *Control Engineering*, Vol. 8, No. 8 (August, 1961), pp. 77–81.

[23]   M. Hanan, "On Steiner's Problem with Rectilinear Distance." *SIAM J. Appl Math.*, Vol. 14, No. 2 (March, 1966), pp. 255–265.

[24]   Hardgrave, *On the Relationship Between the Traveling Salesman and Longest Path Problems*. Bell System Monograph No. 4336.

[25]   C. H. Haspel, "The Automatic Packaging of Computer Circuitry." *IEEE International Conv. Rec.*, Vol. 13, Part 3 (1965), pp. 4–20.

[26]   B. R. Heap, "Permutations by Interchanges." *Computer J.*, Vol. 6 (October, 1963), pp. 293–294.

[27]   U. Kodres, "Formulation and Solution of Circuit Card Design Problems Through Use of Graph Methods," in G. A. Walker, ed., *Advances in Electronics Circuit Packaging*, Vol. 2 (New York, Plenum Press, 1962) pp. 121–142.

[28]   J. B. Kruskal, Jr., "On the Shortest Spanning Subtree of a Graph and the Traveling Salesman Problem." *Proc. Amer. Math. Soc.*, Vol. 7 (1956) pp. 48–50.

[29]   J. Kurtzberg and J. Seward, *Program for Star Clustering Wiring of Backplane*. Burroughs Internal Report, January, 1964.

[30]   J. Kurtzberg, *Background Wiring Algorithms for the Placement and Connection Order Problems*. Burroughs Tech. Report TR 60-40, June 28, 1960.

[31]   S. E. Lass, "Automated Printed Circuit Routing with a Stepping Aperture." *Comm. ACM*, Vol. 12, No. 5, May, 1969.

[32] F. B. Lavering, J. J. Lyons, and P. J. Corey, "Automated Printed Circuit Board Design," *Proc. SHARE Design Automation Workshop*, June 24–26, 1964.

[33] C. Y. Lee, "An Algorithm for Path Connections and Its Applications." *IRE Trans. on Electronic Computers*, Vol. EC-10, No. 3 (September, 1961), pp. 346–365.

[34] C. L. Liu, *Introduction to Combinatorial Mathematics*. New York, McGraw-Hill Book Company, 1968.

[35] H. Loberman and A. Weinberger, "Formal Procedures for Connecting Terminals with a Minimum Total Wire Length." *J. ACM*, Vol. 4 (October, 1957), pp. 428–437.

[36] E. Majorani, "Simplification of Lee's Algorithm for Special Problems." *Calcolo*, Vol. 1 (1964), pp. 247–256.

[37] W. D. Markovitz, "Multilayer Boards by Computer." *Electronic Products*, Vol. 9 (1967), pp. 92–96.

[38] M. E. McLean, "Autodesign Automated Wiring Design System." *Proc. ACM 15th Nat. Meeting*, August, 1960.

[39] W. Miehle, "Link-Length Minimization in Networks." *J. Oper. Res.*, Vol. 6 (1958), pp. 232–243.

[40] K. Mikami and K. Tabuchi, "A Computer Program for Optimal Routing of Printed Circuit Connectors." *IFIPS Proc.*, Vol. H47, 1968.

[41] E. F. Moore, "Shortest Path Through a Maze,", in *Annals of the Computation Laboratory of Harvard University*, Harvard University Press, Cambridge, Mass., Vol. 30 (1959), pp. 285–292; also Bell System Monograph No. 3523.

[42] O. Ore, *The Four Color Problem*. New York, Academic Press, Inc., 1967.

[43] W. E. Pickrell, "An Automated Interconnect Design System." *Proc. FJCC* (1965), pp. 1087–1092.

[44] T. Pomentale, "An Algorithm for Minimizing Backboard Wiring Functions." *Comm. ACM*, Vol. 8, No. 11 (November, 1965), pp. 699–703.

[45] R. C. Prim, "Shortest Connection Networks and Some Generalizations." *Bell System Tech. J.*, Vol. 36 (November, 1957), pp. 1389–1401.

[46] C. V. Ramamoorthy, "Analysis of Graphs by Connectivity Considerations." *J. ACM*, Vol. 13 (April, 1966), pp. 211–222.

[47] D. L. Richards, *SWAP, A Programming System for Automatic Simulation, Wiring, and Placement of Logical Equations*. Presented at the 1967 SHARE Design Automation Workshop.

[48] D. P. Rozenberg and J. S. Rupp, *The Automatic Routing of Multiple Plane Wiring*. IBM Report No. 64-825-1159, Fed. Systems Div., 1964.

[49] N. Schuster and W. Reiman, *Multilayer Etched Laminates in High Density Electronic Equipment*. Litton Industries, presented at the Electronic Circuit Packaging Symposium, University of Colorado (August 18–19, 1960), pp. 437–471.

[50]   M. N. Weindling and S. W. Golomb, *Multilaminar Graphs*. Douglas Missile and Space Systems Division, Paper No. 3594, September, 1965.

[51]   J. Weissman, *The Mathematical Basis of the Autonetics Etched Interconnection Design Program*. Paper No. 7, Department 3341-51, Autonetics, Anaheim, Calif.

[52]   J. Weissman, *Boolean Algebra, Map Coloring, and Interconnections*. Autonetics, Department 3341-51, Anaheim, Calif.

chapter seven

FAULT TEST GENERATION

RALPH J. PREISS

Ralph J. Preiss
IBM Systems Development Division
Poughkeepsie, New York

7.1 INTRODUCTION

Design automation techniques have been applied to hardware testing as an outgrowth of design simulation. Simulation could prove that a good logic design would behave properly (i.e., as expected), and a faulty design would behave improperly and thus be discovered. It is, therefore, no surprise that this technique of simulating a faulty design was applied to simulating a good design with a faulty circuit. In either case, the response will indicate a design fault. Thus, test patterns could be developed by systematically simulating circuit failures in a good design, running a trial test pattern, and observing the response. Later, these test patterns and their outputs could be used to uncover faulty circuits in a machine that has been built. This chapter deals with the problems of automatically finding an appropriate test pattern sequence for fault detection.

Fault test generation has been used to develop tests for individual circuit logic functions, functionally packaged circuits, subassemblies, and full systems. This chapter deals only with system testing for two reasons. First, design automation deals with system design; and second, by implication, system testing encompasses testing of subassemblies, functionally packaged cards, and individual logic circuits.

Although many researchers have worked in the field of diagnostic test generation, only two major schools of thought have successfully developed. These schools might be called the heuristic school, since its work is ad hoc or heuristically derived, and the analytical school, since its work is analytically derived. While it is not really certain who influenced whom in the development of these schools, it is interesting to note that both schools developed in the late 1950s and early 1960s. Both schools treat combinational logic and sequential logic separately. The emphasis is on combinational logic with the justification that a sequential circuit is at least an order of magnitude more difficult to analyze yet it can be treated as combinational at an instant in time. If a solution for one is found, a solution for the other is soon to follow.

In this chapter, we shall take the work of David Muller and Sundaram Seshu as representative of the heuristic school, and the work of J. Paul Roth as representative of the analytic school. We shall spend more time exploring the heuristic school because this school was used first for directly generating tests from data contained on logic design automation master tapes. Here, the power of the digital computer was used on a trial-and-error basis to develop a set of shortcuts in cases where no elegant solution was thought of or available.

One problem that the analytic or algorithmic school tried to solve and which often ran into massive amounts of computer time is the generation of a set of tests with 100% detection capability followed by the selection of a minimal or, at least, an irredundant subset of these. It turns out that, in a modern microsecond computer, we really do not need a minimum or a near-minimum test set. We may spend a lot of computing time trying to reduce 1000 tests obtained in some manner to an optimum, or irredundant, 900 tests; but we could have spent the time doing better things. If these tests were manually applied, the 10% reduction would certainly be worth the effort. However, considering the speed of commercially available computers at present, and assuming that these tests can be applied at computer speeds, 10,000 tests or even 50,000 tests can be executed in a matter of minutes or seconds. (In mobile computers, storage of the test patterns may be a problem; however, high-density storage media are available, and only an engineering decision is required on how the tests are to be stored.)

The problem given to design automation groups is to develop the tests in a reasonable length of time. Early calculations indicated that it might take some 800 hours of IBM 7090 time to generate a set of tests for one of the System/ 360 machines. This frightened even the most stouthearted developers of the programs. But by choosing shortcuts in developing the program specifications, rather than utilizing minimizing algorithms, they reduced the computing time to about 20 hours, a number that could be cost-justified.

Further relaxation of specifications is possible if we realize that a one-to-one correspondence between test results and the malfunctioning replaceable components is not necessary. Manual help is available during the repair operation where it is possible to use the old-fashioned method of swapping components, or probing with an oscilloscope before performing the swap. (It might be said that this technique does not apply to large scale integration (LSI). But in LSI, the failure need only be detected and localized to some gross component since a single replaceable unit represents a large part of the entire circuitry. With good design practices, the LSI problem may therefore be reduced to one that is less severe.)

In Section 7.2 we shall define terms that are generally used by researchers in the field, and which provide background on some of the important-to-understand concepts used in diagnosis. Next, in Section 7.3, we shall review those efforts that historically have spurred diagnostic research. In Sections 7.4 and 7.5, we shall discuss some representative test-generating methods for combinational logic and sequential logic, respectively. In Section 7.6, we shall touch lightly on the matter of test application and the steps still required after the tests have been generated, but before they are useful to the people who apply the tests—the manufacturing test and the field personnel.

7.2 DEFINITIONS

A *machine failure* or error is the detected presence of an unwanted action, or the detected absence of a wanted action [1]. Because of the myriad of redundancies and don't care conditions or states designed into a digital system, a fault may exist in a computer for a long time without being discovered. In other words, a machine failure does not appear as long as a fault does not manifest itself in a CPU miscalculation. (This may explain why computers are much more reliable than calculations based on the reliability of individual components would indicate.) A *fault* may be a missing wire, a shorted diode in a high-resistance path, or an open diode in a low-resistance path. It may be any other physical defect that keeps the circuit from functioning as intended at the time the circuit is needed. A fault may also be present in a digital system because of a flaw in the design or in the building of that system; yet it does not cause a machine or a system failure until the component takes part in a calculation and causes an incorrect result to be generated.

This fault is analogous to an incorrectly coded branch in a program. If the branch condition is not used, the program will run correctly despite the built-in fault.

A classical method [2], [3], of determining whether a machine failure exists in a computing system is to have redundant logic designed into the system to detect failures as they occur. Parity calculations, error-correcting codes, duplicate circuits which calculate and compare the answers, and instruction-repetition circuits are some of the design techniques employed. In some cases, these techniques may also correct an error before the incorrect result is passed to the outside world. Since a fault is as likely to appear in the checking circuitry as in the circuitry being checked, the addition of the checking function increases the probability of failure in the resultant circuit. A design tradeoff is usually made in determining the amount of checking performed in a design.

Machine failures exhibit themselves as solid or intermittent. A *solid failure* is one that always exhibits itself in the same manner to conditions over which the machine user has control. An *intermittent failure*, on the other hand, is one over whose occurrence the machine user has little or no control. In other words, a solid failure is repeatable in a synchronous manner under user control, whereas for an intermittent one, though it may repeat, the exact moment of failure repetition is not predictable because all conditions contributing to its occurrence are not known. Once they are known, the failure may be reclassified as solid under our definition.

In practice, we can simulate an intermittent failure by applying a solid fault to a digital circuit at a repetition rate which is asynchronous to that of the machine clock. Another method of simulating an intermittent failure is to apply a solid fault to a circuit for just a few machine clock cycles anytime in a calculation. From this, we see that a failure may be due to a solid cause, and that the length of time, Δt, during which a fault is activated may be used to characterize it. As Δt becomes small, the failure becomes intermittent [4]. For this reason we shall concentrate on developing techniques to generate test patterns for detecting solid faults only. We shall depend on the design of the hardware or the user of the tests to make the tests fail often enough to permit locating the faulty components.

A diode implementation of a logic function is said to be *energized* or *activated* if the logic function is realized. If the conditions for activation are not realized, the logic function is said to be *inhibited*. Each input to a logic circuit is sometimes referred to as a *leg*. In Fig. 7-1, for example, if a negative voltage is applied to leg 1 or 2 of the OR circuit, output 3 is at $-B$ volts since no current can flow through the almost infinite backward resistance of the diode. If a zero or higher voltage is applied at leg 1 or 2, the voltage at output 3, often called the *source*, is raised to that level (assuming almost no forward resistance in the diode) to energize the next circuit, often called the *sink*.

Sec. 7.2   Definitions   339

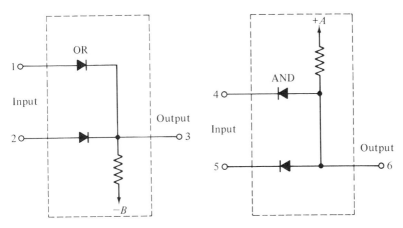

Fig. 7-1. OR circuit implementation, simplified.

Fig. 7-2. AND circuit implementation, simplified.

The AND function is shown in Fig. 7-2. Here current is constantly flowing from $+A$ to the most negative voltage applied at leg 4 or 5. Again assuming almost no forward resistance in the diode, the voltage-divider action of the circuit causes output leg 6 to follow the lowest applied input voltage. If one voltage is zero and the other positive, the AND condition is not met and the circuit is inhibited. However, as soon as both legs 4 and 5 raise their voltages to zero or a positive value, current can no longer flow through the diodes from $+A$ to either leg; thus output point 6 is raised to $+A$ volts. The logical conditions are met and the circuit is activated.

Figure 7-3a illustrates the more complicated AND-OR-INVERT (A-O-I) logical function frequently found in computers. The associated simplified representative circuit is shown in Fig. 7-3b. As long as input legs 1 and 2 or input legs 3 and 4 are not both positive, the base of the transistor is kept essentially at $-B$. This prevents current from flowing through the load resistor from $+C$ to ground and keeps output 5 at essentially $+C$ volts. As soon as one of the conditions is met, the voltage at the transistor base increases, current flows through the load, and the voltage at point 5 swings to ground. The AND condition is met and the output is inverted.

Taking the same circuit, let us determine what happens if the transistor has a collector short-to-ground fault. No matter what the base voltage is, the output voltage at point 5 is ground. If the transistor fault is an open (i.e., cannot conduct), the output voltage always stays at $+C$. Having ascertained this phenomenon, and using an arbitrary notation (1 = activation; 0 = inhibition), we can begin to characterize the failures that may be associate with the A-O-I circuit (Table 7-1). Notice that the AND diode-1 short condition does not logically represent a failure. We take advantage of this condition

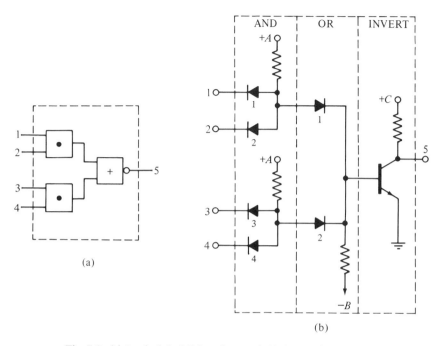

Fig. 7-3. (a) Logical A-O-I function, and (b) A-O-I circuit implementation, simplified.

in certain circuit families and form a "wired" logical function (called a dot-AND). In other circuit families, permitting the current to flow in both directions may sooner or later cause another circuit to overheat and break down. This process will continue until a logical failure is produced.

From the example of Table 7-1, we realize that by applying certain *test patterns*, stimuli, or vectors to the *primary inputs* or *entry points* of circuits, and by observing the *primary outputs* or *exit points*, we can determine whether a circuit is functioning correctly. By analyzing what the output patterns look like, we have a good chance of determining what the cause of failure might be. We can apply the game of "twenty questions" and pinpoint it quite readily. Each of the twenty questions are fault detection tests, and the total constitute the diagnostic tests. A *fault detection test* or a *test* for a fault used in diagnosis is a pattern of setable input signals that causes an incorrect value to appear at the observable output points if a fault is present.

Refering to Table 7-1, we note that in reference line 4, in the transistor-open case, the output is different from that for the no-failure condition. Therefore, this pattern, when applied, is a valid detection test for this failure. We note further that, no matter what input patterns we apply, the output is "stuck-at-1." Similarly, in the transistor-short case, the output is "stuck-at-0."

TABLE 7-1. Failure Characterization of the Circuit in Fig. 7-3b

Reference line	Input 1	Input 2	Input 3	Input 4	Output						
					No failure	Transistor short	Transistor open	AND diode-1 short	AND diode-1 open	OR diode-1 short	OR diode-1 open
1	0	0	0	0	1	0	1	1	1	1	1
2	0	0	0	1	1	0	1	1	1	1	1
3	0	0	1	0	1	0	1	1	1	1	1
4	0	0	1	1	0	0	1	0	0	1	0
5	0	1	0	0	1	0	1	1	0	1	1
6	0	1	0	1	1	0	1	1	0	1	1
7	0	1	1	0	1	0	1	1	0	1	1
8	0	1	1	1	0	0	1	0	0	1	0
9	1	0	0	0	1	0	1	1	1	1	1
10	1	0	0	1	1	0	1	1	1	1	1
11	1	0	1	0	1	0	1	1	1	1	1
12	1	0	1	1	0	0	1	0	0	1	0
13	1	1	0	0	0	0	1	0	0	0	1
14	1	1	0	1	0	0	1	0	0	0	1
15	1	1	1	0	0	0	1	0	0	0	1
16	1	1	1	1	0	0	1	0	0	0	0

Another failure to investigate is the case of AND diode-1 open. The behavior of this diode is to prevent any current from flowing through it; therefore, it has the same effect as a positive voltage at that input, i.e., represents a condition met. In other words, it is stuck-at-1, and, even though only signal 2 is applied in reference lines 5, 6, and 7, the condition for activation is met, and the output is different from that for the no-failure output. Therefore, we may conclude that physical faults result in inputs or outputs that are stuck-at-0 (s-a-0) or stuck-at-1 (s-a-1). This concept is very important and is used by all researchers in diagnosis.

The classical method of diagnosis of a piece of digital hardware is to apply test stimuli for which known results are available or can be calculated in an independent manner. The known results (under non-failing conditions) are referred to as the *good machine*. The test results are then compared with the expected results. If the results are equal, we know that the digital hardware can produce the correct results to the test stimuli. But since we do not know the extent of the logic which the test stimuli cover, we really cannot say much about the condition of the hardware. If the results are unequal, a failure has been detected. But diagnosis is not complete until the faulty component is located. *Diagnosis* is therefore made up of fault detection and fault location.

Pinpointing a failure to a faulty component depends on the addressing resolution and the timing resolution of the test stimuli. For example, if the fault-detection capability of a 72-bit adder is contained in a single check bit, then a failure detected by the circuitry supporting that bit will pinpoint the fault to anywhere within the 72-bit adder. Consequently, the addressing resolution is low and, unless the 72-bit adder resides in one replaceable component, the time to restore the machine to proper working order will be high. For diagnosis, the *addressing resolution* is a way of distinguishing failure symptoms (caused by different failing components) through the differences of observable output patterns (or *bad machines*) that these symptoms generate.

Similarly, if too many operations, e.g., addition, carry propagation, and checking, can take place in one clock cycle (the smallest unit of time in which a machine's state may be frozen), the cause of a detected error might be located in any of the three areas mentioned. Thus, the *timing resolution* is low, and diagnosis is impeded. Therefore, fault location depends not only upon the completeness of a set of tests, but also on how these tests are applied to take advantage of the addressing and timing resolutions designed into the hardware.

7.3 REVIEW OF CLASSICAL APPROACHES

Computing systems are composed of a large variety of circuits. Some convert voltage or impedance levels, others convert digital information into analog form, or vice versa, and still others perform a combination of these functions. Some of the circuits, or subsystems, because of their nature or their use, are easier to diagnose than others. For example, in a read-only control storage, we know exactly what bit patterns to expect (response) for each storage location addressed (stimulus). Furthermore, a control storage is usually small enough and fast enough to permit us to read out each word and compare it against the bit pattern which the original control-storage design specified. The approach is straightforward, and the design automation records can easily be used to develop a set of fault detection tests. Faults, in turn, are readily found by the maintenance engineer through his knowledge of the design and package of the control storage. For example, if every even word fails, then the addressing circuitry may be suspect. If bit 7 of every word fails, then the fault may be a sensor circuit of the storage element associated with bit 7. Of course, more complicated testing techniques may be used to reduce the personnel skill-level required for fault location. For example, the entire control storage may be converted to an equivalent logic circuit by means of the logic synthesis techniques described in Chapter two. The diagnostic techniques described later in this chapter may then be used to generate a set of tests to provide the diagnosis and fault location. This author is

Sec. 7.3   Review of Classical Approaches   343

not aware of a project in which this technique has been used commercially.

In the magnetic core read/write storage, the patterns required for testing are complicated and numerous. The reason for this is that a user may store any combination of bit patterns at any time. The combinations possible, which are a function of addressing size and word length, make it impractical to use an exhaustive test-pattern technique. The designers provide the diagnostic engineer with a series of worst-case test patterns that attempt to induce incorrect outputs either by magnetic flux buildup or cancellation (e.g., record 1's and 0's in alternate planes or sectors of a core stack such that continued reading of the 1's might induce the 0's to switch, or vice versa). These worst-case patterns are then loaded into the read/write storage under test and later retrieved. If the data retrieved do not correspond to the data introduced, a failure is observed. The fault location is deduced by a combination of the number of different patterns that happen to detect the fault and the ingenuity of the diagnostic engineer in interpreting the results. Occasionally, a fault (i.e., a material flaw) in a physical storage box may go undetected for years until, one day, a unique combination of data forces the fault to exhibit itself.

Another type of storage device for which the traditional method of functional testing is employed is the magnetic recording surface. Various problems are diagnosed functionally, depending on the design. In a fixed-head, fixed-speed, and fixed-recording-density device such as a magnetic drum, the read/write and addressing capabilities are checked in a manner similar to that used in the core storage. Again, flux buildup (or creeping) and flux cancellation are induced on the theory that a worst-case pattern combination can be generated and retrieved. The exact worst-case test pattern may be derived from a study of the electrical, mechanical, and magnetic characteristics which might build up or cancel magnetic flux strengths, e.g., by recording and reading of 1's and 0's on alternate-track and alternate-addressable fields.

In the fixed-head, variable-speed, and fixed-recording-density device such as a tape drive, data are written on the tape, the tape is rewound (or read backwards), and then read forward again. A comparison is made between the original data that should have been written and the data that was read back. The object is to determine whether the data conversion (from data to motion and magnetic flux) and its reconversion (from magnetic flux back to data) indeed works. If any of these functions do not perform as expected, suspicion is immediately focused on the circuitry involved in the nonperforming function. The maintenance engineer has the ability to request that one or more functions be performed repetitively so that he may localize the fault to a replaceable or a repairable component. While many tape control functions such as load, unload, and rewind can be neglected at this time, there are two read/write functions that exist in tape testing that do not exist for the drum. These are the problems of skew (since more than one track is recorded in parallel on the tape, all heads must be so aligned that all bits written in one unit of time in parallel will be read again in one unit of time) and recording tolerances (since the tape may be written on one drive and read on another slightly faster or slower drive). Therefore, additional tests and physical movement of the recorded tape from one drive to the other must take place.

In the movable-head, variable-density, fixed-speed recording device such as a magnetic disk, recording tolerances must also be maintained to allow disk interchangeability between drives. While it should be possible to calculate analytically what the worst-case patterns are, and under what conditions the recording tolerances may be relaxed, or for that matter, what design parameters influence the tolerances, this information does not seem to appear in the design automation or diagnostics literature. The reader might find it an interesting project to provide the analytical tools for better test design and fault diagnosis in this area.

When we consider read-only and write-only devices, we lose the capability of achieving a fully automatic mode of testing. In testing card readers, paper tape readers, and film strip readers, for example, we depend on known sequences of prerecorded data patterns as inputs. These must be properly read, encoded, and decoded, and compared against what was prerecorded. The ability to modify the patterns to detect the failing element depends on the availability of a card punch, a paper tape punch, or a filmstrip recorder at the site of the failing device. In testing printers, punches, and displays, the readback capability is not readily provided. For these devices, a character raster that can be visually inspected is presented, and one can easily determine a break in the output raster pattern. The circuits or the mechanical elements associated with the failing raster position would be prime suspects, and fault location may be readily accomplished with a little ingenuity on the part of the maintenance engineer.

However, these functional diagnostic techniques tend to break down when they are used on areas of design devoted to the transformation of data by passage through adders or other logic elements. Not only is this area of the design very large and usually constructed with uniform building blocks, but it is also an area that permits literally billions of legitimate combinations, many of which are don't care subsystem design conditions that, in practice, once the system is assembled, are unrealizable. It is here that the maintenance engineer cannot rely on a seat-of-the-pants approach, but needs the help of an algorithm or systematic methodology. He should not be found in the embarrassing position of having his diagnostic programs work correctly and successfully while his customers' programs do not.

But a systematic approach is not enough to automate the test generation process. The approach has to be applied to the logic design (which, in turn, has to meet certain requirements) to make the process economically possible. These requirements, so often overlooked, include:

1. complete documentation of the design;
2. standard notation used in the documentation;
3. high usage of the components used to build the machine;
4. minimum exceptions to rules.

If these requirements are not met for components, different unique-usage programs have to be written to analyze the logic. The complexity of these programs is proportional to the number of exceptions to the rules or the number of rules that are broken in a design. Furthermore, unless the design is completely documented, additional exceptions have to be written into the analysis programs. As the design proceeds, if the notation changes from one part to another, additional complexities have to be written into the programs. Usually at this point of utter chaos, the programming budget is in danger of being exceeded, the deadline may be missed, and somebody may decide that a new algorithm is required to get this program back on the road. It is therefore wise for the person given the task of implementing a program to generate test patterns, to heed the four conditions above, and to obtain a commitment from the users that these conditions will be met, before he accepts his assignment.

In the sections that follow, we shall review some of the techniques, models, and algorithms that have been successfully applied to the test pattern generation for large logic arrays, and we shall try to place these in historical perspective. We have to realize that a generalized algorithm for testing a large sequential circuit with limited access or observation points has not yet been developed. Each algorithm, or model, or technique, therefore, is designed for and is applicable to a special case only. The more a computing system is designed to follow one of these special cases, the better the chances are of having the design automation system generate the tests for it automatically. This leads to a better design. If the resultant computer can be tested automatically, and if it is built with replaceable and sufficiently reliable components, it can be easily maintained. Therefore, the well-designed computer is the easily maintained computer which can be made to do its job of solving problems for customers with minimal down times.

7.4 TEST GENERATION MODELS FOR COMBINATIONAL LOGIC

In the following sections we shall investigate a number of test generation techniques for combinational logic that have been successfully applied in commercial computing equipment. Each computer design group or manufacturer has his favorite test generation techniques, and some of these turn out to be computer-aided rather than fully automated. The literature is quite extensive [5]. We shall investigate some of these techniques from the point of view of the man in design automation who must furnish a service at a reasonable cost. It is his job to decide which test generation method to use and how to program it on the computer available to him. The emphasis, therefore, is to provide him with a thorough understanding of what is involved, and also to make him wary of techniques that work with lightning speed on small concocted problems, but seldom produce a result for large problems associated with present-day computers.

We also distinguish between combinational logic (discussed in this section) and sequential logic (discussed in Section 7.5). The reasonableness of this approach is evident when one assumes that, from a practical point of view, a sequential circuit may be considered combinational during one clock cycle or between the transitions from one machine state to the next. If we can solve the combinational problem, we can use it to solve the sequential. Alternatively, it can be shown that computers can be designed to take advantage of combinational tests [6].

7.4.1. Eldred's Method of Activating a Failure

R. D. Eldred [7] faced the problem of writing diagnostic programs for the DATAMATIC 1000 computer. This machine was designed mainly with OR-AND-Register logic. In the mid-1950s, test routines for use by the servicing engineer were written by first analyzing the description of the machine and then generating enough tests to check all the functions the machine was designed to perform. Although these tests were indeed useful in determining whether the design engineer faithfully implemented the design specifications, they were found wanting for the purposes of diagnosis. Eldred therefore proposed using the system logic diagrams as a starting point. Here, the conditions of operation of each logic function were completely defined, and a set of tests could be developed to make each logic circuit perform all possible functions. He proposed four simple rules which, when applied to each component, should lead to the complete testing of a system:

1. Let each input buffer diode conduct when activated (i.e., each input to an OR gate).
2. Let each AND gate leg diode inhibit when not activated.
3. The output of the gating structure should not be lost (i.e., inhibited by a signal not under test) when traveling via output components.
4. The output components should not provide a signal when the input conditions are inappropriate.

He discovered that output components such as auxiliary amplifiers, delay circuits, and transmission lines were also tested by the tests applied to the gating structure with which they were associated. He therefore could select an optimal number of test conditions necessary to make each diode (and vacuum tube) conduct and inhibit at least once, and thereby claim complete diagnostic coverage.

While others [8] had been engaged in applying faults to the hardware to generate failure dictionaries, Eldred had written a set of diagnostic routines for his machine using the simple technique he developed from the description of the logic for that machine (which happened to be a series of Boolean equations). His four simple rules recognized that, no matter what the fault is, it will not exhibit itself as a failure until an inappropriate output signal is generated when it should have been "appropriate." In other words, a fault must exhibit itself at an observable output as a stuck-at-1 or stuck-at-0 condition.

7.4.2. The Forbes–Stieglitz–Muller Techniques

In the late 1950s, R. E. Forbes, C. B. Stieglitz, and D. Muller developed a series of diagnostic techniques (FSM techniques) as part of the IBM Systems Error Analysis project [9]. These techniques were then programmed so that an IBM 704 computer could make use of design automation data stored on logic master tapes. They assumed that faults manifest themselves as stuck-at-1 or stuck-at-0 conditions. They defined the "good machine" as a design without a faulty circuit and a "bad machine" as a design with one faulty circuit. Furthermore, they assumed that only one fault could occur at any given time, mainly because the permutations and combinations of multiple faults would make the task of computation an unending one. They argued that the probability of such multiple faults is quite low and the chances are quite high that one fault masks out the consequences of another fault. Also, after having located and fixed one fault, they could reapply the tests to locate the next fault. This was tried out and demonstrated as valid by this author in a large number of cases on an actual system.

Forbes, Stieglitz, and Muller set out to develop, for any large combinational circuit, a nonexhaustive set of test patterns which would permit them to pinpoint a fault with very high resolution by observing the input/output patterns of the circuit. Using basically Eldred's method, they developed a few computer-use rules, or theories.

In one theory, called the *failing net theory*, the researchers make no distinction between the erroneous output of a logic block and the erroneous inputs to other blocks, provided the blocks are all part of the same *net*, i.e., they are electrically common. In this theory, the entire fan-out path from one logical output, or source, to any and all inputs to other logical functions, or sinks, assumes the same logical value. If the sources and sinks appear on different replaceable components, then all would be named as suspects, provided that the error appearance propagates to an observable output point.

Note that in Fig. 7-4 the output of block *m* should be 0 in both applied tests; but because of the failure, it is a 1. Applying the failing net theory, the inputs to blocks *j* and *k* are therefore affected; in test 1, observable output *E* is wrong and observable output *F* is correct. (The reader may want to trace the correct values through.) In test 2, observable output *E* is correct, while observable output *F* is wrong. Since the net propagates over physical package 2 and physical package 3, by this theory the two are not distinguished, and therefore both are suspects.

In another theory, called the *failing input theory*, the output of a logic block still performs its true logic function provided the proper input conditions are met. Some of these input conditions are met inadvertently because of a fault. In this theory, an input fault does not drag with it the whole net

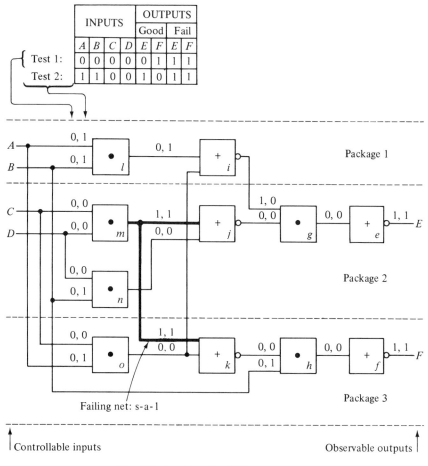

Fig. 7-4. Applying the failing net theory.

Sec. 7.4   Test Generation Models for Combinational Logic   349

to which it is connected. Thus every sink in a multiple fan-out network behaves independently. This theory is equivalent to Eldred's. However, the researchers found that, by applying this theory, a test is never generated for the output of an AND block stuck at 0 or for the out of an OR block stuck at 1.

Applying this theory to the same circuit (Fig. 7-4), we observe this time, in Fig. 7-5, that only the input to block k is stuck-at-1, and block j behaves normally. Thus value E passes both tests, but value F passes only the first. Package 3 is therefore the sole suspect. A combination of the foregoing two theories was then proposed and programmed in what is called the ANA-LYZER. This program produces some superfluous tests; but at least it is possible to distinguish between input and output failures, except in the case

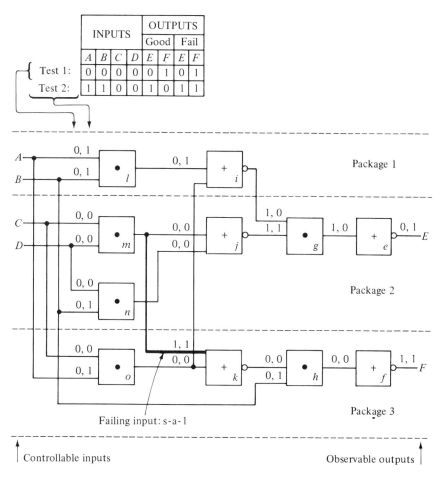

Fig. 7-5. Applying the failing input theory.

where single-input, single-output logic or physical elements (i.e., amplifiers, drivers, and inverters) are involved.

The programs developed for Forbes, Stieglitz, and Muller are applicable only to combinational logic. The inputs to this logic, as well as the outputs, must be clearly identified after all latches and triggers are removed or after the feedback loops are cut by the removal of one or more connecting lines. The first step of the calculation is to assign a level to each logic block. The first level of logic blocks is composed of those blocks that are fed only by input signals. The next level of logic blocks is obtained by finding those blocks that are fed only by inputs and the first level. This leveling is continued until the last level is obtained. This nth level is fed by any of the other levels ($n - 1, n - 2, \ldots, 2, 1$) and feeds only the output.

The second step of the calculation is to take these identified logic levels and compile them into a logic simulator.

The third step is to subject the identified logic levels to the ANALYZER program to calculate a series of input (test) patterns by utilizing the failing-net and failing-input theories. This step for the E-signal trace of Fig. 7-4 is shown in Table 7-2. Whenever one of the variables is essential to producing the desired output, its value is fixed, and only the remaining unassigned values in the same term need be varied from there on. The fixed, nonconflicting values at the primary input constitute a test.

In the fourth step, each input pattern is applied in turn to the logic circuit, and the value of the output pattern is calculated with a simulator program that assumes every circuit performs properly. Then each leg of the logic circuit is simulated to fail, in turn. The same input patterns are applied, as before, and the output patterns are calculated for each assumed fault.

In the fifth step a list is produced of all faults which produce identical output patterns, but which are different from those produced by the nonfaulty machine simulation. Referring again to Fig. 7-4, note that the m-input leg of logic block j stuck-at-1 is tested by the same input pattern and may be indistinguishable from the j output stuck-at-0, or the j-input leg to block g stuck-at-0, or the output of g stuck-at-0, or the input of e stuck-at-0, or the output of e stuck-at-1. Another test, however, utilizing the n-input leg to j as activating signal, may be able to make the distinction. Since all of these indistinguishables are found in the same package, and if that package is the basic replaceable unit, then it is not necessary to compute any further. Steps 4 and 5 are continued until every input pattern has been exhausted.

In step 6, the tests are put into a decision-tree sequence. The list of fault-producing suspects for each output pattern is examined to determine if the number of supects is within the addressing resolution set by parameter into the calculation. If so, the test is ready for use as a final test in the decision tree. If not, secondary tests are chosen from the previously produced set and are used to increase the resolution of the diagnosis until every circuit has been

TABLE 7-2. Step 3 of FSM Test Generation Procedure

For Circuit	Output	Input	E-Trace Test 1	E-Trace Test 2
e	1	g'		1
	0	g	1	
g	1	i, j	1	
	0	$i', (j + j')$		1
	0	$(i + i'), j'$		
i	1	l', o'	1	
	0	$l, (o + o')$		1
	0	$(l + l'), o$		
j	1	m', n'	1	1
	0	$m, (n + n')$		
	0	$(m + m'), n$		
l	1	A, B		1 fix A, B
	0	$(A + A'), B'$	1 fix B'	
	0	$A', (B + B')$		
m	1	C, D		
	0	$(C + C'), D'$	1 fix D'	1 fix D'
	0	$C', (D + D')$		
n	1	B, D		
	0	$(B + B'), D'$	1 fix D'	1 fix D'
	0	$B', (D + D')$		
f	1	h'		
	0	h		
h	1	B, k		
	0	$B', (k + k')$		
	0	$(B + B'), k'$		
k	1	o', m'		
	0	$(o + o'), m$		
	0	$o, (m + m')$		
o	1	A, C		
	0	$(A + A'), C'$	1 fix C'	1 fix C'
	0	$A', (C + C')$		
Final Pattern Selected			$(A + A'), B', C', D'$	A, B, C', D'

diagnosed within the addressing resolution limit. The secondary-test generation problem will be discussed in Section 7.5.1.

By making use of the 36-bit word structure of the IBM 704, the logic simulator was able to calculate the outputs for 36 different input combinations simultaneously. These programs were used very successfully for generating tests for the diagnosis of the character-recognition circuits of the IBM 1418, IBM 1419, and IBM 1428 Magnetic-Character and Optical-Character Reader Sorters. The tests were automatically applied to the 1418, 1419, and 1428 by the IBM 1401 data processor which, in turn, printed out the failure suspect list when a test produced an incorrect result.

7.4.3. The Maling–Evans Model

K. Maling and M. W. Evans at IBM were given the Forbes–Stieglitz–Muller (FSM) model and three plain and simple objectives:
1. Increase the size of the logic to be tested in a 32K-word IBM 7090 from 1000 to 4000 logic blocks or connectives.
2. Reduce the IBM 7090 computing time by at least one order of magnitude.
3. Base all diagnostic decisions on the status of a single observable signal at a time.

A thorough analysis of the FSM model brought to light a few ideas for substantially saving time in generating the tests. For example, about a third of the time was spent in the ANALYZER developing patterns for the simulator to use. Why not use random numbers to both save that time and reduce the size of the program needed? The concept was great; but unfortunately, random numbers were not the answer. In disbelief the researchers observed thousands of numbers run through the simulator and test no new logic. The random numbers were valid for testing adder designs and data paths, but they were much less efficient than the ANALYZER-produced test patterns for testing control circuitry. The researchers realized that random luck was not on their side and that only certain combinations of input-pattern values would reveal a bad machine by an inversion of the correct value at an output.

A thorough examination and understanding of Fig. 7-6a-d may bring this lesson home. The input pattern 0, 0, 0, 0 applied in Fig. 7-6a and b does not exhibit any difference at the output whether a failure exists in the circuit or not. The input pattern 1, 1, 0, 0, as shown in Fig. 7-6c and d, on the other hand, does. The problem is to find enough tests to distinguish all possible (i.e., assumed) bad machines from the good machine (fault detection) and, if possible, to distinguish all bad machines from each other (fault location).

Maling showed that any method of solution, including the exhaustive, may look good if the circuits considered are composed of less than sixty logic connectives or blocks. Beyond that number, an algorithmic procedure should be superior. It is for this reason that Evans and Maling may have been led astray with the random-number generator, hand-applied to a small number of logic blocks. They proceeded to analyze further what was happening and arrived at certain very interesting concepts, which ultimately caused them to develop a new program that tried to overcome the anomalies they had discovered.

Let us now assume that a net is stuck at a value, say 1, in block e of Fig. 7-7. A test for this immediately implies that the signal values on the input nets (H and I) to the logic connective (block) producing the output value E should be such as to produce a zero output. The values of these inputs are called *implied values* and are chosen such that more than one input signal

Sec. 7.4   Test Generation Models for Combinational Logic   353

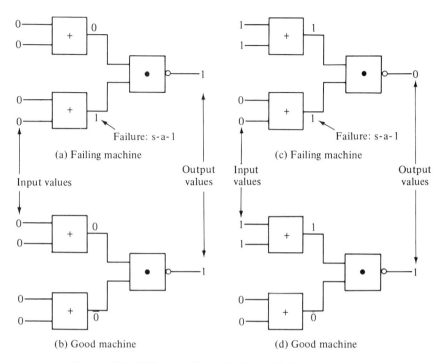

Fig. 7-6. The failing machine is indistinguishable from the good machine at the output of (a) and (b); but can be distinguished as shown in (c) and (d).

(if possible) will provide the desired output value. In the example, both values must be 1's. This in turn implies that all the other blocks feeding the input nets are set to fixed values, etc. (The inputs to blocks h and i should all be 1's.)

A backward trace from the failure level (level 4) to the input or entry level has to be made to determine all the entry values. One net feeding block h is an entry net, net S, which can be set to 1. The other is fed by block k; therefore the implied values must be traced further back. Similarly, block i is fed by an entry net, net R, which can be set to 1. The other input is net L, which must be traced further back. The program, having exhausted its required calculations in level 3, now proceeds to level 2 and determines that, for block k to have an output of 1, both inputs must be 0. Since neither is an entry net, a further trace is required. Similarly, block l has to be set at 1, which implies that nets O and P must be set to 0. Since neither is an entry net, the trace must proceed to level 1. Block m is fed by an entry net, net Q; therefore this trace ends with Q set to 0. For block n to have the value 0, both inputs R and S must have the value 0. But both are entry nets and already have implied values of 1. This causes an implication conflict, marked X

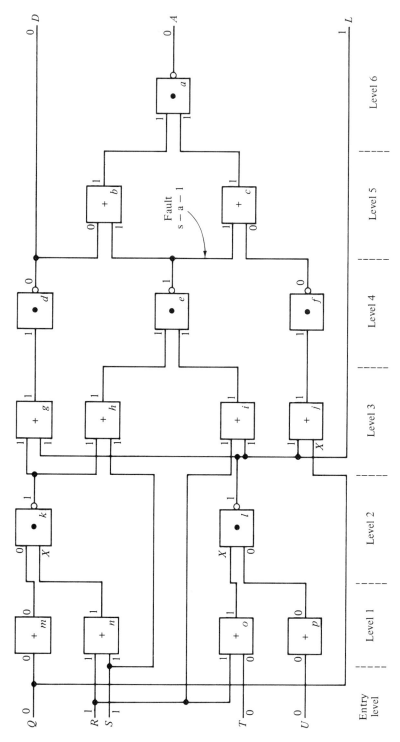

Fig. 7-7. Multiple-output logic network with net E stuck-at-1.

Sec. 7.4   Test Generation Models for Combinational Logic   355

at the input to k. Similarly, block o must be fed by two 0's; but one of the nets already has an implied 1. This causes another implication conflict, again marked X at the input to l. The second input feeding that block, net T, may be set to 0 since it is still an unassigned entry net. Block p must be fed by a 0, which is no problem to net U, another entry net.

Now the forward trace has to begin. Does net E feed an exit? No, but it feeds blocks b and c. To have the signal of net E propagate forward, net D must be an implied 0. Similarly, net F must be an implied 0. Since b and c are the only blocks that feed a, the exit block, no further forward trace is neccessary. However, another backward trace is necessary, to ensure the implied values of D and F. Block d is fed by g, which must be a 1. Block f is fed by j, which must be a 1. Block g is fed by nets K and L and, since both have an implied value of 1, this trace is completed without conflict. Block j is fed by block l and signal Q. Both already have implied values of 1 and 0, respectively. The latter is in conflict and is also marked X.

Now the marked conflicts must be analyzed and made consistent. This means that one of the values must be chosen to produce the required result or the candidate test pattern must be discarded.

Two rules are applied:

1. If another input to an AND block exist whose implied value is 0, the conflict may be resolved by assigning the implied value of 1 to the "conflict" input.
2. If another input to an OR block exists with the implied input value 1, the conflict may be resolved by assigning the implied value of 0 to the "conflict" input.

Rule 1 may be applied to blocks k and l, and both X's change to 1's. Rule 2 may be applied to block j, and the X is changed to 0. This results in a test for net E stuck-at-1 of Q, R, S, T, and U equal to 0, 1, 1, 0, 0 as shown.

If the conflicts had not been resolved, a test for the bad machine involving net E stuck-at-1 may not have been found, and we would have no idea whether one exists. We therefore try to resolve the conflicts with other combinations by changing conflict-producing entry values and calculating the consequent implied values. In this network, however, many solutions exist. For example, the reader might want to prove that another solution exists with U set to 1. In that case, net F is an implied 1; therefore, the fault at E will propagate only through block b. Other solutions exist with Q set to 1, or R to 0, or S to 0, or T to 1.

The propagation path of the error through the logic is known as the *sensitive path*. This is the path for which Roth searches in his *D*-algorithm (to be discussed in Section 7.4.5). It is also the path independently described by Armstrong [10]. The heuristic researchers noted that, if any net on the sensitive path switches values, the output value at the observable exit will also switch. Evans also discovered that in some designs destructive and constructive interference took place. *Destructive interference* is present when the implied values cannot be resolved because it is discovered that a sensitive path branched out into two (or more) parallel paths at some level and then reconverged with its value inverted at a subsequent level. Consequently, the apparent solution has to be discarded. To minimize destructive interference, the researchers developed a strategy to use constructive interference when a net fans out from the sensitive path. *Constructive interference* is present when a failure is masked by another net such that no sensitive path exists. For example, if net F in Fig. 7-7 were set at 1, then the sensitive path through block c would have been eliminated.

From exercises similar to that just described, and from additional observations of the concept of interference, Maling and Evans came to the conclusion that the computation time could be kept in check if the analyzer and simulator portions of the FSM model were intertwined and calculated simultaneously. Computation could then be discontinued as soon as interference was discovered. Furthermore, a test of one bad machine is a test of many bad machines tracing the entire sensitive path. Also, since the sensitive path reverses values because of a fault, it follows that the inverse fault would also invert the values at the exit (by definition). Therefore, once a sensitive path has been found, the program can take full advantage of it and identify every fault associated with that path, so as not to go through all that tracing again for each separate fault detectable along that path. On the other hand, if no sensitive path can be found for a fault because of constructive interference, then all potential faults along the sensitive path traced up to there share the same fate.

Additional observations indicated that, in general, the fewer the levels of logic through which a sensitive path propagates, the fewer the implied values that have to be set. And if fewer implied values are required, then fewer conflicts exist. Therefore, the design of the program should develop tests first for the logic trees with the fewest number of levels between exit and entry. A *logic tree* is comprised of all logic blocks from its apex, a single exit, to its roots, the entry level. For example, tree D in Fig. 7-7 is comprised of blocks d, g, k, m, n, l, o, p, and exits at level 4. Tree L has only two levels, while tree A has six levels. The three trees, in this example, constitute the entire logic to be analyzed at one time and are termed a *forest*. One should bear in mind the difference between a tree and the sensitive path. A sensitive path may at times include an entire tree; but this is not generally so. The proof of this is left as an exercise to the reader.

One final but important observation on methods: So far, our discussion of Maling's and Evans' work has concentrated on distinguishing the good machines from the bad. How does one distinguish the bad machines from each other? The set of bad machines that can be distinguished from the good machines at one exit will be called a *class*. The researchers observed that the test generation solution time is approximately proportional to the square of the size of the class. Therefore, a substantial improvement in computer time can be obtained if a bad machine is diagnosed within the smallest class to which it belongs. This also accomplishes a diagnostic objective, namely, the addressing resolution, without additional computation.

Now let us concentrate on the computer and the coding techniques. The researchers observed that high speed storage is usually available as a fixed number of words; hence, the largest forest that can be analyzed is fixed after the size of the program and its working storage have been determined. A careful trade-off must therefore be made between execution time for obtaining solutions and execution time for packing the representation of the model and working lists into the least amount of storage. The trade, they concluded, is usually in favor of storage. Improvements in coding technique for speed turns out to be less significant than improvements in method; and the penalty for exceeding a high speed storage boundary is extremely high, both in programming effort and in waiting and execution time.

Another observation they made ruled out a compiled simulator, i.e., one which simulates the logic directly with a string of computer instructions. The researchers determined that, if evaluation and test development had to be intertwined, both had to use the same logic representation if one were to stay within the storage limitation. An interpretive simulator therefore had to be developed. An interpretive simulator is slower in execution than a compiled simulator, but it does have the advantage that it can be terminated at any level, i.e., whenever the effect of a fault ceases to propagate. Also, to run efficiently, a compiled simulator has to simulate a whole forest at once, even if the results of only a single tree are of interest. A significant reduction in simulation results, therefore, when only the sensitive path is simulated.

With those observations implemented, the objectives Maling and Evans set out to accomplish were realized. The resulting model can be summarized by the following procedures:

1. Given all exit names in a forest, extract each tree from the forest and discard all remaining unconnected logic blocks.

2. Divide each tree into levels.

3. Sequence the trees for test development by the number of levels in ascending order.

4. Make a list of faults to be tested in each tree. For example, use the failing net theory.

5. Start at the fault (e.g., s-a-1) closest to the entry level, i.e., lowest level (not previously tested), in a tree with the least number of levels, and find the sensitive path to the exit.

6. Assign implied values to every logic block contributing to the activation of that path, and calculate the entry values which now become the test candidate input pattern.
7. If a solution is blocked by interference, try to resolve the interference by other implied assignments. The maximum number of retries (i.e., discarding of candidate patterns) may be set by parameter before the program concludes that a sensitive path cannot be found.
8. Check off all faults (in the list developed in procedure 4) that the sensitive path tests, and assign these checked-off faults to a class.
9. Do procedures 5-8 using the alternate fault (e.g., s-a-0) at the point just completed.
10. Find the next remaining untested fault closest to the entry level, and repeat procedures 6-10 until no faults are left untested in the list or until conflicts prevent a test from being found for the remaining untested faults.
11. Go to the next tree and repeat procedures 5-11 until all trees are exhausted.
12. Sequence the resulting test patterns so that the pattern with the smallest number of bad machines in the class comes first. In this manner, those blocks closest to the exit level will be diagnosed before others further back. (Sequencing will be further discussed in Section 7.6.3).

The Maling-Evans model has been successfully utilized by IBM on a commercial basis as the central computer algorithm of the Fault Locating Test generator program complex [1] for generating tests for System/360 Models 50, 65, 75 and a number of subsystems.

7.4.4. The Galey-Norby-Roth Technique

As a by-product of his logic minimization effort using cubical complexes, J. P. Roth became interested in diagnosis. With the help of a programmer, R. E. Norby, and a diagnostic engineer, J. M. Galey, he developed the first systematic procedure to compute all patterns that can distinguish the behavior of a machine with a failure from that of a good machine [11]. The beauty of his technique was that it could first develop all input combinations which may be used to test the circuit, and then reduce these to a minimum set so that very few tests would be required to check out a circuit. This technique constituted the basis for his later work on the D-algorithm, to be covered in the next section. It is interesting to review this early work to get a better understanding for what followed.

The example Roth shows in his first report is an 8-bit parity check circuit with 43 logic functions that had 102 possible failure modes (the conventional stuck-at-1 or stuck-at-0 faults). The program developed four input combinations that tested all 102 failures (i.e., distinguished between the good machine and any of the 102 bad machines). Soon thereafter, however, the technique fell into disfavor because of the unpredictable amount of computer

Sec. 7.4   Test Generation Models for Combinational Logic   359

time it required. For example, after one of the problems presented to the program, involving about 65 logic blocks with 115 variables, did not come to a speedy solution, and a program bug was suspected, someone calculated that $(10)^8$ reels of magnetic tape, each 2000 feet in length, and recorded at 556 bits per inch, were required to record a minimal normal form expression for the off-array, a necessary ingredient to the calculation. Needless to say, more work was required to tame the calculation.

In this chapter, we shall not go into the calculus of the cubical complex that Roth developed; instead, we shall delve right into its application. Papers are available [12], [13], in which the notation, the calculus, and the proofs are described in great detail. Some of this information also appears in Chapter two of this book. We shall use the example of Fig. 7-7 to illustrate the technique. The output, A, in Roth's terminology, constitutes a Boolean function of five variables, Q, R, S, T, and U, expressed by a series of injection words.

In Table 7-3

$$\prod_A^{\&'} B, C$$

is the *injection word* which states that the ON- and OFF-arrays in the rows above the line have the ON- and OFF-arrays of the $\&'$ (AND-INVERT)

TABLE 7-3. ON-OFF MATRIX FOR TREE A OF THE CIRCUIT IN FIG. 7-7

Line		U	T	S	R	Q	P	O	N	M	L	K	J	I	H	G	F	E	D	C	B	A	
1																					1		on
2																					0		off	$\prod_A^{\&'} B, C$
3																			X	0			
4																			0	X			on
5																			1	1			off	$\prod_B^{v} D, E$
6																	0	0	X				
7																	0	0	0				
8																	X	1	0				
9																	1	X	0				on
10																	X	1	1				off
11																	1	X	1					$\prod_C^{v} E, F$
12																	0	0	0				
13	Discard (see text)																	X	Φ	0				
14																	1	0	0				
15																	0	0	0				
16																	0	0	1				
17	Discard (see text)																	0	Φ	1				
18																	0	0	X				on
19																	X	1	1				off
20																	1	X	1				
21																	X	1	X				
22																	1	1	X				

function in terms of the variables B and C substituted for the variable A. The arrays are in positional notation, where each column stands for a different variable.

The *ON-array* represents the Boolean function when it is a 1, and the *OFF-array* represents the Boolean function when it is a 0. The X is used here to indicate a don't care value (i.e., either 1 or 0). Similarly

$$\prod_{B}^{v} D, E$$

stands for the substitution of the ON- and OFF-arrays of the variables D and E for the variable B. This kind of substitution can go on until the last ON-array and OFF-array are purely in terms of the input varables, Q, R, S, T, and U. The reader may wish to carry this calculation to the bitter end to gain a little feel for the meaning. The first step of the Galey–Norby–Roth (GNR) technique is to state the problem to be solved in both the ON- and the OFF-array form.

Actually, there is nothing mysterious about what is being accomplished. In Boolean algebra, the Boolean tree can be represented by the equation for the ON-array

$$A = 1$$

and the equation for the OFF-array

$$A = 0$$

The injection word

$$\prod_{A}^{\&'} B, C$$

represents the equation

$$A = (B \cdot C)' = B' + C'$$

Therefore, the ON-array is

$$A = 1 = B' + C'$$

and the OFF-array is

$$A = 0 = B \cdot C$$

Placing these into positional notation yields the data in Table 7-3. (No attempt at simplification has been made in the table.) Again, the reader may want to carry this calculation out to the end.

Notice that the substitution of

$$\prod_{C}^{v} E, F$$

in line 6 of Table 7-3, results in an impossible condition in line 13, where variable E has to be both 1 (because of the OR function) and 0 (because of the input condition in line 6) simultaneously to satisfy the overlay of the substitute array on the base array. Consequently, line 13 has to be dropped. This is different from line 11's expansion into line 22. Here the X of the substitute array was assigned the value 1 from the base array.

Table 7-3 has been carried far enough to allow us to proceed to the next step of the GNR technique, namely, the introduction of a fault via the concept of pruning. *Pruning*, here, means cutting a branch in a Boolean tree and inserting the failure, i.e., the incorrect value, at the cut. At this point, we must realize that a *branch* is nothing but the variable, or column, of the array. We must also realize that four possible conditions have to be accounted for:

1. The on-array may contain a value of 1 at the cut.
2. The on-array may contain a value of 0 at the cut.
3. The off-array may contain a value of 1 at the cut.
4. The off-array may contain a value of 0 at the cut.

We must therefore keep track of these four conditions in four arrays, a, b, c, and d; but fortunately, each is smaller since it is only a part of the whole.

The second step of the technique is therefore to segregate the four arrays, using column E as the cut point, as shown here.

	F E D	Remarks
$a^{on}_{E=1} =$	Null	ON-array with the value of $E = 1$ at the cup point
$b^{on}_{E=0} =$	$\begin{bmatrix} X & 0 & 0 \\ 0 & 0 & X \end{bmatrix}$	On-array with the value of $E = 0$ at the cup point
$c^{off}_{E=1} =$	X 1 X	Off-array with the value of $E = 1$ at the cup point
$d^{off}_{E=0} =$	1 0 1	Off-array with the value of $E = 0$ at the cup point

Note that no term exist for $E = 1$; therefore array a is empty. Replacing $X00$ for lines 12 and 14 of Table 7-3 and using line 18 instead of lines 15 and 16, we obtain array b. Line 21 provides us with array c. And substituting 0 for X in line 20 gives us a term for array d.

Since the "stuck-at" value must appear as a value inversion at the output, the third step is to set the cut point to the inverse value and expect a term of the inverse array as a result, if a test exists. To implement this, substitute 0 for the cut value in array a and form a new array e by intersecting the new a with array d as shown below.

	F E D	Remarks
$e^{on}_{(0,1)} = a_{E=0} \cap d_{E=0} =$	Null	Assume failure is s-a-1 at cut point

This step obtains all possible patterns to distinguish the good from the failing circuit by setting the cup point (E) to the inverse value and expecting, therefore, a term of the inverse array for each existing test.

	F E D	Remarks
$f^{on}_{(1,0)} = b_{E=1} \cap c_{E=1} =$	$\begin{bmatrix} X & 1 & 0 \\ 0 & 1 & X \end{bmatrix} \cap \begin{bmatrix} X & 1 & X \end{bmatrix} = \begin{bmatrix} X & 1 & 0 \\ 0 & 1 & X \end{bmatrix}$	Assume failure is s-a-0 at cut point
$g^{off}_{(0,1)} = c_{E=0} \cap b_{E=0} =$	$\begin{bmatrix} X & 0 & X \end{bmatrix} \cap \begin{bmatrix} X & 0 & 0 \\ 0 & 0 & X \end{bmatrix} = \begin{bmatrix} X & 0 & 0 \\ 0 & 0 & X \end{bmatrix}$	Assume failure is s-a-1 at cut point
$h^{off}_{(1,0)} = d_{E=1} \cap a_{E=1} =$	Null	Assume failure is s-a-0 at cut point

Similarly, form array *f* by substituting 1 for the cut point in array *b* and intersecting the new *b* with array *c*. Form array *g* by substituting 0 for the cut point in array *c* and intersecting the new *c* with array *b*. Form array *h* by substituting 1 for the cut point in array *d* and intersecting the new *d* with array *a*. The reader may wish to verify the result, realizing that an array intersected with a null array produces a null array. Similarly, a 1 and a 0 intersected in the same column produces a null term, not an *X*.

We have now obtained all possible values which will distinguish this failing circuit from a good circuit. Array *e*, which is nonexistent, nevertheless states that, if *E* is stuck-at-1, the output *A* will not be a 1. Array *g* comes to the rescue and states the conditions under which *E* stuck-at-1 can be tested and will produce a 0 at the output, *A*. Similarly, array *f* states that, if *E* is stuck-at-0, the output, *A*, will be 1 if any row of the array is applied. (Note that the Boolean sum of all terms gathered from the intersection of the foregoing matrices represents

$$A(F, E, D) \oplus A(F, E', D) = [E \cdot A(F, 1, D) \oplus E' \cdot A(F, 0, D)]$$
$$\oplus [E' \cdot A(F, 1, D) \oplus E \cdot A(F, 0, D)]$$

which is called the *Boolean difference* and will be discussed in Section 7.4.6.)

Step 4 is now a matter of performing all the injection operations to the non-null arrays in order to obtain the values of the input variables (Q, R, S, T, U) which test the given failures.

Step 5 requires that steps 2–4 be repeated for every column, proceeding from right to left, until the first input variable is reached. This step then has produced the totality of tests which differentiate the good from the bad machines. Choosing an irredundant set of tests to test all failures in the tree requires another step, namely, the solution to a covering problem of the type discussed in Chapter two of this book. It is, therefore, not repeated here.

As one can see, the technique is elegant, simple, iterative, and can be handled in parts. It lends itself easily to computer solutions and seems perfect for design automation implementation. The technique has just one drawback: both the ON-array and OFF-array are required for the solution; and usually one array is small and the other's size is unpredictable.

7.4.5. Roth's D-Algorithm

So far we have seen some representative techniques which have been used to generate diagnostic tests for acyclic or combinational circuits. Eldred's method (Section 7.4.1) was manual and represented the first systematic technique published. The Forbes–Stieglitz–Muller (FSM) technique (Section 7.4.2) may have been the first large-scale, commercially used automated method. The method suffered from two drawbacks:

1. It finds a candidate test pattern which supposedly tests for a particular failure and then attempts to prove, by simulation of the whole logic network, whether the pattern indeed is a test. This proof is time-consuming and frequently ends in failure due to masking or blocking before the failure symptom can propagate to the primary output and be observed.
2. Even if it can be proven that a test for a particular failure exists, the method could not guarantee to find it, since it depends on trial and error.

The Maling–Evans (ME) model (Section 7.4.3) improved on the FSM technique by localizing the simulation to the sensitive path and stopping it as soon as masking or blocking is detected. Thus, while disadvantage 1 was eliminated, disadvantage 2 still remained.

The Galey–Norby–Roth technique was the first systematic analytic attempt at diagnosis. However, since it required both the true and the complement functions (ON- and OFF-arrays) as part of its solution, the amount of computer time involved, even in relatively small commercial problems, was unpredictable and often exorbitant. A more "localized," analytic method with the following characteristics was required:

1. If a test for a fault exists, it will indeed be found.
2. An extant test will be discovered in a small amount of computer time.

Instead of tracing a single sensitive path as in the ME model, it was necessary to trace all sensitive paths simultaneously. Such a technique was proposed by Roth [14] and was named the D-algorithm. The number of terms needed in its calculation is proportional to the number and complexity of the logic functions under test. Proponents of the D-algorithm, therefore, claim that it is as efficient as the best of the previously known heuristic techniques and, further, if a test exists, the D-algorithm will find it. In fact, it utilizes the ME sensitive-path concept, here called the *activity vector*, to localize calculations as much as possible. With DALG-II [15], the TEST-DETECT algorithm is advanced, which mimics the ME heuristic by identifying all faults along the D-path once a test pattern has been found. This approach reduces significantly the number of calculations needed, since it does not compute failures on a one-at-a-time basis.

Only the highlights of the D-algorithm will be discussed in this book. The reader is referred to reference [15] which discusses DALG-II and its various strategies in detail and provides the necessary background for any reader who plans to program this method. Another excellent discussion is contained in reference [16]. We shall treat only the method here, and we show how it fits into an automated test scheme. We shall start by the unorthodox method of stating the algorithm, and then explaining, where needed, the new terms utilized in its statements.

The D-Algorithm Stated. The D-algorithm is used to generate a test pattern or test cube for a particular stuck-at-1 or stuck-at-0 failure in a logic net whose primary input coordinates may be set to any combination of 0's and 1's, and whose primary output coordinates are observable:

1. Select a primitive D-cube of the failure ($pdcf$).
2. D-intersect it with the primitive D-cubes (pdc) of the logic net so as to form a connected chain of D-coordinates to a primary output. (This operation is also called D-drive.)
3. Complete the connected chain from primary input to primary output by D-intersecting the resultant cube obtained in step 2 with the singular cover of the logic net. In this step, a set of primary input values is obtained which will imply all 0's and 1's required in forming the D-chain. (This operation is also called the *consistency operation*.)

A systematic application of these three steps leads to the generation of a test for each testable failure in the logic net. As a by-product of this algorithm, a list of untested failures may be produced. The logic designer may then be alerted to change his design or to introduce more observable test points (primary outputs) to provide adequate maintenance for his design.

The pdc and $pdcf$. Going back to our ON- and OFF-arrays of Section 7.4.4, we find that we can simplify our notation slightly by assigning a value of 1 to the coordinate position of the output of a logic function of the ON-array and a value of 0 to the coordinate position of the OFF-array. Thus, for example, the injection operator

$$\prod_A^{\&'} B, C$$

represented in lines 3–5 of Table 7-3 and lines 1–3 of Table 7-4 is now changed to the three 3-coordinate singular cubes shown in Table 7-5. We call this the *singular cover*, a truth-table description of the function of one logic block, here shown with output A, and inputs B and C.

TABLE 7-4. CUBICAL COVER OF THE INJECTION OPERATOR $\prod_A^{\&'} B, C$

Line	C	B	A
1	X	0	
2	0	X	On
3	1	1	Off

TABLE 7-5. SINGULAR COVER FOR OUTPUT COORDINATE A

Line	C	B	A
1	X	0	1
2	0	X	1
3	1	1	0

We are no longer substituting for the coordinate A. Instead, we keep the coordinate explicitly. The saving becomes more meaningful when we go to the next injection operator

$$\prod_B^v D, E$$

Sec. 7.4   Test Generation Models for Combinational Logic   365

represented in lines 6–11 of Table 7-3 and lines 1–6 of Table 7-6, which is now changed to the three 5-coordinate singular cubes of Table 7-7, with coordinates C and A unfilled (or assigned as don't cares). What warrants the change and the corresponding simplification? We have to realize that, in the injection operator matrix, we substituted one set of variable coordinates for

TABLE 7-6. CUBICAL COVER OF THE INJECTION OPERATOR $\prod_{B}^{v} D, E$ IN THE ACYCLIC NETWORK A

Line	E	D	C	B	A
1	0	0	X		
2	0	0	X		
3	X	1	0		
4	1	X	0		On
5	X	1	1		Off
6	1	X	1		

TABLE 7-7. SINGULAR COVER FOR THE OR FUNCTION WITH OUTPUT COORDINATE B

Line	E	D	C	B	A
1	1	X		1	
2	X	1		1	
3	0	0		0	

another. We know immediately by looking at line 1 of Table 7-6 that, if E and D are set to 0, no matter what value C has, A will result in a value 1 (assuming a failure-free logic circuit). This relationship is lost to us in the simplification we undertook in Table 7-7. All we can tell from the first line, for example, is that, if E is 1, no matter what D is, B will be 1. We don't know how that affects A (or C). To win back that relationship, Roth introduced the D symbol as a dependent value having certain properties. The properties of D are:

1. D may take on the value of 0 or 1 in a coordinate of a cube.
2. Once assigned a value in one coordinate of a cube, the dependency must be such that every appearance of D in that cube is forced to assume the same value in an error-free circuit and the opposite value if an error exists.
3. The complement of D ($\bar{D}$ in Roth's notation; D' in the present chapter) appearing in a cube has the inverse value of D and must maintain that inverse value in every appearance.

These D's have to be placed in a particular manner into the singular cover of all the logic blocks making up the circuit to be tested. This process is akin to the intersection of the four arrays in the third step of the GNR technique, except that the intersection is performed on a local level, namely, on individual logic blocks, producing what Roth has named *primitive D-cubes* or *pdc*'s.

To construct a *pdc* of a singular cover:

1. Pick a subset of coordinate positions in a given singular cube.

2. Complement the values of the subset picked, together with the output coordinate position.
3. Intersect the new cube with each singular cube of the prime implicant cover of the function.
4. The nonempty resultant cubes may be used to form a *pdc* after D or D' is substituted for the picked coordinates whose original values were 1, and D' or D is substituted for the picked coordinates whose original values were 0.

(*Note:* The alternate D-values may be used to generate the dual *pdc*'s.)

Returning to the second step of the GNR technique, note that we segregated the prime implicant cover of the logic function into four arrays:

1. Array a had the output function value 1 and the picked coordinate E value 1.
2. Array b had the output function value 1 and the picked coordinate E value 0.
3. Array c had the output function value 0 and the picked coordinate E value 1.
4. Array d had the output function value 0 and the picked coordinate E value 0.

We could apply the same procedure to the cover of the function shown in Table 7-7, noting that the ON-array is now synonymous with $B = 1$, and the OFF-array, with $B = 0$

$$\text{Array } a = \begin{matrix} E & D & B \\ \begin{bmatrix} 1 & 0 & 1 \\ 1 & 1 & 1 \end{bmatrix} \end{matrix} \tag{7-1}$$

$$\text{Array } b = [0 \quad 1 \quad 1] \tag{7-2}$$

$$\text{Array } c = \text{Empty} \tag{7-3}$$

$$\text{Array } d = [0 \quad 0 \quad 0] \tag{7-4}$$

Next, as in the third step of the GNR technique, we perform four separate intersections to obtain groups $e, f, g,$ and h. The first intersection is

$$e = a_{E=0} \cap d \tag{7-5}$$

which states that array e consists of those terms whose output is inverted if a single failure, E stuck-at-1, is to be tested with terms from array a. Array d represents the output inversion, $B = 0$, under failure-free conditions.

Applying this calculation to the function of Table 7-7, we obtain

$$e = \begin{matrix} E & D & B \\ \begin{bmatrix} 0 & 0 & 0 \\ 0 & 1 & 0 \end{bmatrix} \end{matrix} \cap [0 \quad 0 \quad 0] = [0 \quad 0 \quad 0] \tag{7-6}$$

Now substituting D' for the picked coordinate E and the output B, we are left with a pdc

$$e = \begin{array}{ccc} E & D & B \\ \hline D' & 0 & D' \end{array} \tag{7-7}$$

Similarly, we perform the second intersection

$$f = b_{E=1} \cap c \tag{7-8}$$

which states that array f consists of those terms whose output is inverted if a single failure, E stuck-at-0, is to be tested with terms from array b. Array c represents the output inversion, $B = 0$, under failure-free conditions. But c is empty; therefore f is empty.

The third intersection

$$g = c_{E=0} \cap b \tag{7-9}$$

is again empty because c is empty. It would have represented an inversion at the output if E stuck-at-1 is to be tested with terms from array c such that the output value inverts and looks like one caused by terms of array b.

The fouth intersection

$$h = d_{E=1} \cap a \tag{7-10}$$

states that array h consists of those terms whose output is inverted if a single failure, E stuck-at-0, is to be tested with terms from array d. Array a, of course, represents the inversion of the output value ($B = 1$) and assumes no other failures. Applying this calculation to the function of Table 7-7 gives

$$\begin{array}{c} E \quad D \quad B \\ \hline h = [1 \ 0 \ 1] \cap \begin{bmatrix} 1 & 0 & 1 \\ 1 & 1 & 1 \end{bmatrix} = 1 \ 0 \ 1 \end{array} \tag{7-11}$$

and substituting D for the picked coordinate, we obtain another *pdc*

$$h = \begin{array}{ccc} E & D & B \\ \hline D & 0 & D \end{array} \tag{7-12}$$

We immediately note that Equation (7-12) is the dual of Equation (7-7). We interpret Equation (7-12) to state: if coordinate value D is held at 0, a test for the failure (s-a-0 at coordinate E) would be detected as a value 0 on the output B if the failure exists, a value 1 otherwise. Similarly, Equation (7-7) states that, if coordinate value D is held at 0, a test for the failure (s-a-1 at coordinate E) would be detected as a value 1 on output B if it exists, a value 0 otherwise.

We can perform the same calculations for a failure at coordinate D, and obtain

$$e = 0 \quad D \quad D \qquad (7\text{-}13)$$

$$h = 0 \quad D' \quad D' \qquad (7\text{-}14)$$

Since we could have had a choice of how many coordinates to pick, we may pick both E and D simultaneously, thus getting the additional terms DDD and $D'D'D'$. Since the inverse value is implied by the notation itself, we may discard all the dual terms and come up with a set of *pdc*'s for Table 7-7 as shown in Table 7-8.

TABLE 7-8. D-Cube Representation of the OR Function with Output Coordinate B

Line	E	D	C	B	A
1	D	0		D	
2	0	D		D	
3	D	D		D	

We may also give an interpretation to arrays f and g (in this case found with the empty array c), namely, that holding coordinate D to the value 1 causes a failure at E *not to propagate* through to the output ($DD'1$). This concept is made use of in reference [15] but is not needed for this discussion.

Another way of looking at the results of Table 7-8 follows. For example, the second D-cube states that, if E is held at 0, the value of coordinate D controls the value of coordinate B in a failure-free logic circuit. This value would be the reverse of one found in a failing circuit. We must remember that a test for a signal stuck-at-0 is one in which we try to set the conditions of the logic function to realize the complement, i.e., the value 1. Thus, applying a 1 in coordinate D of the first D-cube in Table 7-8 would result in a 1 at coordinate B if neither coordinate D nor B is stuck-at-0. Similarly, applying a 0 at coordinate D of the second D-cube in Table 7-8 would result in a 0 at coordinate B if neither coordinate D nor B is stuck-at-1. We thus find that, by introducing the D symbol in the D-cube, $0DD$ we realize a shorthand notation for the two cubes, 011 and 000, which could possible test for four failure conditions.

We are now in a position to define a primitive D-cube, abbreviated in the literature as *pdc*. A pdc is a singular D-cube associated with an individual logic block or circuit; it does not necessarily constitute a test when that block is embedded in a larger circuit.

A primitive *D-cube of a failure*, abbreviated *pdcf*, on the other hand, is

Sec. 7.4   *Test Generation Models for Combinational Logic*   369

a *D*-cube that identifies a test for a failure at the output of a logic function in terms of the inputs to it. Thus, the shorthand notation for testing a stuck-at-0 condition for the output of a two-legged AND function is $11D$. This states that, if the failure were not present, the output would be a 1. With a failure, it is 0. $0XD'$ or $X0D'$ would have been used if the failure-free output is 0 and if the output with failure is a 1.

D-Intersection. Now let us investigate how we might chain the *pdc*'s together to assign values to primary or setable inputs to detect faults at the observable outputs of the logic net. This is done by *D*-intersection, a calculus discovered and defined by Roth [14].

The following table defines the *D*-intersection operator. The operator is applied, coordinate by coordinate, with one *D*-cube on another at a time.

D-Intersection	0	1	*X*	*D*	*D'*
0	0	Empty/null	0	Undefined	Undefined
1	Empty/null	1	1	Undefined	Undefined
X	0	1	*X*	*D*	*D'*
D	Undefined	Undefined	*D*	Special-case 1	Special-case 2
D'	Undefined	Undefined	*D'*	Special-case 2	Special-case 1

Rule 1. If a single coordinate of a cube (term) encounters an empty/null or an undefined condition after intersection, the resultant cube is discarded.

Rule 2. If one or more coordinates in a cube encounters special-case 1 only (and no other), then the *D*-intersection is the value of *D* or *D'*. We thus chain two *D*-cubes together.

Rule 3. If one or more coordinates in a cube encounters special-case 2 only (and no other), then convert special-case 2 into special-case 1 by changing *D* to *D'* or *D'* to *D* in every coordinate of the second factor, and then chain the *D*-cubes together as before.

For example, to *D*-intersect the following two *D*-cubes

$$\begin{array}{ccccc} E & D & C & B & A \\ 0 & D & X & D & X \\ X & X & 1 & D & D' \end{array}$$

take the values of coordinate *E* (0 and *X*) and from the *D*-intersection table, get the resultant 0. Take the values of coordinate *D* (*D* and *X*) and, from the *D*-intersection table, get the resultant *D*. Continue this process until every coordinate has been changed to the resultant value or until the cube is discarded according to the rules established earlier. With the D-intersection completed, a new cube is formed as follows

$$\begin{array}{ccccc} E & D & C & B & A \\ \hline 0 & D & 1 & D & D' \end{array}$$

This new D-cube establishes a connected D-chain from the lowest input, E, to the output, A. It states that, if E and C are held at 0 and 1, respectively, a failure at coordinate D (say, stuck-at-1) would be discovered as a 1-output at coordinate A. A stuck-at-0 failure at the same coordinate would be discovered as a 0-output at coordinate A. This connected D-chain now brings back the relationship between output and input that the substitution inherent in the injection operator implied, but which was temporarily lost by the new notation.

In performing the D-intersection, there are two qualities that we look for:

1. A D-cube intersection which provides a connected D-chain driving closer to the output. This means that the output coordinate of the working (or driving) D-cube must intersect an input coordinate containing a D or D'. Preferably there should be only one D or D' input coordinate in the chosen D-cube.
2. The implied assigned values of the other participating coordinates of the resultant D-cube must be consistent.

The Consistency Operation. Now that we have shown the necessary definitions and usages of the D-notation, let us apply the D-algorithm to the logic circuit of Fig. 7-7. Table 7-9 shows its singular cover. Note that the output is in the top line and the input, in the bottom. This is consistent with the injection operator procedure, but not with Roth's present method. To make up for the difference, we shall number our cubes in the reverse order and thus obtain the same results. The first thing we must do is to pick a *pdcf*. We shall use the same one we used earlier, namely, a fault at coordinate E. The *pdcf* is

$$\begin{array}{ccc} I & H & E \\ \hline 1 & 1 & D' \end{array}$$

We must now intersect this *pdcf* with the *pdc*'s which may be used to form a connected chain between coordinates E, the fault, and A, the primary output. Table 7-10 shows the *pdc*'s in question. It remains an exercise for the reader to work out all *pdc*'s for this circuit.

Selecting D-cube 1 in Table 7-10 as the driving *pdcf*, we make a trial intersection with *pdc* 2. Since the result does not produce a connected chain—fails quality-1 test of the preceding section (D-intersection)—we

TABLE 7-9. Singular Cover for the Circuit of Fig. 7-7

Cube	U	T	S	R	Q	P	O	N	M	L	K	J	I	H	G	F	E	D	C	B	A
44	.	.	.	.	.	.	.	.	.	.	.	.	.	.	.	.	.	.	1	1	0
43	.	.	.	.	.	.	.	.	.	.	.	.	.	.	.	.	.	.	X	0	1
42	.	.	.	.	.	.	.	.	.	.	.	.	.	.	.	.	.	.	0	X	1
41	.	.	.	.	.	.	.	.	.	.	.	.	.	.	.	X	1	.	1	.	.
40	.	.	.	.	.	.	.	.	.	.	.	.	.	.	.	1	X	.	1	.	.
39	.	.	.	.	.	.	.	.	.	.	.	.	.	.	.	0	0	.	0	.	.
38	.	.	.	.	.	.	.	.	.	.	.	.	.	.	.	X	1	.	1	.	.
37	.	.	.	.	.	.	.	.	.	.	.	.	.	.	.	1	X	.	1	.	.
36	.	.	.	.	.	.	.	.	.	.	.	.	.	.	.	0	0	.	0	.	.
35	.	.	.	.	.	.	.	.	.	.	.	.	.	0	.	.	.	.	1	.	.
34	.	.	.	.	.	.	.	.	.	.	.	.	.	1	.	.	.	.	0	.	.
33	.	.	.	.	.	.	.	.	.	.	.	1	1	.	.	0	.	.	0	.	.
32	.	.	.	.	.	.	.	.	.	.	.	X	0	.	.	1	.	.	1	.	.
31	.	.	.	.	.	.	.	.	.	.	.	0	X	.	.	1	.	.	1	.	.
30	.	.	.	.	.	.	.	.	.	X	1	.	.	.	.	1	.	.	.	.	.
29	.	.	.	.	.	.	.	.	.	1	X	.	.	.	.	1	.	.	.	.	.
28	.	.	.	.	.	.	.	.	.	0	0	.	.	.	.	0	.	.	.	.	.
27	.	.	X	.	.	.	.	.	.	1	.	.	1	.	.	.	.	.	.	.	.
26	.	.	1	.	.	.	.	.	.	X	.	.	1	.	.	.	.	.	.	.	.
25	.	.	0	.	.	.	.	.	.	0	.	.	0	.	.	.	.	.	.	.	.
24	.	.	.	X	.	.	.	.	.	1	.	.	1	.	.	.	.	.	.	.	.
23	.	.	.	1	.	.	.	.	.	X	.	.	1	.	.	.	.	.	.	.	.
22	.	.	.	0	.	.	.	.	.	0	.	.	0	.	.	.	.	.	.	.	.
21	.	.	.	.	.	.	.	.	.	.	.	0	.	.	.	.	1	.	.	.	.
20	.	.	.	.	.	.	.	.	.	.	.	1	.	.	.	.	0	.	.	.	.
19	.	.	.	.	X	.	.	.	.	1	.	1	.	.	.	.	.	.	.	.	.
18	.	.	.	.	1	.	.	.	.	X	.	1	.	.	.	.	.	.	.	.	.
17	.	.	.	.	0	.	.	.	.	0	.	0	.	.	.	.	.	.	.	.	.
16	.	.	.	.	.	.	.	1	1	.	0	.	.	.	.	.	.	.	.	.	.
15	.	.	.	.	.	.	.	X	0	.	1	.	.	.	.	.	.	.	.	.	.
14	.	.	.	.	.	.	.	0	X	.	1	.	.	.	.	.	.	.	.	.	.
13	.	.	.	.	1	1	.	.	0	.	.	.	.	.	.	.	.	.	.	.	.
12	.	.	.	.	X	0	.	.	1	.	.	.	.	.	.	.	.	.	.	.	.
11	.	.	.	.	0	X	.	.	1	.	.	.	.	.	.	.	.	.	.	.	.
10	.	.	.	1	.	.	.	1	.	.	.	.	.	.	.	.	.	.	.	.	.
9	.	.	.	0	.	.	.	0	.	.	.	.	.	.	.	.	.	.	.	.	.
8	.	.	X	1	.	.	.	1	.	.	.	.	.	.	.	.	.	.	.	.	.
7	.	.	1	X	.	.	.	1	.	.	.	.	.	.	.	.	.	.	.	.	.
6	.	.	0	0	.	.	.	0	.	.	.	.	.	.	.	.	.	.	.	.	.
5	.	X	.	1	.	.	1	.	.	.	.	.	.	.	.	.	.	.	.	.	.
4	.	1	.	X	.	.	1	.	.	.	.	.	.	.	.	.	.	.	.	.	.
3	.	0	.	0	.	.	0	.	.	.	.	.	.	.	.	.	.	.	.	.	.
2	1	.	.	.	1	.	.	.	.	.	.	.	.	.	.	.	.	.	.	.	.
1	0	.	.	.	0	.	.	.	.	.	.	.	.	.	.	.	.	.	.	.	.

TABLE 7-10. D-Cube for D-Drive of $pdcf = \dfrac{IHE}{11D'}$

D-Cube	I	H	G	F	E	D	C	B	A
11	.	.	.	.	.	.	D'	1	D
10	.	.	.	.	.	.	1	D'	D
9	.	.	.	.	.	.	D'	D'	D
8	.	.	.	.	D	0	.	D	
7	.	.	.	.	0	D	.	D	
6	.	.	.	.	D	D	.	D	
5	.	.	.	D	0	.	D		
4	.	.	.	0	D	.	D		
3	.	.	.	D	D	.	D		
2	.	.	D'	.	.	D			
1	1	1	.	.	D'				

discard the result. Next we perform a trial intersection with pdc 3. After applying special-case-2 rule to pdc 3, the result produces a chained D-cube

$$\frac{I\ H\ G\ F\ E\ D\ C\ B\ A}{1\ 1\ X\ D'\ D'\ X\ D'\ X\ X} \tag{7-15a}$$

However, the quality-1 test states that, although this is an acceptable D-cube, a preferable D-cube might be found. We therefore try to intersect the $pdcf$ with pdc 4. Again, after applying special-case-2 rule to pdc 4, we obtain a chained D-cube

$$\frac{I\ H\ G\ F\ E\ D\ C\ B\ A}{1\ 1\ X\ 0\ D'\ X\ D'\ X\ X} \tag{7-15b}$$

This time we pass the quality-1 test and choose to continue our chain search using pdc (7-15b) as the driving D-cube. We intersect this with pdc's 5, 6, 7, and 8, each time discarding the result because we could not produce a connected chain. However, with pdc 9, we finally produced a chain-connected D-cube

$$\frac{I\ H\ G\ F\ E\ D\ C\ B\ A}{1\ 1\ X\ 0\ D'\ X\ D'\ D'\ D} \tag{7-16a}$$

The quality-1 test warns us to try again before choosing this D-cube. We find that intersecting a D-cube (7-15b) with pdc 10 is not defined. We try it with pdc 11 and obtain a solution that passes the quality-1 test

$$\frac{I\ H\ G\ F\ E\ D\ C\ B\ A}{1\ 1\ X\ 0\ D'\ X\ D'\ 1\ D} \tag{7-16b}$$

Sec. 7.4   Test Generation Models for Combinational Logic   373

Since the output coordinate is the primary output coordinate A, no further chain searching is required. We now delve into step 3 of the D-algorithm, that of the consistency operation, where we are trying to find an attainable nonconflicting value assignment of all primary input values to obtain the pdc (7-16b).

The first thing we must now do is determine the implied values of the other coordinates of the resultant D-cube if values are already assigned as dependencies for the test to succeed. If coordinate B is a 1, then intersection of the singular cover of the logic block whose output is coordinate B gives us this value. We therefore intersect the D-cube (7-16b) with singular-cube 41 of the singular cover of the whole circuit as given in Table 7-9 and get an answer

$$\frac{I \ H \ G \ F \ E \ D \ C \ B \ A}{1 \ 1 \ X \ 0 \ D' \ 1 \ D' \ 1 \ D} \qquad (7\text{-}17)$$

Since coordinate D is now assigned a value 1, we intersect the D-cube (7-17) with singular-cube 35 in that table, resulting in a new D-cube

$$\frac{I \ H \ G \ F \ E \ D \ C \ B \ A}{1 \ 1 \ 0 \ 0 \ D' \ 1 \ D' \ 1 \ D} \qquad (7\text{-}18)$$

We follow this procedure of intersection until we reach the primary inputs in a consistent manner. In case of a conflict, we back up to the previous singular-cube intersection with a D-cube, e.g., (7-18), and determine if an alternate singular cube is available, which may give us this consistency. If not, we try another, until no alternates are available, and we back up another step in the intersection process by finding a previous D-cube, e.g., (7-17), and finding another singular cube to come up with a consistent value assignment. Should we run through all members of each singular cover of each logic block, and should all of these efforts fail to produce a consistent value assignment, then no possible test for this failure exists. This is where the D-algorithm proves its worth over a heuristic approach.

The reader may calculate a resulting test by this procedure

$$\frac{U \ T \ S \ R \ Q \ P \ O \ N \ M \ L \ K \ J \ I \ H \ G \ F \ E \ D \ C \ B \ A}{1 \ X \ 1 \ 1 \ 1 \ 1 \ 1 \ 1 \ 1 \ 0 \ 0 \ 1 \ 1 \ 1 \ 0 \ 0 \ D' \ 1 \ D' \ 1 \ D} \qquad (7\text{-}19)$$

Since coordinate T remained X, its value does not influence the test, and it may be arbitrarily set to 0 or 1. Alternate tests may result because of the sequence in which the cubes in the singular cover were arranged to begin with, or because of the choice of alternatives used during the consistency operation.

Improvements in the D-algorithm [15] take advantage of all D's appearing in the resultant D-cube and therefore state that the same test finds a stuck-at-1 failure at coordinate C and a stuck-at-0 failure at output coordinate A. Furthermore, the consistency operation may be performed within the algorithm at a more strategic time. This new procedure prevents large numbers of inconsistent calculations from being made before the inconsistency is discovered. It was the secret that made the Maling–Evans model so very useful. The same kind of shortcuts are being applied in the D-algorithm. The realized improvement in speed and the guarantee of finding a test (if one exists) make the D-algorithm, and its subsequent improvements, a very strong candidate for becoming the workhorse of the diagnostic test generating community.

7.4.6. The Method of Boolean Differences

This section is meant as an introduction to another analytic approach to diagnostic test generation, namely, Boolean differences. Since this approach does have some following in the literature, it is included here. However this author does not believe that it offers any advantages to design automation over the D-algorithm. The reader will have to judge for himself.

Given a function $F(X)$, where X represents a number of Boolean input variables, $x_1, x_2, \ldots, x_n$, we note that, in order to observe a failure in that function, a valid term of that function with one of its variables (x_i) inverted must cause the complement of $F(X)$ to be produced. By defining the boundary at which these input variables exist, similar to the injection word of Roth's (Section 7.4.4), we should be in a position to find tests for that boundary without any trouble. The existence of a test for a fault affecting variable x_i is proven with the simple rule

$$F(x_1, x_2, \ldots, x_i, \ldots, x_n) \oplus F(x_1, x_2, \ldots, x'_i, \ldots, x_n) = 1$$

The left-hand side of the equation is called the Boolean difference [17]. For convenience of notation, the derivative operator

$$\frac{d}{dx_i}$$

is used as a representation of the Boolean difference operator. A test for a fault affecting the variable x_i exists if

$$\frac{d}{dx_i}(F(X)) = 1$$

and does not exist if

$$\frac{d}{dx_i}(F(X)) = 0$$

A proof of this statement is left to the reader. A shorter and more convenient notation for the Boolean difference of the function $F(X)$ is possible and is used hereafter, namely, $D_i F$.

Since a Boolean function can only be a 1 or a 0, a simplification can be made in the process of finding the Boolean difference, namely, substituting a 1 for the variable x_i on one side of the EXCLUSIVE-OR Operator and substituting a 0 for the same variable on the other side. In other words

$$D_i F = F(x_1, x_2, \ldots, 1, \ldots, x_n) \oplus F(x_1, x_2, \ldots, 0, \ldots, x_n)$$

A proof of this simplification is left to the reader.

Taking the example of the previous section (but using lowercase letters for the signal names in Fig. 7-7) in which we attempt to find a test for variable e, let us write the expressions in Boolean form

(1) $\quad a' = b' + c'$
(2) $\quad b' = d'e'$
(3) $\quad c' = e'f'$

Substituting (2) and (3) into (1) produces

(4) $\quad a = d'e' + e'f' = e'(d' + f')$
(5) $\quad D_e a = 1(d' + f') \oplus 0(d' + f') = d' + f'$

Now we have to find $D_e a$ in terms of the primary input variables $q, r, s, t,$ and u. The equation for each of the dependent variables follows:

(6) $\quad d' = g$ (11) $\quad l = o' + p'$
(7) $\quad f' = j$ (12) $\quad m' = q'$
(8) $\quad g = k + l$ (13) $\quad n' = r's'$
(9) $\quad j = l + q$ (14) $\quad o' = r't'$
(10) $\quad k = m' + n'$ (15) $\quad p' = u'$

The substitution (left as an exercise to the reader) results in:

(16) $\quad D_e a = q' + r's' + r't' + u' + q$

Equation (16) states that the five logical terms will provide an incorrect value at a if an error exists at e. This is a fantastically fast and simple answer. But when we realize that each circuit is represented by a Boolean equation, that the same substitution computation has to be repeated for each fault, consuming massive amounts of storage, and that some logic networks may have to test for a few thousand faults, then this method does not turn out to be so attractive. The reader is referred to Marinos [18] for obtaining complete minimal test input sequences.

7.5 TEST GENERATION MODELS FOR SEQUENTIAL LOGIC

Though we can assume that every primary input for combinational logic networks is accessible to the tester and, therefore, the logic may be set to any state in one step, we cannot assume this to be true for sequential logic networks. All feedback signals (which are part of the state of the machine under test but embedded within it) are not available to the tester. A failure, therefore, may never permit us a test pattern that we are interested in applying. Furthermore, a failure may cause us to think we are applying one test pattern, while we may actually be applying a completely different one. This problem has worried researchers for a long time, and some are still trying to discover an answer. But engineers have designed machines with a special reset condition that is supposed to provide a well-defined starting state—at least a state from which we may proceed to test the whole machine. If that starting state cannot be reached in a hardware model under test, then we must resort to drastic manual techniques for diagnosis.

We must also know how many "blind" states there are between the time we are sure of a good-machine state and the next time we get to look at it. For example, if we can observe the state of a three-bit counter only when it resets to zero, we have seven blind states between observable states. Thus, we have to assume that the counter is working properly during the blind period. In the test-generating models that follow, we will assume that no blind states exist and that every next-state is available to the tester for decision-making. (This assumption is necessary if we are to continue with a single-failure model. If we allow blind states, we may be using the same fault to produce several failure conditions and, thus, be subject to multiple failures.)

Having thus simplified the problem, we must now consider in detail how the tests are applied and how their results are gathered. If we use a delayed-analysis tester, i.e., one that supplies the input patterns in a fixed sequence without regard to the outcome of each test and which reads output patterns and stores them for later analysis, we have to generate one type of test. We can run through a whole sequence and determine at which point the output of the model under test differs from the good machine as it was simulated. We can repeat this for the next sequence, etc. Then, knowing what logic blocks have been tested up to the point of failure detection in each test sequence, we can calculate a suspect list from the remaining untested logic blocks. Of course, the various test sequences must be well chosen to obtain good fault-locating resolution. For each failure, some sequences must work without failure long enough to test a significant part of the machine. This method of test, which is in use at Bell Telephone Laboratories for its electronic switching system, will be further discussed in Section 7.6.3. However, the Bell engineers did not originally use good-machine simulation as stated here but, rather, a method of fault insertion in a physical model of the machine under test [8].

If we use an instant-analysis tester, i.e., one that makes decisions while reading the output patterns, we have to generate a different type of test. This latter method is extensively used for integrated-circuit testing and for computer-system testing. Here the test sequences may be modified, depending on the outcome of each test cycle. The generation of tests for this application, using heuristic models attributable to Seshu, will be taken up in the next two sections.

Although the analytic school always maintained that a sequential diagnostic test generator was only an extension of the combinational one, no reasonably fast algorithm has been reported in the literature. It is still too early to comment on Kriz [19] who has used the *D*-algorithm technique on special types of sequential circuits designed with clocked flip-flops.

7.5.1. Seshu–Freeman Model

S. Seshu and D. N. Freeman worked on a technique similar to the Forbes–Stieglitz–Muller (FSM) model, however this time aimed at sequential circuits [20]. They implemented a compiled simulator which included a time-delay capability, based on the Huffman model of the sequential circuit [21]. This simulator also had the ability to use the output pattern for the good machine at time t as the feedback part of the input pattern for the next simulation at time $t + 1$. The simulator does not, however, depend on synchronous clock cycles but may be used to generate any event-driven asynchronous circuit. Therefore, time $t + 1$ simulation may start only if a stable state has been reached between the output and the feedback signals in the time t calculation. (If no stable state develops, the test pattern is disqualified.) Another reason for this method of simulation is that even a synchronous design may behave asynchronously under failure conditions. On the other hand, a restriction in the definition of an asynchronous machine, which makes simulation simpler, is that a test sequence can differ only one bit per input pattern at a time.

The researchers also point out a fundamental difference between combinational and sequential procedures for test organization. These procedures are independent of the organization of the machine under test. Under the *combinational test procedure*, which is the method of FSM, a fixed sequence of tests is applied and the results observed. Based on the particular tests passed and failed, and knowing what each test pattern checks, it is possible to deduce the failing circuit. Under the *sequential test procedure*, the set of tests applied to the machine under test is not fixed. The result of each test is used as a basis for deciding on the next test to be performed. The Seshu–Freeman (SF) model adopts the sequential test procedure on the assumption that it is more efficient and, therefore, fewer tests need be executed for diagnosis.

After the circuit has been assigned levels and the inputs, outputs, and feedback lines are clearly identified, step 1 of the SF method is to place the circuit under test in its reset state. Unless this state can actually be set and verified in the real circuit, the simulated stable state and the real circuit do not have a common starting point and, therefore, are not representative of each other.

Step 2 is to change arbitrarily one of the input bits and to calculate the next stable state for the good machine, as well as for each failing machine.

In step 3, the output patterns are grouped into equivalence classes, i.e., all failing machines producing the same new stable state are gouped together.

Step 4 takes the stable state of one of the equivalence classes and arbitrarily changes one of the input signals for this state to calculate the next stable state for each of the failing machines in this class.

Step 5 continues step 4 for the next input signal, until all possible next-inputs, one bit at a time, have been simulated. We also eliminate all patterns causing unstable next-states.

In step 6, we repeat step 3 for each of the patterns and group the subset of failing machines into equivalence classes. We then choose the patterns which provide us with partitioned equivalence classes equal to the resolution we desire, or which partition these classes as evenly as possible.

In step 7, we repeat steps 4 through 6 until each subset of equivalence classes is reduced within the resolution parameter given, or until no further partitioning is possible.

In step 8, we reset the circuit and arbitrarily change the next input bit as in step 2. We then repeat all the calculations in steps 2 through 8 until every one of the bad machines has been placed into partitioned groups within the specified resolution parameter, or until no further partitioning is possible.

Let us apply this technique to an example in order to get a better feel for the steps involved and their consequences. For our example, let us use the general model of a 4-input, 3-output circuit, with two feedback lines, as shown in Fig. 7-8. We need one more piece of information, namely, a failure-free starting state. We can arbitrarily state that the good machine is reset in such a way that, whenever all four inputs are set at 0, the outputs are 101, and the feedback is set at 00. This is entered in line 1 of Table 7-11. The computer also knows from the details of the circuits in M_1 that there are M_2 through M_{10} bad machines associated with that circuit.

Sec. 7.5 Test Generation Models for Sequential Logic 379

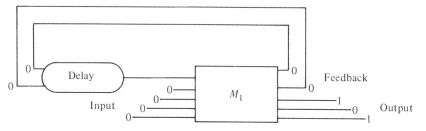

Fig. 7-8. Simulation model.

We now apply step 2 and calculate all stable states corresponding to the machines M_1 through M_{10}. This is shown in line 2 of Table 7-11, where we have arbitrarily changed the value of the first input bit from 0 to 1.

We now go to step 3 and group together M_1, M_4, M_5, M_9, and M_{10} into one class since the output patterns are identical. Similarly, M_2 and M_3 are grouped into a second class, and M_6, M_7, and M_8 are each in a class by itself. Let us further assume that the resolution parameter is 1, i.e., each bad machine must be distinguishable from any other. The result of this step is shown in Fig. 7-9.

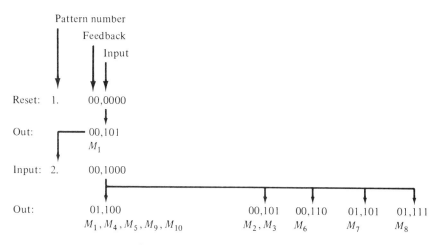

Fig. 7-9. Sequential test development.

The application of step 4 to our example is illustrated in line 3 of Table 7-11. We choose the feedback from the class containing the good machine, M_1, and vary another bit of the input pattern. This time, instead of calculating the stable states for all the machines, we calculate only the stable states for the machines in this class.

We go next to step 5. The results are illustrated in lines 3, 4, 5, and 6. We

TABLE 7-11. Step-by-Step Calculation

	Line	Fb, Input	Fb, Output M_1	M_2	M_3	M_4	M_5	M_6	M_7	M_8	M_9	M_{10}
	1	00,0000	00,101	—	—	—	—	—	—	—	—	—
	2	00,1000	01,100	00,101	00,101	01,100	01,100	00,110	01,101	01,111	01,100	01,100
X	3	01,1100	11,100	—	—	11,100	11,100	—	—	—	11,100	11,100
X	4	01,1010	11,100	—	—	U	11,100	—	—	—	01,100	10,101
	5	01,1001	11,101	—	—	11,101	01,100	—	—	—	01,100	01,100
X	6	01,0000	10,100	—	—	10,101	U	—	—	—	10,101	10,101
X	7	11,1101	U	—	—	U	—	—	—	—	—	—
X	8	11,1011	U	—	—	—	—	—	—	—	—	—
	9	11,0001	01,110	—	—	01,100	—	—	—	—	—	—
X	10	11,1000	11,110	—	—	11,110	—	—	—	—	—	—
	11	01,1101	—	—	—	—	11,101	—	—	—	11,100	11,000
	12	01,1011	—	—	—	—	00,101	—	—	—	00,101	10,100
X	13	01,1000	—	—	—	—	U	—	—	—	U	11,000
X	14	01,0001	—	—	—	—	00,101	—	—	—	U	U
X	15	00,1100	—	10,101	10,101	—	—	—	—	—	—	—
	16	00,1010	—	01,100	01,101	—	—	—	—	—	—	—
X	17	00,1001	—	01,100	U	—	—	—	—	—	—	—
X	18	00,0000	—	U	01,101	—	—	—	—	—	—	—

Note: Fb = feedback, U = unstable, and X = discard.

immediately discard the calculations for lines 4 and 6, since they contain unstable solutions. Applying step 6, we note that the calculation of line 3 did not partition the bad machines into further classes. Therefore, we discard it too, and are left with only the solution of line 5. Our present status is illustrated in Fig. 7-10.

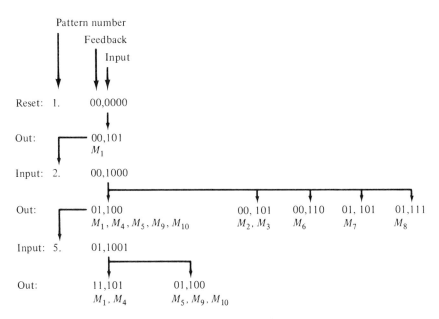

Fig. 7-10. Sequential test development.

Lines 9 and 10 in the table illustrate the results of trying step 7 on the class containing M_1. Figure 7-11 then illustrates the status at completion. We did not use the pattern from line 10 because it did not partition the machines further. Also, since we met our resolution requirements at this stage, we apply the technique to generating secondary tests for the class involving M_5, M_9, and M_{10}.

Lines 11 to 14 illustrate the calculations. Calculations in lines 13 and 14 are discarded because of the unstable states. A choice has to be made between the patterns of lines 11 and 12. The pattern in line 11 is chosen because the machines M_5, M_9, and M_{10} are resolved uniquely, i.e., within the resolution parameter as illustrated in Fig. 7-12.

Similarly, we try to resolve machines M_2 and M_3 by means of the secondary test calculations in lines 15–18. Again using the reasoning developed earlier, the pattern of line 16 is finally chosen. Figure 7-13 illustrates the final sequential testing tree. We see that we chose five test patterns plus the reset

382 *Fault Test Generation* *Chap. 7*

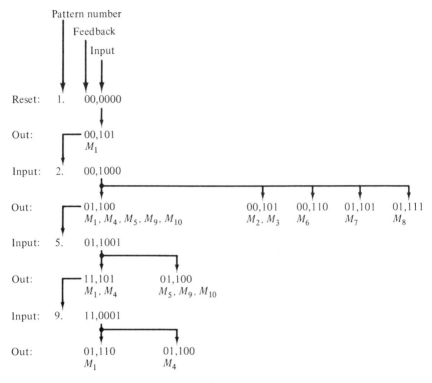

Fig. 7-11. Sequential test development.

pattern to isolate each of the bad machines. However, we had to calculate eighteen patterns in the process.

Measurements of a program testing out this model showed that it took about a minute of computer running time to generate tests for every ten faults in designs with several hundred possible faults. A number of unresolved problems to consider before this model may be generally applied are pointed out by the researchers:

1. In general, the computer cannot determine the reset conditions of a sequential circuit. (This information is usually known by the designer and should not be hard to obtain as part of the input conditions.)

2. The states that the computer calculates as desirable next-states with very high resolutions may not be realizable in a large circuit unless special design considerations are given to all primary inputs.

3. The program requires more information to choose the best patterns when large combinational circuits are interspersed with the sequential. (Presumably this is the same kind of problem that Maling and Evans noted.)

Sec. 7.5   Test Generation Models for Sequential Logic   383

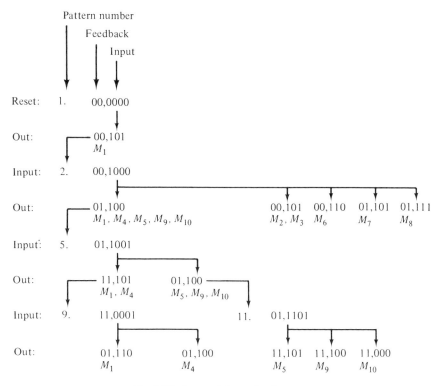

Fig. 7-12. Sequential test development.

The reader is referred to reference [22] for an excellent treatment of the sequential analyzer (fault simulator), written for 96 primary inputs and 48 feedback loops, which represents the heart of this test-generation technique. It also discusses some innovations that permit manual insertion of test patterns. The effective use of manual intervention will be described in the next section.

7.5.2. Auch–Cheng Model

A. G. Auch and D. D. Cheng at IBM were given the Seshu–Freeman (SF) model and some program design objectives:

1. Handle a sequential circuit of up to 3800 logic blocks in a 32K-word IBM 7094 (later changed to 64K words).
2. Handle 200 primary input bits and 260 primary output bits (as opposed to 96 in and 96 out [22]).

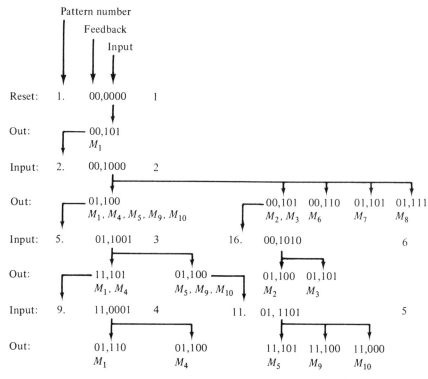

Note — Test numbers are circled.

Fig. 7-13. Sequential test development.

3. Provide for up to 760 feedback loops (as opposed to 48 [22]).
4. Discard input patterns that produce more than 512 output patterns. (This restriction turned out not to be important; they managed to handle 760 output patterns.)
5. Test at observable scoping points only.
6. Use only legitimate test sequences and provide manual intervention capability at each step of the simulation.
7. Reduce computing time to a reasonable amount and maintain a linear relationship between computer time and size of logic being tested, as opposed to the greater than quadratic relationship then observed.

What were the time-consuming calculations in the SF model? For one, the number of simulated failures. This number was reduced, certainly, when the objectives called for detecting failures at observable scoping points only. This change of the specifications is reasonable since a failure within a package (but not at an observable scoping point) sooner or later must affect an observable scoping point. For example, a machine with 3000 logic blocks can exhibit perhaps 10,000 failures, but only 2000 might be observed at scoping points. Even this 5-to-1 reduction in observable failures still leaves 2000 failures to be simulated.

Another major time-consuming calculation was the exact-simulation technique required for asynchronous logic. If some simplification were possible here, such as an introduction of a pseudosynchronous simulation, much computer time might be saved. The requirement calls for the simulated logic to be logically equivalent at observation (i.e., log-out) time, even though at other, in-between, times it is not necessary to be that precise.

A third major time-consuming calculation was the determination of the sequence of patterns to be generated. Under the SF model, individual input bits would be varied and used to determine the next pattern. Since one of the design criteria was that only legitimate states could be used for testing, it seemed that, with so many primary input bits available, the test analyzer (pattern generator) might come up with myriads of test patterns, few of which would be legitimate. Therefore, a new approach had to be taken. The diagnostic engineers would have to supply the legitimate sequences and the test analyzer would be used to evaluate these pattern sequences for their fault-detection capability.

It might be argued that, if the first two computer-reduction techniques have not taken this problem out of the design automation realm and placed it squarely into the computer-aided design realm, the third one certainly did. We must realize, however, that we are still using the same logic design data base that is used for cabling and wiring; only this time we are calculating tests. Until a fully automatic technique can be found, a partial technique must suffice. Heuristic design automation does not have to wait for the ultimate algorithm. It can attack a problem from many sides and evolve a solution which at first may not have been obvious. The advances discussed in this section are based on unpublished reports by the researchers and this author [23], [24], [25].

The researchers took this problem to task and came up with the following additional program requirements:

(a) Provide a language the diagnostic engineer can use to express the input patterns, their progression from state to state, and the observation times.

(b) Provide a trace of all feedback paths for possible manual analysis of feedback race conditions and for assistance in editing decisions.

(c) Provide a manual input path into the simulated logic to permit editing, i.e., pseudodelay block and pseudoclocking signal insertion.

(d) Provide the simulator with the capability of being interrupted at any stable-state time, of saving all feedback conditions and failure-list tables in computer- and man-readable form, and of being restarted from that interrupted condition.

(e) Provide for editing of the interrupted saved data.

(f) Provide a program for comparing the test-analyzer-produced log-out data of the good machine with the data produced by the physical hardware for purposes of checking the interactive simulator model.

These requirements were implemented and successfully used on two large-scale sequential circuits—one involving 1800 logic blocks and 300 feedback loops, and another involving 2800 logic blocks and 600 feedback loops. Further enhancements to simplify diagnostic engineering use, along with reprogramming for an IBM System/360, could make it applicable to other, even larger-scale designs.

Now let us see how the foregoing series of programs may be used to produce fault-locating tests for a sequential logic design. The diagnostic engineers supply the reset signal pattern for the logic design, followed by various sequences of signal changes. Since engineers are familiar with timing diagrams, this method of input is most desirable [item (a) above]. The timing diagrams can then be transcribed for use by the test-analyzer simulator (TAS). The timing diagrams represent the states of the primary input lines. Unlike the method used in the SF model, here multiple input signals could be changed at one time. In addition, whenever a prespecified pseudoline is set to 1, the TAS interprets it to mean that observation time is here (log-out commences) and that TAS should go through its paces of saving all conditions, etc., for further use and analysis [item (d) above].

When the test patterns are applied to the real hardware, the sequence should be as follows:

1. Reset
2. Variation 1
3. Settle
4. Log-out
5. Reset
6. Variation 1
7. Variation 2
8. Settle
9. Log-out
10. Reset
11. Variation 1
12. Variation 2
13. Variation 3
14. Settle
15. Log-out
16. Etc.

TAS, on the other hand, only has to simulate:

1. Reset
2. Variation 1

Sec. 7.5   Test Generation Models for Sequential Logic   387

    3. Settle and log-out
    4. Variation 2
    5. Settle and log-out
    6. Variation 3
    7. Settle and log-out
    8. Etc.

It is then up to some other procedure to make up the test-pattern application sequences and merge these with the TAS bad machine patterns in the production of the final test library.

As was hinted earlier, an attempt had to be made to use a zero-time-delay logic simulator to simulate the stable states of asynchronous logic at discrete time intervals. An order of magnitude of computer time saving was foreseen if it were successful. This meant that the asynchronous logic taken from the design automation logic master tape had to be "prepared" through editing in various ways so that at log-out times, at least, the "good" hardware machine should be equivalent to the simulated "good" design [item (c) above]. The researchers worked out a technique utilizing a compiled simulator as the base, whereby each machine state was divided into distinct simulation intervals. These intervals and the logic simulated in each were so chosen that a zero-time delay adequately represented the real hardware response. By manually inserting some pseudotime delays in critical feedback paths [discovered by the program in item (b) above], it was possible to move some of the logic to be simulated into the next simulation interval. By this means, the researchers could eliminate feedback race conditions that exist in the simulator but not in the hardware. The first clue to the feasibility of this approach is to apply the reset condition to the real hardware and the TAS logic, and to compare the two conditions. (If an asynchronous simulator for that large a circuit existed, it could have been used to provide this verification.) Once the reset conditions matched, some test sequences could be tried and further adjustments made to the "prepared" logic. The program that compares TAS-produced results with hardware-produced log-outs [item (f) above] is invaluable in speeding up the process of making the two logically equivalent for all contemplated test series. Then some faults could be simulated utilizing program item (e) above and compared against the same faults introduced into the actual hardware. When this checked out, the simulation model was ready for production use.

One more computer time-reducing by-product of the program in items (d) and (e) above is that it was possible to edit out faults which cannot be located automatically, such as hard faults in the reset condition, and those controlled through manual switches only. A further option, called *bug elimination* [23], could be introduced into the TAS once the exact sequence of test application had been fixed. If the maintenance procedure is to use the first failure indication for diagnosis [24], then those faults located in an earlier step in the simulation need not be further simulated. This could be accomplished by the removal of these bugs from the list of faults to be simulated.

Since a logic design is subject to many minor changes before it finally reaches the customer's location (and even then, some changes may still be made), an attempt to find a way to diagnose as much of the logic with as little expense in computer time does pay off. This feat can be performed with a *test series appreciation program* [25], which can provide:

1. the number of faults (bugs) detected vs. the number of bugs simulated in a sequence (or series);
2. sorted lists of detected and undetected bugs;
3. sorted lists of unique faults detected in each sequence (i.e., not detected in any other sequence).

With this information, it is possible to eliminate whole sequences of presumably redundant patterns that do not detect unique faults. It is also possible to bring to the attention of the diagnostic engineers some long test sequences, which detect only a few faults, with the idea of shortening these sequences and still retaining the fault detecting coverage; or to approach the testing for these faults in a different manner, so as to cover these faults and others not yet detected.

The heuristic tools supplied by a design automation organization (working in conjunction with an engineering design group) and a diagnostic engineering group permitted the reduction of computer time by almost two orders of magnitude from that estimated for the "fully automatic" approach. Yet the approach is general enough to be used on a number of different designs, provided that the testing method stays the same. The close cooperation between the three groups permitted the reduction to practice of a very powerful diagnostic tool, many years before even a limited sequential test algorithm had been developed.

7.6 AUTOMATED TEST APPLICATION

In this section, we shall review some dependencies that have to be met before a useful set of diagnostic test patterns can be automatically generated for a digital system and used by the field and manufacturing engineers. It is necessary, for instance, to know what the testing plan for the system hardware will be:

1. On what medium are the test data to be stored?
2. What is hard-core logic (i.e., must be working before the automated tests may be applied)?
3. What troubleshooting techniques are to be used, and what documents are required for them to be effective?

It is also necessary to know the characteristics of the logic to be tested:

1. Is the logic sequential or combinational?

2. For input states, are we limited to legal states only, or can illegal states also be used?
3. Do the test patterns contain control information for the tester, or is control information not embedded?

And it is necessary to know what the tests are expected to accomplish:
1. Is fault location or fault detection desired?
2. Is diagnosis based on the order of test application, or is diagnosis based on a post-test analysis of results?

Of course, it is assumed here that the data which describe the design, together with the parameters needed for accomplishing the test objectives, are available to the test-generating programs.

7.6.1. System Test Plan

An example of a system test plan involving IBM System/360 automated tests is described by Preiss [26]. Here the core storage is the starting point for testing the machine. This storage can be tested from a built-in test-pattern generator that permits all 1's and all 0's to be stored in various combinations of addresses throughout the core storage and then read out again and compared. The circuitry involved is quite simple and effective. Thus, some test patterns for this part of the machine are always readily available, being built-in. But sooner or later, the need for an external source of test patterns arises. In the example described, a tape provides the rest of the patterns. This tape fills up a part of core with test patterns, utilizing the same addressing and data registers that were used in the built-in storage tests. Then, with some additional "tester" circuits, the new data are applied to other parts of the machine, being directed or programmed by the contents of the data.

As part of the test plan, the first set of data that comes off the tape functionally checks the tester circuits, and then checks every bit of the read-only control store. Once these are checked out, the next series of tests functionally check out an additional part of the tester circuitry, this time controlled by the read-only storage (ROS). Next, the tests are applied to the combinational logic under ROS control to check out the arithmetic and logic unit and all the registers and control triggers associated with that unit. This is followed by the application of another set of storage patterns (of the type initially applied to the core storage) to the local storage, a high speed core used as intermediate-results storage registers of the machine.

Successfully passing this battery of tests, the machine is in a position to execute its first full instruction. It does this by reading in a control program which makes extensive use of a special DIAGNOSE instruction, which has the capability of addressing one or a series of ROS control words. It is with this instruction that the sequential tests for the I/O channels are applied.

Since the I/O channels are the last to be tested, they cannot, and should not, be relied upon to print out any diagnostic messages. Part of the test plan, therefore, requires that all tests run prior to the completion of the channel tests must communicate to the outside world via lights on the display console. Since four bytes (eight hexadecimal digits) are displayed on the IBM System/360 console, all failures could be communicated through a numbering convention. Some numbers are direct-indicating, stating, for example, that bit 31 of ROS word 204 is stuck-at-0; others refer to tables or preprinted directories produced by the design automation programs. These references list either points to be probed by oscilloscope or the most likely suspect circuits to be replaced.

The Achilles' heel of the entire test plan is getting the tests into storage from an external medium such as tape. This must be accomplished through an I/O channel of some type. In the machine described, more than one independent I/O channel was always available to read in the data. If that had not been the case, a redundant or hardened single channel, or a built-in procedure for testing the channel, independent of external data, would have been required [27].

The test plan just described indicates that a design (keeping maintenance in mind) can begin with a very small (2–5%) hard core of hardware and test equipment which, once functioning, can bootstrap additional, more complex, tests into the system up to the point where the full power of the computer may be used to test itself. The test plan does not necessarily depend on the ability to generate tests automatically, although that certainly helps and guarantees as complete a test coverage as possible. Bock and Toth [28] describe a test plan that was developed after most of the hardware was designed. They estimate that the essential (i.e., hard-core) logic required before the testing scheme could be implemented involved only 15–20% of the processor logic. This should be compared with the 65–75% of the logic that needed to be operational in first generation computers to enter and execute the first conditional-branch instruction. The tests used by Bock and Toth were not automatically generated; but the design automation system was used extensively to aid in the overall design of the tests, their verification, and certainly their updating after engineering changes.

We should note that it does not follow that, unless the machine is designed with bootstrapping and self-testing in mind, it will be difficult to test or diagnose. Some systems or partial systems are tested with external testers (possibly other computers) to which they have access [29]. This is a common technique employed in manufacturing plants, where one computer may check out one or more others [30], or in field-installed duplex systems, where one processor checks out the other [31]. Special designs may also be employed to ensure that a computer is capable of self-test, even if a portion of it is incapacitated [32]. More and more testing is also being done on subsystem components. With the advent of integrated circuits, more and more logic can be placed into one pluggable subassembly. These subassemblies must be tested before they are assembled into the next-level package. It is therefore not uncommon to see computers test these subassemblies [33]. Computers provide the flexibility through program changes to apply the controls to electrical robots that are built to hold the subassemblies and provide the necessary signals (sequence, duration, voltage levels) which permit the tests to be performed. This aspect of testing, of necessity, is beyond the scope of this chapter. However, the tests used may be generated by the schemes described earlier. Also note that, through robot action, it may be possible to apply both valid and invalid test patterns to a subassembly that could never be applied once the subassembly is mounted in the system. For example, a subassembly may be tested as combinational logic in a robot and as sequential logic in the system.

7.6.2. Designing Logic for Simplified Testing

Any digital design consists of storage elements (flip-flops or triggers that hold the state of the machine for a period of time), combinational logic elements, clocking gates (if synchronous design), manually activated switch inputs, and observable switch or light outputs. The earliest method of test was to enter data into the designed hardware by means of manual switch settings and to observe the light outputs to determine whether the data have been entered properly. Then, by means of a push button that initiates a clock pulse (or other propagating gate signal), the data are propagated through the design (or to the next observable machine state). The person performing the tests knows what the next state should be, and he checks the lights to see whether that state has been reached before he presses the push button again for the following state, etc. This method proved very effective for fault-locating since very little added logic was used in every subsequent clock cycle. But since it was manual in nature, it was time-consuming and expensive.

By redesigning the data entry and exits from the storage elements so that they interfaced directly with the storage (memory) bus, it was possible to pass data into and out of the design at computer speeds [6]. This represents a start towards automated testing. The degree to which automation of test-pattern generation is possible depends on the uniformity of the design used in the entry and exit operation. If, for example, every storage element of the design has a test-pattern entry and exit point (called scan-in and scan-out or log-out path), then combinational tests can be automatically generated and applied without trouble. A design of this kind is illustrated in Fig. 7-14. If, for example, 50% of the storage elements have both paths and 45% of the remaining logic elements have log-out paths only, then test generation becomes more complex. Some logic may require combinational tests, other logic may

392   Fault Test Generation   Chap. 7

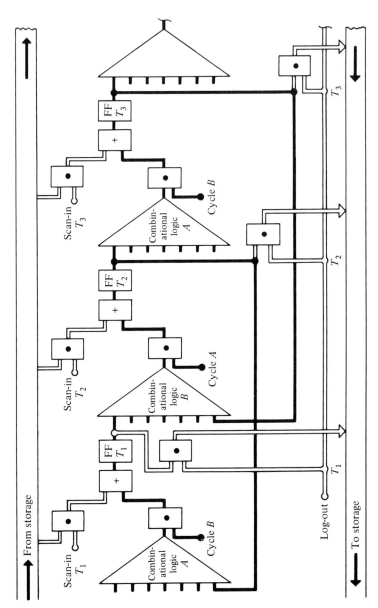

Fig. 7-14. Scan-in and log-out gating design.

require sequential tests, and the rest of the logic may require hand-produced functional tests.

If we design every storage element with a scan-in and log-out path, the application of the tests can be done in three stages. The first stage is to test every scan-in/log-out path, together with its storage element (flip-flop). This is called a 0-cycle test, because no clock cycle is provided to permit the scanned-in data to propagate. Figure 7-15 shows the testing path of flip-flop T_1 from scan-in at the top left to log-out at the bottom left of the figure. Note that the data do not enter T_2 because the cycle-A signal is not provided.

The second application stage is the DC, or static, test stage with a 1-cycle advance. Here the data are permitted to propagate to the immediately next state (when an A cycle is given; see Fig. 7-16) and are then logged out. The test is a DC test because the time permitted to propagate to the next state while waiting for all storage elements to be set from external storage is long compared to a machine cycle (time between A and B cycles). Realize that, as soon as data are scanned into the storage elements, the signals propagate through the combinational logic to the point at which the next clock gate stops them. In the 1-cycle tests, the clock gate is opened just long enough to permit the signal to set the immediately adjancet flip-flop or storage element. It is this immediately adjacent storage element that is sampled during log-out to determine its state. Since it was already tested by the 0-cycle tests, the 1-cycle tests provide a DC test of the combinational logic that precedes the sampled storage element.

The third stage of testing is the 2-cycle or dynamic test. Here two clock pulses are gated into the logic under test: pulse A and pulse B in Fig. 7-17. This test is dynamic because the data enter the logic under test at A-cycle time and must propagate through the already statically tested logic by B-cycle time so that they can set the B-cycle storage elements whose status will be logged out. Unless this action is completed between the two cycles, a dynamic failure is detected. By these three stages of testing, the circuits of a whole system may be fully checked out.

However, in most designs, some storage elements cannot be directly attached to the memory bus. In some designs it becomes difficult to provide the necessary gating to set a storage element to any arbitrary state. In that case, the designers can be coaxed into providing a reset state, at least, and thus permit subsequent states to be logged out on a cycle-by-cycle basis. The tests required for this design are the sequential kind. If some storage elements cannot be logged out, some blind states appear in the design and fault-locating becomes quite difficult; but this does not usually interfere with fault detection. A fault is detected as soon as the precalculated expected response does not agree with the actual results provided by the hardware under test. This sometimes is enough information with which to make an accept/reject decision for a single part. However, if one of many parts is

394   *Fault Test Generation*   *Chap. 7*

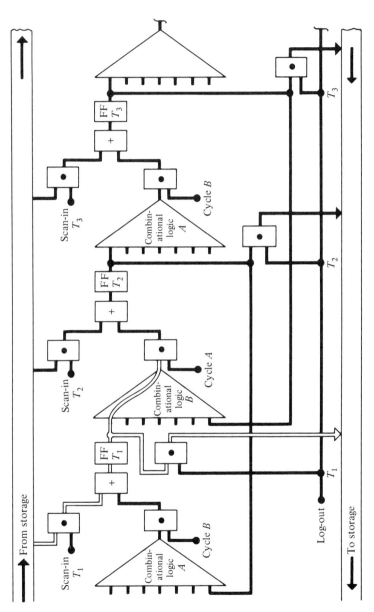

Fig. 7-15. O-cycle path.

Sec. 7.6 Automated Test Application 395

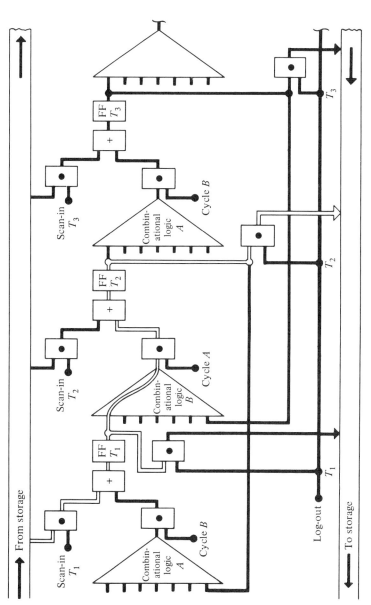

Fig. 7-16. 1-cycle path.

396 *Fault Test Genertaion* Chap. 7

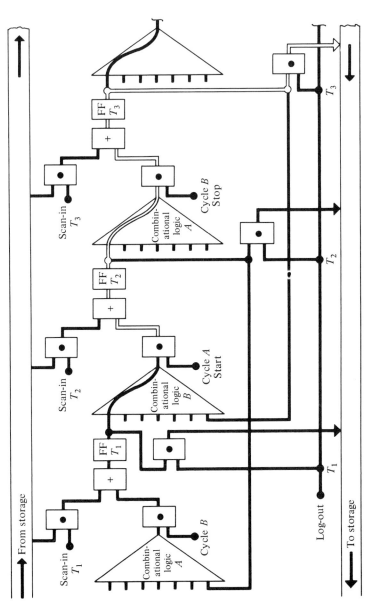

Fig. 7-17. 2-cycle path.

involved, it may be more desirable to identify the failing part than to accept/reject the whole system.

Another area where the designer needs some convincing follows: A good way to verify the simulator and the logic data it works on is to design the tester in such a way that the hardware log-out patterns may be accumulated and compared against simulation-produced good-machine patterns. This should be accomplished before the faulty machine patterns are produced for production usage. Much computer time and untold hours of engineering investigation (and of course incorrect error-pattern usage) may be saved if this precaution is taken. It takes just a little more design ingenuity, and that pays off in the long run.

7.6.3. Test-Pattern Arrangement for Detection or Location

To be effective, the test patterns produced by a test generator-analyzer must be placed in some order for use, and documentation must be produced. For combinational circuits, input patterns may be used in any order in a test sequence. For sequential circuits, the order is fixed for any one series (from a reset value to an end-of-sequence value). This is illustrated in Fig. 7-13. Patterns 1, 2, 5, and 9 are used as test numbers 1, 2, 3, and 4. Pattern 11 becomes test numbers 5 and is used if test number 3 fails. Also, pattern 16 becomes test number 6 and is used if test number 2 fails. However, the order of the series may be arranged in various ways, depending on the objectives to be accomplished. Since tests are redundant, in general, their optional selection and their sequencing can be considered an art or can be reduced to a covering problem with additional constraints. For fault detection only, it is desirable to pick an irredundant set of tests which covers all possible failures, (e.g., Fig. 7-13, patterns 1, 2, 5, and 9). For fault location, the ideal cover consists of a sequence of redundant tests, each of which tests one more condition than the previous one. Thus, fault location is guaranteed if the previous test passed and the present test failed.

Test-pattern arrangement also depends on the test application method. An external tester may subject the test logic to a battery of tests, and the logic consisting of those elements tested by the failing tests is saved for later intersection with those saved from other failing tests [6]. The components suspected of failing are produced from this intersection or from an analysis of how often the same component is suspect when a test fails. A second application method is that of changing the sequence of test execution based on the passing or failing of any executed test [1]. For example, one might first execute irredundant fault-detecting tests, followed by redundant fault-locating tests as soon as a failure is detected. A third method of test application is to test in sequence an ascending number of failures, using the assumption that any previously tested logic cannot fail when tested again. Thus, the incremental failures tested for are suspect when a test fails, and diagnosis is accomplished with the first test failure [24].

Generated test patterns may also be used more than once for failure location. If the tester can observe only one output bit at a time, the same test may be used; but different outputs must be observed to subdivide the logic into small tree segments as illustrated symbolically in Fig. 7-18. If the tester can observe all bits simulated simultaneously, the tests may still be arranged incrementally as discussed earlier, with previously tested portions being removed from the suspect list after redundant tests have been reapplied. Since the foregoing test arrangement methods lead to so many practical solutions, they will not be taken up here in detail. Let an example suffice.

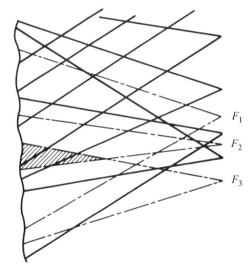

Fig. 7-18. Grouping trees—failure resolution.

Suppose the machine used in Figure 7-8 can be tested by cutting all feedback paths (thus reducing it to a combinational circuit) and by applying the desired feedback and input signals in any sequence (as opposed to the one sequence illustrated in Figure 7-13). Suppose, further, that the tester used is capable of testing the status of only one of the three output bits at a time (hereafter referred to as bits A, B, and C, respectively). Furthermore, suppose the component packaging is such that faults M_2, M_3, and M_4 reside in Replaceable Unit 1 (RU1), faults M_5, M_6, and M_7 in RU2, and faults M_8, M_9, and M_{10} in RU3. The problem is to produce and arrange a set of tests that will isolate a fault to one RU.

Since we can test only one output bit at a time, we transform Table 7-11 into Table 7-12 by choosing only those output bits whose values differ from the good machine and yield stable output patterns. Pattern 2, for example,

TABLE 7-12. SINGLE-OUTPUT TEST PATTERNS FROM TABLE 7-11

Line	Table 7–11 Pattern #	Input	Output A B C	Simulated Fault
1	2	00, 1000	0 · ·	An impossible condition to halt simulation.
2	2	00, 1000	· 1 ·	M_6, M_8
3	2	00, 1000	· · 1	M_2, M_3, M_7, M_8
4	5	01, 1001	· · 1	M_5, M_9, M_{10}
5	9	11, 0001	· 0 ·	M_4
6	11	01, 1101	1 · ·	M_5, M_9
7	11	01, 1101	· · 0	M_9, M_{10}
8	12	01, 1011	· · 1	M_5, M_9
9	12	01, 1011	· · 0	M_{10}
10	16	00, 1010	· 0 ·	M_2, M_3

differs in both bits B and C from the good machine. Bit A, on the other hand does not change values. This condition may be used as a means of verifying the test hardware, and halting the test application, by requiring a 0 value result, an impossible condition. Having generated Table 7-12, we may now arrange the patterns for the one-RU resolution.

In section 7.5.2, we discussed the advantages of testing only for the remaining bugs in a bug list. We can make use of this idea in the fault-locating and test sequencing phase. Assuming that only one fault occurs at one time, then if more than one test pattern tests for a particular fault, either all of them fail if the fault is present, or none of them fail if it is absent. Since each test pattern presumably tests for a different combination of faults, the theory states that only those faults which had not been tested for earlier can now fail. We may, therefore, incrementally remove bugs from the list and obtain excellent fault resolution with a very simple sequencing routine.

The first step is to merge the geographic data (RU) with the simulated fault data for each candidate test pattern.

The next step is to order the candidate test patterns by the number of logical faults and RU's identified. The result of these two steps is shown in Table 7-13.

The third step is to take each fault in turn from the simulated fault list associated with each test pattern, and determine if that fault still appears on the bug list. If it does, we remove it from the bug list on a trial basis. If it does not appear, it must have been removed earlier and we repeat this process with the next fault associated with the test pattern. In the process of removing the fault from the bug list, we flag the fault tested and count the number of RU's removed. Upon exhausting the simulated fault list, we check whether the RU count is non-zero and smaller or equal to the

TABLE 7-13. TEST PATTERNS SORTED AND MERGED WITH GEOGRAPHICAL DATA

Line	Pattern #	Output	Simulated Fault	Replaceable Unit
1	9	$B = 0$	M_4	RU1
2	12	$C = 0$	M_{10}	RU3
3	16	$B = 0$	M_2, M_3	RU1
4	11	$C = 0$	M_9, M_{10}	RU3
5	2	$B = 1$	M_6, M_8	RU2, RU3
6	11	$A = 1$	M_5, M_9	RU2, RU3
7	12	$C = 1$	M_5, M_9	RU2, RU3
8	5	$C = 1$	M_5, M_9, M_{10}	RU2, RU3
9	2	$C = 1$	M_2, M_3, M_7, M_8	RU1, RU2, RU3

permitted resolution. If that condition exists, the pattern becomes the next test in sequence and the faults are permanently removed from the bug list. If not, this pattern remains on the candidate test pattern list, the faults are not removed from the bug list, and step 3 is repeated on the next pattern.

This process is continued until all patterns are exhausted, until the bug list is exhausted, or until no change takes place in a complete pass through the test pattern list.

Let us apply this step to Table 7-13. Lines 1-4 immediately become tests 1-4 as shown in Table 7-14, because each pattern meets the one-RU criterion.

TABLE 7-14. FINAL TEST SEQUENCE FOR MACHINE OF FIG. 7-8

Test	Pattern #	Output	Simulated Fault	Replaceable Unit
1	9	$B = 0$	*M_4	*RU1
2	12	$C = 0$	*M_{10}	*RU3
3	16	$B = 0$	*M_2, *M_3	*RU1
4	11	$C = 0$	*M_9, *M_{10}	*$RU3$
5	11	$A = 1$	*M_5, M_9	*RU2, RU3
6	2	$B = 1$	(Dummy Test—See Text)	
7	9	$C = 1$	$M_2, M_3, M_7,$ *M_8	RU1, RU2, *RU3
8	2	$B = 1$	*M_6, M_8	*RU2, RU3
9	9	$C = 1$	$M_2, M_3,$ *M_7, M_8	RU1, *RU2, RU3
10	2	$A = 0$	(Final Stop)	

Note: *used above if primary suspect.

Line 5, however, does not meet the criterion, and is left behind. Line 6 removes bug M_5, and since M_9 had been removed earlier, the one-RU criterion is met and test 5 is produced. Line 7 is discarded also since the new RU count remains zero. The same fate befalls line 8. Line 9 does not pass the one RU criterion and remains. Having reached the bottom of the

Sec. 7.6   Automated Test Application   401

pattern list, we repeat the third step with the remaining test patterns beginning again on the top of the list. This time only lines 5 and 9 remain on the pattern list. No change in the list is produced in the second pass. We, therefore, proceed to step four.

In this step, we determine if the RU criterion can be met by using two or more test patterns (called secondary tests) to break the tested logic into smaller parts, assuming that only a single fault can cause a test to fail. By using the listed faults tested on a trial basis as the full bug list, we can apply step 3 and some pass/fail reasoning.

Applying step four to the example, we use line 5 (the first line remaining on the pattern list) and force M_6 and M_8 to substitute for the remaining bug list. Taking line 9 next, (the next line on the pattern list) we note that M_2, M_3, and M_7 do not appear in the forced bug list, and, therefore, M_8 is the only bug removable, and our one-RU criterion is met. Line 5 becomes test 6, and line 9 becomes test 7. If both tests 6 (a dummy test) and 7 fail, the suspect is M_8. We now remove M_8 from the full bug list and return to step 3. We immediately find that only M_6 remains untested with line 5, which meets its own one-RU criterion and thus becomes test 8. Line 9, in turn has only M_7 untested and becomes test 9. All faults tested for the first time are flagged to be easily identifiable in the fault-location dictionary.

For any test arrangement method, then a failure dictionary or a series of failure suspect lists may be produced. These lists require massive computations to associate logic data with physical data and to edit the results into a man-readable form or machine-readable form so that, when a diagnostic conclusion has been reached, a printout of the physical circuit suspect list may be consulted. While the test plan of Section 7.6.1 was designed for an exact-match log-out, the documentation provided enough cross-referencing information so that any failing test, when used with a manually controlled tester and an oscilloscope, could find the failing component, even though it may not have been included explicitly in the suspect list. For each combinational test, a dictionary lookup number (e.g., 03-2CD) displayed by the tester yields a printout of the probing points along the sensitive path, or D-chain (Fig. 7-19). Both the physical (geographic) location in the hardware and the logic location in the design blueprint, together with the error-free scoping value to be observed while the test is being executed, are used in practice. Some of the suspects are further identified as being tested in this test for the first time and are therefore prime suspects (marked G/F in Fig. 7-19). For sequential tests, the suspect list for each error log-out is printed, together with the assumed stuck-at-x value. Should the value observed on an oscilloscope change from the stuck-at-x value at any time during the execution of a repetition of the test, the suspected circuit is immediately removed from suspicion. The machine for which this method has been applied never had more than 760 error log-out patterns for each test.

```
        V       PIN            NET#    REF  FED BY

       ***     03             2CD     ***

        1   01A-E1M6B10    PA011BG4   AA   AB
        0   01A-E1M6D05    PA011BD4   AB   AC
        1   01A-E1M6D07    PA011BC4   AC   AD,AE
        0   01A-E5G2B07    SA011BA4   AD   AF,AG
        1   01A-E1F2D03    PB031AJ4   AE
  G/F   0   01A-E5G1D13    RA721BB4   AF
  G/F   1   01A-E3K2B10    SA011EJ4   AG   AH,AJ
  G/F   0   01A-E5G1B09    RA721AA4   AH
  G/F   1   01A-E3K2D05    PA101BA4   AJ
```

Fig. 7-19. Failure dictionary printout example.
Note: G/F = tested for the first time
V = logical value
Pin = geographical location of test point
Net = logical reference for systems page, etc.
Ref = reference line for "Fed By" connection

Another approach to test documentation (the cell dictionary) has been tried by Bell System engineers when a much higher number of log-out possibilities exists [34]. Here, a diagnostic conclusion is reached by accumulating the test numbers that failed in a fixed battery of tests and then deriving a number that may be used as a failure dictionary entry number. Since this number could consist of some 5000 bits, and was quite sparsely occupied, mapping of the binary number consisting of these 5000 bits into a number consisting of only 12 decimal digits was produced. As field usage of this scheme increased, it was found that many numbers were encountered that had no dictionary entries. Thus, a method of interpolation was required, and the cell dictionary was the result. The requirement for an exact match for test failure sequences was eliminated and a near match, with the Hamming distance being used as a measure of nearness, was introduced. In subsequent field usage, if an exact match was not found in the dictionary, a near match with a lesser failure resolution could be found.

While this approach is a way around the exact-match problem, it is this author's contention that design automation should not be made to cover up hardware design deficiencies—that the design engineer can design for a high degree of fault resolution. It should be part of his design objectives.

7.6.4. DA System for Test Generation

In this section we review the requirements of an automated test-generation system. To make visualization simpler, we shall make extensive use of Fig. 7-20. Just as with the placement, routing, and wiring problem, the physical information of the logic is required. But unlike these, we also need the logic part of the DA file. Also, since we need concern ourselves with only a part of the design at a time, we select only that logic that we need for calculations. The diagnostic engineer is responsible for subdividing the design into testable portions and assigning a sequence in which the portions will be tested in the field (this sequence may be different from that used in the engineering laboratory or in the manufacturing plant). This selection is illustrated in process 1, Fig. 7-20. The resultant extracted logic is held in two separate data sets, physical and logical, to permit separate manual interventions on each. However, to provide a tie between the files, a common index is program-generated.

In process 2, the logic data are divided into levels; the feedback paths, the primary inputs, and primary outputs are identified; and manually edited last-minute changes are checked and updated. The physical (geographical) data are checked and a bug-list (a list of potential faults to be tested) data set is optionally generated (and cross-indexed) and may be further edited in process 3. A separate print program identified in process 4 provides the options to print all data desired about these three separate data sets. When the diagnostic engineer is satisfied that the data needed by the next process are all contained on the "level" logic data set, he can subject this logic to the test pattern candidate-generation routines of process 5, utilizing the edited bug list.

As discussed in Sections 7.4 and 7.5, not every test-pattern candidate generated is useful in testing for faults. Thus, many candidates may be discarded before the final test-pattern list is given a stamp of approval by the test-evaluation algorithms or heuristics. "Fault simulator" and "test analyzer" are other names given to process 6. In some cases, manually generated test-pattern candidates may also be added to complement the test-generating algorithms. In others, the manually generated pattern is the sole source of input patterns (as discussed in Section 7.5.2). The editing, checking, and updating of the test-pattern candidate file is thus performed in process 7.

The output of process 6 consists of three files needed for further processing. The test file at this point is a compact, tester-independent data set that is cross-indexed to a second data set containing the sensitive path for each test pattern. This sensitive path contains a list of faults for which each pattern is a test. It is later utilized in the fault-isolation calculation. Optionally, the test patterns may be compared against the entries on the bug list, and a bug list entry for a fault may be removed when a pattern is found that tests for the listed fault. The reduced bug list is then fed back into process 5 for additional test-pattern candidate generation. Ultimately, the bug list could be made to disappear.

Process 8 is a rather complicated procedure and is tester-dependent. First,

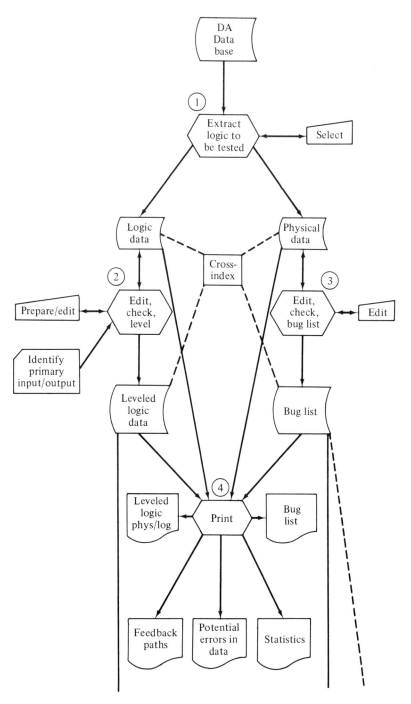

Fig. 7-20. Design automation system for test generation.

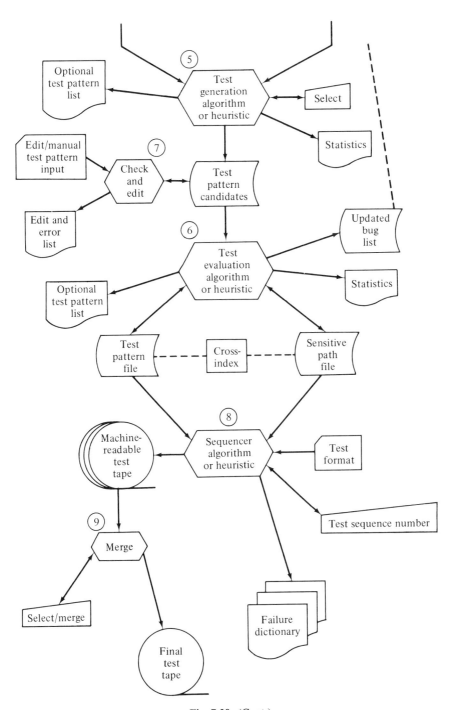

Fig. 7-20. (Cont.)

the exact test format must be spelled out. Also, the method of identifying faults must be defined and the method of enumeration given. The format of the fault dictionary must be specified. The method of test, the expected resolution, and the hardware dependencies must be known. The chances are very high that a special program will have to be written for each tester and each uniform piece of hardware to be tested (see Section 7.6.3). But since the amount of computer time to be utilized in this phase is quite negligible when compared to the generate-evaluate phase, a higher-level language could be utilized effectively for the unique parts of this program.

The final process is that of merging all the independently generated sections of tests onto one test library for use by the maintenance personnel in the field. This library should contain tests for the hard core, as well as tests for all parts of the machine in an ascending order of complexity. For instilling confidence and guarding against runaway tests, the maintenance personnel should have positive confirmation that various sections of tests have been executed successfully. This confirmation may also have to be merged in with process 9.

While the algorithms or techniques discussed in Sections 7.4 and 7.5 play a major role in actually producing tests for use on systems, this section points out that they are only a part of the total test-generating design automation system. While the algorithms consume most of the computer time, the analyzing, editing, merging, etc., require a significant amount of programming. But without the convenience of automation for the latter, the best and fastest test-generating algorithm may end up in disuse.

ACKNOWLEDGMENTS

The author wishes to express his appreciation to Raymond H. Luiggi for assistance with the original manuscript and to Samuel B. Lee for his editorial guidance, critical review, partial rewrite of the final manuscript, and his fine-combed proofreading of the manuscript and galley proofs. Linda M. Vouloukos was given the painstaking task of typing various portions of various versions of the manuscript, and also the final version. Her accuracy and extreme care for details reduced the author's burden considerably, and this help is hereby acknowledged with gratitude. Critical guidance from, and discussions with the editor, Dr. Melvin A. Breuer, and also J. Russell Washburne, Jr., Dr. Ginofranco R. Putzolu, Klim Maling, and Alfred G. Auch were appreciated, and the author's thanks are hereby added.

PROBLEMS

1. Using the circuit of Fig. 7-3b, construct a table similar to Table 7-1 for the following faults:
   (a) a short in AND diode 3;
   (b) a short in OR diode 2.
2. Using the results of Table 7-1, show that a broken wire feeding AND diode 1 is logically equivalent to AND diode 1 being open.
3. Using Table 7-1, if the wire feeding AND diode 1 shorts with the wire feeding AND diode 3, show the resultant output values and compare them against the good circuit.
4. Given a read-only control store containing n-bits per word, and r-bits to address each:
   (a) Find the number of tests needed to test for each bit of each word.
   (b) If, on the average, one-third of the bits in a word are 1's, find the approximate number of tests that would be generated by a program such as the D-algorithm. Assume that the control store has been converted into an equivalent logic circuit, and that the resultant tests checked one bit at a time, with an average resolution of four logic blocks. (Make additional assumptions as needed.)
   (c) What size control store represents the crossover point for the number of tests produced by the methods used in Problems 4(a) and 4(b)? (Use the same assumptions as in problem 4(b) to arrive at this answer.)
5. Show that a test for the output of an AND block stuck-at-0, or the output of an OR block stuck-at-1, is not generated by means of the failing input theory alone.
6. Using the circuit of Fig. 7-4, construct the F-trace in a table similar to Table 7-2, and find a pattern to test for the indicated failure.
7. (a) Show that an algorithmic procedure is superior to an exhaustive procedure for test-pattern generation on circuits consisting of more than sixty logic blocks. (Assume ten times as many logic blocks as primary inputs.)
   (b) Repeat this problem, this time comparing against a random-number procedure.
8. Using the Maling–Evans model, develop one or more tests to distinguish between an output failure at block k (K stuck-at-0 in Fig. 7-7) from another output failure at block e (E stuck-at-1).
9. Show that

$$D_i F = F(x_1, x_2, \ldots, x_i, \ldots, x_n) \oplus F(x_1, x_2, \ldots, x_i', \ldots, x_n)$$
$$= F(x_1, x_2, \ldots, 1, \ldots, x_n) \oplus F(x_1, x_2, \ldots, 0, \ldots, x_n)$$

10. Show that the results of the third step of the GNR technique in Section 7.4.4 are identical to the results of Equation (5) in Section 7.4.6.

11. Using the *D*-Algorithm, find the other tests discovered by Maling and Evans in the example of Section 7.4.3.

REFERENCES

[1] W. C. Carter, H. C. Montgomery, R. J. Preiss, and H. J. Reinheimer, "Design of Serviceability Features for the IBM System/360." *IBM J. Res. and Devel.*, Vol. 8 (April, 1964), pp. 115–126.

[2] J. Von Neumann, "Probabilistic Logics and the Synthesis of Reliable Organisms From Unreliable Components" *Annals of Mathematics: Automata Studies*, Vol. 34 (Princeton University Press, 1956), pp. 43–98.

[3] W. H. Kautz, "Automatic Fault Detection in Combinational Switching Networks." *Proc. AIEE 2nd Annual Symposium on Switching Circuit Theory and Logical Design* (October, 1961), pp. 195–214.

[4] M. Ball and F. Hardie, "Effects and Detection of Intermittent Failures in Digital Systems." *Proc. 1969 FJCC*, Vol. 35, pp. 329–335.

[5] M. A. Breuer, "General Survey of Design Automation of Digital Computers." *Proc. IEEE*, Vol. 54, No. 12 (December, 1966), pp. 1708–1721.

[6] K. Maling and E. L. Allen, Jr., "A Computer Organization and Programming System for Automated Maintenance." *IEEE Trans. on Electronic Computers*, Vol. EC12, No. 6 (December, 1963), pp. 887–895.

[7] R. D. Eldred, "Test Routines Based on Symbolic Logic Statements." *J. ACM*, Vol. 6, No. 1 (January, 1959), pp. 33–36.

[8] S. H. Tsiang and W. Ulrich, "Automatic Trouble Diagnosis of Complex Logic Circuits." *Bell Sys. Tech. J.*, Vol. 41 (July, 1962), pp. 1177–1200.

[9] R. E. Forbes, C. B. Stieglitz, and D. Muller, "Automated Fault Diagnosis." Unpublished report presented at the 1961 AIEE Conference on Diagnosis of Failures in Switching Circuits, University of Michigan, May, 1961.

[10] D. B. Armstrong, "On Finding a Nearly Minimal Set of Fault Detection Tests For Combinational Logic Nets." *IEEE Trans. on Electronic Computers*, Vol. EC15 (February, 1966), pp. 66–73.

[11] J. M. Galey, R. E. Norby and J. P. Roth, "Techniques for the Diagnosis of Switching Circuit Failures," *AIEE Trans. of Communications and Electronics*, Vol. 83, No. 74 (September, 1964), pp. 509–514.

[12] J. P. Roth, "Algebraic Topological Methods for the Synthesis of Switching Systems, Part I." *Trans. Amer. Math. Soc.*, Vol. 88 (July, 1958), pp. 301–326.

[13] J. P. Roth and E. G. Wagner, "Algebraic Topological Methods for the Synthesis of Switching Systems, Part III." *IBM Journal of Research and Development*, Vol. 3 (October, 1959), pp. 326–344.

[14] J. P. Roth, "Diagnosis of Automata Failures: A Calculus and a Method." *IBM Journal of Research and Development*, Vol. 10 (July, 1966), pp. 278–291.

[15] J. P. Roth, W. G. Bouricius, and P. R. Schneider, "Programmed Algorithms to Compute Tests to Detect and Distinguish Between Failures in Logic Circuits." *IEEE Trans. on Electronic Computers*, Vol. EC16, No. 5 (October, 1967), pp. 567–580.

[16] H. Y. Chang, E. C. Manning, and G. Metze, *Fault Diagnosis of Digital Systems*. (New York, Wiley-Interscience, 1970), pp. 29–46.

[17] F. F. Sellers, Jr., M. Y. Hsiao, and L. W. Bearnson, "Analyzing Errors With the Boolean Difference." *Trans. IEEE Computer Group*, Vol. C17, No. 7 (July, 1968), pp. 676–683.

[18] P. N. Marinos, "A Method of Deriving Minimal Complete Sets of Test-Input Sequences Using Boolean Differences." *Proc. 1970 IEEE Computer Group Conference*, Vol. 70-C-23C (June, 1970), pp. 240–246.

[19] T. A. Kriz, "A Path Sensitizing Algorithm For Diagnosis of Binary Sequential Logic." *Proc. 1970 IEEE Computer Group Conference*, Vol. 70-C-23C (June, 1970), pp. 250–259.

[20] S. Seshu and D. N. Freeman, "The Diagnosis of Asynchronous Sequential Switching Systems." *IRE Trans. on Electronic Computers*, Vol. EC11, No. 4 (August, 1962), pp. 459–465.

[21] D. A. Huffman, "The Synthesis of Sequential Switching Circuits." *J. Franklin Institute*, Vol. 257 (March, 1954), pp. 161–190; Vol. 257, (April, 1954), pp. 257–303.

[22] H. Y. Chang, E. G. Manning, and G. Metze, *Fault Diagnosis of Digital Systems*. (New York Wiley-Interscience, 1970), pp. 90–125.

[23] D. D. Cheng, *A Method to Optimize the Selection of a Bug List For Failure Simulation*. Informal paper presented at the 1968 IEEE Computer Group Design Automation Workshop, Michigan State University, East Lansing, Michigan, October 2, 1968.

[24] R. J. Preiss, *The IBM Scopex—A Document Used in Fault Locating*. Informal paper presented at the 1966 IEEE Computer Group Design Automation Workshop, Michigan State University, East Lansing, Michigan, September 28, 1966.

[25] A. G. Auch, *The Simulation Systems Statistical Analysis Program*. Informal paper presented at the 1966 IEEE Computer Group Design Automation Workshop, Michigan State University, East Lansing, Michigan, September 28, 1966.

[26] R. J. Preiss, "The Use of Fault-Location Tests in Prototype Bring-up." *Proc. IFIP Congress 65*, Vol. 2 (Washington, D.C., Spartan Books), pp. 511–512.

[27] F. J. Hackl and R. W. Shirk, "An Integrated Approach to Automated Computer Maintenance." *IEEE Conf. Rec. on Switching Circuit Theory and Logical Design*, Vol. 16-C-13 (October, 1965), pp. 289–302.

[28] R. V. Bock and A. P. Toth, "Hardware and Software for Maintenance in the B5500 Processor." *1965 IEEE International Conv. Rec.* Part 3, pp. 21–27.

[29] P. S. Bottorff, M. E. Schwass, and F. J. Villante, "An Automatic System Approach to the Problem of Memory Circuit Testing and Fault Diagnosis." *Proc. SHARE-ACM-IEEE 7th Annual Design Automation Workshop* (June, 1970), pp. 95–99.

[30] M. S. Horovitz, "Automatic Checkout of Small Computers." *Proc. SJCC*, Vol. 34 (1969), pp. 359–365.

[31] H. Y. Chang and J. M. Scanlon, "Design Principles for Processor Maintainability in Real-Time System." *Proc. FJCC*, Vol. 35 pp. 319–328.

[32] R. E. Forbes, D. H. Rutherford, C. B. Stieglitz, and L. H. Tung, "A Self-Diagnosable Computer." *Proc. FJCC*, Vol. 27 (1969), pp. 1073–1086.

[33] F. B. Cole, "Automatic Generation of Functional Logic Test Programs Through Simulation." *Proc. SHARE-ACM-IEEE 7th Annual Design Automation Workshop*, (June, 1970), pp. 116–127.

[34] H. Y. Chang and W. Thomis, "Method of Interpreting Diagnostic Data for Locating Faults in Digital Machines." *Bell Sys. Tech. J.*, Vol. 46, No. 2 (February, 1967), pp. 289–317.

INDEX

Activated function, 338
Activity vectors, 363
Addressing resolution, 342
ALGOL, 177
ANALYZER, 349–50, 352
AND, 24, 29, 53, 60, 64, 65, 68–70, 73, 78, 104, 106, 113, 118 121 129, 145–46, 153, 177, 181–83, 190, 205, 295, 339–40, 346
AND-AND, 24
AND-INVERT, 359–60
AND-OR, 21, 29, 59, 79
AND-OR-AND, 59

AND-OR-INVERT, 339
AND-OR (NAND-NAND), 28, 53
APL, 22, 111
Assignment algorithms, 280-82
Assignment problem, 221
  quadratic, 222–25
Auch-Cheng model, 383–88
Automated test application, 388–406
  arrangement for detection or location, 397–402
  DA system for, 402–6
  designing logic for, 391–97
  system test plan, 389–91

412 *Index*

AWOB, 130
AZR, 130

Backplanes, 215
Bad machines, 342
Basis nodes, 182
Bi-digraphs, 178–80, 191
Bigraphs, 217
  partial, 232
  signal-set, 218
Bipartite graphs, 217
Bit time, 107
Black-box expression, 115
Boards, 215
Boolean differences, 362
  method of, 374-76
Boolean equations, 22, 25–26, 114–15
Boolean functions, 21, 22
  single-output, 26
  switching, 34
Boolean product term, 28
Branch, 361
  improper, 304–5
  proper, 305
Branch-and-bound methods, 40, 264–71
  exact, 264–69
  Gilmore's approximation methods, 269–71
Branching, 39–41
Bug elimination, 387
Building material influences, 174–76

Cabling, 16
Candidate set, 229
Candidate slots, 232
Canonical normal form, 28
Card selection, 173–212
  building material influences, 174–76
  graph theoretic representation of the logic design, 176–81
  problems, 202–10
  replaceable units, 181–87
  technique influence, 174–76
Cards, 215
Care complex, 29
CAUT, 129–31
Chains, 220, 291
Chromatic numbers, 302–3
Circuit analysis, simulation and, 107–11

Circuit elements, 214
Circuit modification techniques, 77–85
  functional packages, 81–85
  gate type conversion, 78–81
Class, 357
CLOCK, 132, 163–64
Cluster-development method, 240–45
  timing for, 243–45
Coding procedure, 317–18
Column dominance, 44
Combinational logic, models for, 345–76
  Eldred's method of activating a failure, 346–47
  Forbes-Stieglitz-Muller techniques, 347–51
  Galey-Norby-Roth technique, 358–62
  Maling-Evans model, 352–58
  method of Boolean differences, 374–76
  Roth's D-algorithm, 362–74
Combinational test procedure, 377–78
Common factors, 61
Compilation techniques, 126–27
Compiled-code model, 136–37
Compiled-code modeling technique, 107
Complement-factored circuits, 67
Components, 214
Computer system, designing, 10–17
  from concept to defined subsystem, 13–14
  detailing the subsystem, 14–15
  document maintenance, 17
  reduction to hardware, 15–16
  testing, 16–17
Configuration, 215
Conjunction, the, 190
Connection cover:
  generation of, 56–58
  reduction of, 55–56
Connection matrix, 217, 224, 225, 267
Connection pins, 215
Connection tubes, 53
Consensus, 32–33
Consistency operation, 364, 370–74
Constructive algorithms, 214
Constructive initial placement methods, 226–45
Consructive interference, 356
Controls automation system, 9
Critical-length wires, 310

Cube deletion, 45–46
Current-activity stack, 137

DA system for test generation, 402–6
Data base, relationship between simulation and, 161–64
Data structures, 85–87
Data-dependent code 122–24
D-cube of failure, 368–69
DDL, 111
Decision trees, 265
Decomposition, 88
Delay considerations, 117
DeMorgan's theorem, 31–32
Design automation:
  designing a system, 10–17
    from concept to defined subsystem, 13–14
    detailing the subsystem, 14–15
    document maintenance, 17
    reduction to hardware, 15–16
    testing, 16–17
  evolution of, 4–10
    advent of standardization, 8–9
    efforts to automate design procedures, 4–5
    first uses of, 5–7
    technology changes, 9–10
    use of simulation, 7–8
  objectives, 2–3
  usage, 2–3
Design procedures, efforts to automate, 4–5
Destructive interference, 356
DIAGNOSE, 389
Diagnosis, 341
Diagnostic generation, 147–48
Digraphs, 178
Discrete-wire routing, 326–27
Disjunctions, 192
Distance matrix, 267
Document maintenance, 17
Don't care complex, 29
Don't care vertices, 26, 27
Double fan-out technique, 317

ECAP, 108
Eldred's method of activating a failure, 346–47
Electrically common, 215

Element deletion operation, 45
Elements, 214
Energized function, 338
Entry points, 340
Equation simulation, 122–26
Equivalent circuits, 82, 83
EXCLUSIVE-OR, 375
Exercising the model, 131–43
  compiled-code model, 136–37
  next-event synchronous simulation, 137–39
  output descriptions, 141–43
  simulation algorithms, 133–41
  simulation commands, 132–33
  three-value simulation, 133–36
  time-mapping asynchronous simulation model, 139–41
Exit points, 340
Expression descriptions, 113–17
Extraction algorithm, 39–42
  local, 42
Extremals, 39, 44

Factorization, synthesis of constrained NAND networks via, 59–77
  construction of factors, 61–64
  multiple-output fan-in limited, 70–72
  multiple-output fan-out limited, 73–76
  single-output fan-in limited, 64–70
Factors:
  common, 61
  construction of, 61–64
  proper, 75
Failing input theory, 348–49
Failing net theory, 347–48
Failure:
  D-cube of, 368–69
  intermittent, 338
  machine, 337
  solid, 338
False literals, 28
False vertices, 26, 27
Fault, 337–38
Fault detection test, 340
Fault insertion, 144–47
Fault simulation, 143–50
  diagnostic generation, 147–48
  insertion, 144–47
  logic testing, 148–50

414 *Index*

Fault test generation, 335–410
  automated test application, 388–406
    arrangement for detection or location, 397–402
    DA system for, 402–6
    designing logic for, 391–97
    system test plan, 389–91
  classical approaches, 342–45
  combinational logic, models for, 345–76
    Eldred's method of activating a failure, 346–47
    Forbes-Stieglitz-Muller techniques, 347–51
    Galey-Norby-Roth technique, 358–62
    Maling-Evans model, 352–58
    method of Boolean differences, 374–76
    Roth's D-algorithm, 362–74
  definitions, 337–42
  problems, 407–8
  sequential logic, models for, 376–88
    Auch-Cheng model, 383–88
    Seshu-Freeman model, 377–83
Feasible partitions, 188
Feedthrough pin, 285
Feedthroughs, 284–85
Figure of merit, 60–61
Flip-flops, 114–15
Forbes-Stieglitz-Muller techniques, 347–51
Force directed interchange, 259
Force directed pairwise relaxation, 259
Forest, 356
FORTRAN, 24, 113, 151, 177
Four color conjecture, 305
Framing, 317
Function:
  activated, 338
  energized, 338
  inhibited, 338
  multiple-output, 100
  partially specified, 26, 27
  totally specified, 26
Functional multiple-output cube, 70
Functional packages, 81–85
Future-activity stack, 137

Galey-Norby-Roth technique, 358–62
Gate type conversion, 78–81
Gates, 113–15
  logic, 214
Gilmore's approximation methods, 269–71
Good machine, 341
GPSS, 150–51
Graph models, 216–19
Graph theoretic representation of the logic design, 176–81
Graphs:
  bipartite, 217
  complete, 217
  intersection, 302–3
  module-adjacency, 218
  module-connection, 224
Grid segments, routing to preserve, 319–21
Grow factor, 62–64

Hardware, reduction to, 15–16
Hazard cycles, 133–36
HIGH, 117

ICP, 161
ILP, 209
Implicants, 45
  multiple-output, 50, 52
  prime, 29, 34
Implied values, 352–53
IN, 118, 121, 146
Independent modules, 217–18, 247
Inhibited function, 338
INI, 129
Initial connection cover, generation of, 56–58
Injection words, 359
Input-pin permutation, 294–95
Inputs:
  failing theory, 348–49
  primary, 340
Interconnection lists, 287–88
Interconnection matrix, 289–90
Interconnection rules, 219–20
Interference:
  constructive, 356
  destructive, 356

Intermittent failure, 338
Intersect, the, 31
Intersection, 30–31
Intersection calculation, 299–301
Intersection graphs, 302–3
Intersection matrix, 299–300
INVERT, 78, 177, 181–83, 190, 205
Inverter insertion, 78–81
Inverter simulation, 110
I/O channels, 389–90
Irredundant covers, 34, 44–49
Iterative algorithms, 214
Iterative placement-improvement methods, 245–64

Koenig's theorem, 303–5

Language, 113
LATCH, 24
Layering, 295–308
   chromatic numbers, 302–3
   four color conjecture, 305
   intersection calculation, 299–301
   intersection graph, 302–3
   Koenig's theorem, 303–5
   two-color problem, 303–5
   two-layer boards, 305–8
Layout, 215
   See also Wire layout
Lee's algorithm, 312–13
   with arbitrary weights, 322–24
Leg, 338
Level assignment, 78–81
Literals:
   false, 28
   true, 28
Local optimization techniques, 247
Location, 215
LOCS, 111
Logic elements, 113–17, 214
Logic gates, 214
Logic for simplified testing, 391–97
Logic simulation, gate-level, 101–72
   data base and simulation algorithm relationship, 161–64
   developing the model, 113–31
      black-box expressions, 115
      compilation techniques, 126–27

Logic simulation (*cont.*)
   developing the model (*cont.*)
      delay considerations, 117
      equation simulation, 122–26
      expression descriptions, 113–17
      flip-flops, 114–15
      gates, 113–15
      logic element, 113–17
      macro expressions, 115
      model construction, 122–29
      output, 129–31
      parallel simulation, 128–29
      reference catalogues, 115–17
      table-driven techniques, 127–28
      three-value simulation, 119–22
      time algorithms, 117–19
      time mechanization, 117–22
   exercising the model, 131–43
      compiled-code model, 137–39
      next-event synchronous simulation, 137–39
      output descriptions, 141–43
      simulation algorithms, 133–41
      simulation commands, 132–33
      three-value simulation, 133–36
      time-mapping asynchronous simulation model, 139–41
   fault simulation, 143–50
      diagnostic generation, 147–48
      insertion, 144–47
      logic testing, 148–50
   problems, 167–69
   process of, 103–11
      comparative roles, 107–11
      registers and, 111–12
   simulation examples, 157–61
   simulation selection criteria, 164–67
   simulator characteristics, 155–57
   special software techniques, 150–54
      general purpose system simulators, 150–52
      PERT, 152–54
      specialized computer techniques, 154–55
Logic string, 324
Logic synthesis, 21–100
   basic results, 26–33
   circuit modification techniques, 77–85

416 *Index*

Logic synthesis (*cont.*)
  circuit modification techniques (*cont.*)
    functional packages, 81–85
    gate type conversion, 78–81
    data structures, 85–87
    decomposition, 88
    definitions, 26–33
    multiple-output switching function, 50–58
      synthesis of, 51
    problems, 89–93
    rotation, 26–33
    sequential machine, 89
    single-outputs switching function, 33–49
      generation of set Z, 34–38
      irredundant covers, 44–49
      selection of minimum cost cover, 38–44
    synthesis of constrained NAND network via factorization, 59–77
      construction of factors, 61–64
      multiple-output fan-in limited, 70–72
      multiple-output fan-out limited, 73–76
      single-output fan-in limited, 64–70
Logic testing, 148–50
Logic tree, 356
Log-out path, 392, 393
Loops, 290
LOTIS, 111
LSI, 10, 103, 175

Machine failure, 337
Machine translation, 25
Macro expression, 115
Maling-Evans model, 352–58
Manhattan distance, 216, 286
Manhattan geometry, 286
MASK, 146
Matrix:
  adjacency, 217
  algorithm, 42–44
  connection, 217, 224, 225, 267
  distance, 267
  interconnection, 289–90
  intersection, 299–300
  module-adjacency, 218
  module-slot assignment, 269
Maximal independent modules, 248–49

Maximal independent sets, 249–54
Minimal cover, 34
Minimization, 15
Minimum cost cover, selection of, 38–44
Minimum cost tagged cover, 52
Minimum spanning tree, 219
Minimum tree algorithms, 288–89
Minterms, 28, 34
Model construction, 113, 122–29
Modules, 214–15
  adjacency graphs, 218
  adjacency matrix, 218
  adjacent, 217
  connection graphs, 224
  independent, 217–18, 247
  maximal independent, 248–49
  placement problems, 215
  positioning, 232–33
  selection, 226–31
  slot assignment matrix, 269
Monte Carlo method, adaptive, 263
Multilayer boards, 284–85
Multiple-output fan-in limited networks, 70–72
Multiple-output fan-out limited networks, 73–76
Multiple-output functions, 100
Multiple-output implicants, 50, 52
Multiple-output prime implicant, 50, 52
Multiple-output switching function, 50–58
  synthesis of, 51
Munkre's algorithm, 280–82

NAND, 21, 24, 26, 78, 80–81, 104, 106, 107, 113, 115, 122–23, 127, 129, 139
  synthesis of constrained, via factorization, 59–77
    construction of factors, 61–64
    multiple-output fan-in limited, 70–72
    multiple-output fan-out limited, 73–76
    single-output fan-in limited, 64–70
NAND-NAND, 136
NAND-NOR, 78, 79
Nets, 215
  complete, layout of, 324–26
  failing theory, 347–48

Nets (*cont.*)
  single, 324
Network, 324
Next-clock-time, 137
Next-event, 137–39
Next-event synchronous simulation, 137–39
Nodes, 324
  subsets of, 190
NOR, 24, 78, 80–81, 104, 295
Normal form, 28
  canonical, 28

OFF, 107, 146, 153, 359–60
ON, 107, 117, 146, 153, 359–60
ONE, 107
OR, 24, 29, 53, 78, 104, 113, 136, 146, 153, 177, 181–83, 190, 205, 295, 339–40, 346
Ordering, 308–11
  critical-length wires, 310
  rule for, 310
OUT, 118, 121
OUTPUT, 146
Outputs, 129–31
  descriptions, 141–43
  primary, 340

Pair-linking method, 233–40
  timing for, 238–39
Pairwise interchange, 257–62
Parallel simulation, 128–29
Parameter sets, 246
Partial bigraphs, 232
Partial order, 181
Partial permutations, 266
Partitioned list, 36
Partitioning, 173–212
  building material influence, 174–76
  graph theoretic representation of the logic design, 176–81
  problems, 187–202, 209–10
  replaceable units, 181–87
  technique influence, 174–76
Permutation:
  input-pin, 294–95
  partial, 266
PERT, 152–54
Pin-to-connector assignment, 293–94

Placement, 215
Placement configuration, 215
Placement techniques, 213–82
  assignment algorithms, 280–82
  the assignment problem, 221
  branch-and-bound, 264–71
    exact, 264–69
    Gilmore's approximation methods, 269–71
  cluster-development method, 240–45
    timing for, 243–45
  comparison of methods, 271–73
  constructive initial-placement methods, 226–45
  graph models, 216–19
  interconnection rules, 219–20
  iterative placement-improvement methods, 245–64
  Munkre's algorithm, 280–82
  pair-linking method, 233–40
    timing for, 238–39
  pairwise interchange, 257–62
  problem definition, 214–16
  problems, 275–77
  relaxation, 254–57
    force directed, 257
  Steinberg assignment method, 247–54
  stochastic methods, 263–64
  the traveling salesman problem, 221–22
Planes, 215
POP, 138–39
Position, 215
Post-factored circuits, 67
Primary inputs, 340
Primary outputs, 340
Prime implicants, 29, 34
Proper factors, 75
Pruning, 361
PUSH, 138–39

Quadratic assignment problem, 222–25
Quine-McCluskey procedure, 36

Recognized circuits, 182
Rectangular grids, 285–86
Rectilinear distance, 216
Reduced circuits, 82
Redundant, 55
Reference catalogues, 115–17

418   *Index*

Register, simulation and, 111–12
Relaxation, 254–57
    force directed, 257
    force directed pairwise, 259
Replaceable units, 10, 181–87
Reset cycles, 133–36
Residue, the, 60
Residue class, 298
Retracing, 313–14
ROS, 389
Rotation, 26–33
Roth's D-algorithm, 362–74
Routing, 283–333
    basic approach, 286–87
    feedthroughs, 284–85
    layering, 295–308
        chromatic numbers, 302–3
        four color conjecture, 305
        intersection calculation, 299–301
        intersection graph, 302–3
        Koenig's theorem, 303–5
        two-color problem, 303–5
        two-layer boards, 305–8
    Manhattan distance, 286
    multilayer boards, 284–85
    ordering, 308–11
        critical-length wires, 310
        rule for, 310
    problems, 329–30
    the rectangular grid, 285–86
    wire layout, 312–27
        coding procedure, 317–18
        of complete nets, 324–26
        discrete-wire, 326–27
        Lee's algorithm, 312–13
        Lee's algorithm with arbitrary weights, 322–24
        to preserve grid segments, 319–21
        retracing, 313–14
        speed-up technique, 315–17
        storage problems, 317–18
        three-dimensional paths, 314–15
        on a two-layer board, 321–22
        useful variations, 318–19
    wire-list determination, 287–95
        alternate approaches, 291–92
        input-pin permutation, 294–95
        interconnection matrix, 289–90

Routing (*cont.*)
    wire-list determination (*cont.*)
        minimum tree algorithm, 288–89
        pin-to-connector assignment, 293–94
        Steiner's problem, 292
        traveling salesman problem, 290–91
Row dominance, 44
RSFF, 115, 161

Scan-in path, 392, 393
Sensitive path, 355
Sequential logic, models for, 376–88
    Auch-Cheng model, 383–88
    Seshu-Freeman model, 377–83
Sequential machine synthesis, 89
Sequential test procedure, 378
Seshu-Freeman model, 377–83
Set covering, 27
Set Z, generation of 34–38
Sets:
    candidate, 229
    independent, 247
    maximal independent, 249–54
    parameter, 246
    signal, 215
        bigraphs, 218
Sharp-product, 31–32
Sharp-product algorithm, modified, 99–100
SHIFT, 161
Signal, 324
SIGNAL, 132
Signal sets, 215
    bigraphs, 218
SIMSCRIPT, 150–51
Simulation:
    algorithms, 133–41
    circuit analysis and, 107–11
    commands, 132–33
    comparative roles, 107–11
    equation, 122–26
    examples, 157–61
    faults, 143–50
        diagnostic generation, 147–48
        insertion, 144–47
        logic testing, 148–50
    inverter, 110
    next-event synchronous, 137–39

Simulation (*cont.*)
  parallel, 128–29
  register and, 111–12
  relationship between the data base and, 161–64
  table-driven interpretive mode, 107
  three-value, 119–22, 133–36
  time-mapping asynchronous, 139–41
  use of, 7–8
  See also Logic simulation, gate-level
Simulators:
  characteristics, 155–57, 166
  general purpose system, 150–52
  selection criteria, 164–67
Single-output fan-in limited networks, 64–70
Single-output irredundant cover, 46–49
Single-outputs switching function, 33–49
  generation of set Z, 34–38
  irredundant covers, 44–49
  selection of minimum cost cover, 38–44
Sink, the, 338
Slots, 215
Slow-down ratio, 165
Software techniques, special, 150–54
  general purpose system simulators, 150–52
  PERT, 152–54
Solid failure, 338
Source, the, 338
Specialized computer techniques, 154–55
Speed-up techniques, 315–17
Spike condition, 119
Stack, 78
  current-activity, 137
  future-activity, 137
Standardization, advent of, 8–9
State splitting, 61–62
Static hazard, 118
Steinberg assignment method, 247–54
Steiner tree, 219
Steiner's problem, 292
Stochastic methods, 263–64
Stopping rule, 246
Storage problems, 317–18
Strip, 235
Subsets of nodes, 190
Subsuming, 27, 30

Subsystem:
  from concept to defined, 13–14
  detailing the, 14–15
Switch cycles, 133–36
System test plan, 389–91

Table-driven interpretive mode simulation, 107
Table-driven techniques, 127–28
  modeling, 107
Tagged cube, 50
Target location, 255–57
Target point, 255–57
Technology:
  changes in, 9–10
  influences of, 174–76
Test-generation algorithms, 148
Test-patterns, 340
  arrangement of, 397–402
Test series appreciation program, 388
Three-dimensional paths, 314–15
Three-value simulation, 119–22, 133–36
Tie-breaking rule, 230
Time-mapping asynchronous simulation, 139–41
Time mechanization, 117–22
Timing algorithms, 117–19
Timing resolution, 342
Transitivity, 310
Traveling salesman problem, 221–22, 290–91
Trees, 219–20, 324
  decision, 265
  logic, 356
  minimum algorithms, 288–89
  minimum spanning, 219
  Steiner, 219
True literals, 28
True vertices, 26, 27
Two-color problem, 303–5
Two-layer boards:
  layering for, 305–8
  routing on, 321–22

Unfactored circuits, 67

Vertices:
  don't care, 26, 27

Vertices (*cont.*)
   false, 26, 27
   true, 26, 27

Wire layout, 312–27
   coding procedure, 317–18
   of complete nets, 324–26
   discrete-wire, 326–27
   Lee's algorithm, 312–13
   Lee's algorithm with arbitrary weights, 322–24
   to preserve grid segments, 319–21
   retracing, 313–14
   speed-up techniques, 315–17
   storage problems, 317–18
   three-dimensional paths, 314–15
   on a two-layer board, 321–22

Wire layout (*cont.*)
   useful variations, 318–19
Wire routing, 219
Wire-list determinations, 287–95
   alternate approaches, 291–92
   input-pin permutation, 294–95
   interconnection matrix, 289–90
   minimum tree algorithm, 288–89
   pin-to-connector assignment, 293–94
   Steiner's problem, 292
   traveling salesman problem, 290–91

Zero, 281
ZERO, 107
Zero bar, 281
Zero star, 281
Zero-delay networks, 118